Springer Series in Vision Research

The Springer Series in Vision Research is a comprehensive update and overview of cutting edge vision research, exploring, in depth, current breakthroughs at a conceptual level. It details the whole visual system, from molecular processes to anatomy, physiology and behavior and covers both invertebrate and vertebrate organisms from terrestrial and aquatic habitats. Each book in the Series is aimed at all individuals with interests in vision including advanced graduate students, post-doctoral researchers, established vision scientists and clinical investigators. The series editors are N. Justin Marshall, Queensland Brain Institute, The University of Queensland, Australia and Shaun P. Collin, Neuroecology Group within the School of Animal Biology and the Oceans Institute at the University of Western Australia.

Elke Buschbeck • Michael Bok
Editors

Distributed Vision

From Simple Sensors to Sophisticated Combination Eyes

Springer

Editors
Elke Buschbeck
Department of Biological Sciences
University of Cincinnati
Cincinnati, OH, USA

Michael Bok
Department of Biology
Lund University
Lund, Sweden

ISSN 2625-2635 ISSN 2625-2643 (electronic)
Springer Series in Vision Research
ISBN 978-3-031-23215-2 ISBN 978-3-031-23216-9 (eBook)
https://doi.org/10.1007/978-3-031-23216-9

This Springer imprint is published by the registered company Springer Nature Switzerland AG
The registered company address is: Gewerbestrasse 11, 6330 Cham, Switzerland

Preface

This book explores a diversity of distributed eyes and other unusual visual systems in nature. It begins with an overview chapter that introduces theoretical considerations as to how eyes can be distributed in animals, summarizing advantages and disadvantages of distributed, many-eyed visual systems versus centralized eye organizations. The introductory chapter also provides an overview of eye organizations in diverse groups and highlights key themes that are presented in the following chapters. These include in-depth reviews of the extraordinarily widely distributed visual systems of cnidaria and echinoderms, the unorthodoxly organized eyes of bivalves and chitons, and the somewhat more centralized systems of myriapods, insects, crustacea, and spiders. Jointly, these chapters explore unique themes of optics, neural processing, and behavioral control that emerge from these relatively unorthodox visual systems, specifically exploring questions about distributed visual systems. What are distributed visual systems good for? How do they function, and how do such systems that have arisen independently in so many phyla compare? Individual chapters include overviews of the visual systems that exist in specific group of animals, relate vision to each animal group's ecology, and allow readers to compare the most unusual visual systems in the animal kingdom.

Cincinnati, OH, USA — Elke Buschbeck
Lund, Sweden — Michael Bok

Contents

Chapter 1
On Distributed Visual Systems

Michael J. Bok and Elke K. Buschbeck

Abstract Many vision scientists have been drawn to study the remarkable diversity of animal eyes, ranging from very simple light sensors to highly sophisticated image-forming eyes with specializations for color or polarization vision among others. However, relatively few studies exist that specifically draw attention to how multi-eyed visual systems (having three or more eyes) are structured, evolved, and function. Such systems, nearly all of them found among invertebrates, may be centralized, whereas others are completely distributed, spanning across most of the body. Some distributed systems consist of a set of sophisticated visual sensors, providing input to the animal's primary visual system. Other systems consist of very simple organs that, in some cases, are auxiliary to their primary visual system. In this chapter, we provide a theoretical framework on the limits and benefits of distributing vision into multiple organs. We first discuss limitations, as well as benefits, of different organizations in a set of imaginary organisms and then summarize how specific distributed systems are actually organized and how they function throughout major invertebrate groups. This summary includes highlights of the many insightful chapters that authors have contributed to this volume.

Keywords Eye evolution · Vision · Visual ecology · Optics · Visually guided behaviors · Eye design · Distributed vision

M. J. Bok (✉)
Lund Vision Group, Department of Biology, Lund University, Lund, Sweden

E. K. Buschbeck
Department of Biological Sciences, University of Cincinnati, Cincinnati, OH, USA
e-mail: elke.buschbeck@uc.edu

E. Buschbeck, M. Bok (eds.), *Distributed Vision*, Springer Series in Vision Research, https://doi.org/10.1007/978-3-031-23216-9_1

1.1 Introduction

There are many books on the amazing diversity of visual systems invertebrate eyes exhibit (Land 1981; Cronin et al. 2014; Land and Nilsson 2012), but few are devoted to an in-depth exploration of the most unusual groups with distributed visual systems. These include several taxa that are presumably independently evolved visual systems that sample the world through many sensors, which are often distributed broadly over the body, and can be characterized by the absence of a clear anterior-posterior polarity. Others do have such polarity but rely on multiple eyes that acquire different kinds of information from different directions. Before delving into the details of such systems, it seems fruitful to put some thought into a conceptual framework that explores benefits and constraints of different ways to sample light from the environment.

In principle, there are two ways in which visual systems can increase the information content from sampling the environment. The first relates to the level of sophistication that is reached by each individual eye, and the second results from the number of eyes that are present as well as from the way in which they are positioned on the organism and oriented relative to the environment. These principal arrangements can affect many different aspects of vision, including spatial resolution, which is arguably the most important property that defines a high-functioning visual system. Other visual parameters that would be affected by the way eyes are positioned include motion vision, sensitivity at low-light levels (Warrant 2017), the discrimination of colors (van der Kooi et al. 2021) and polarization (Marshall and Cronin 2011; Horváth 2014), as well as depth perception, for example, through stereopsis (Nityananda et al. 2016). The necessary evolutionary steps that allow for the transition from a simple light sensor to a highly advanced image-forming eye already have been well-defined as arising through an evolutionary cascade generating ever-increasing information content that may drive increasingly sophisticated visually guided behaviors in a surprisingly short space of time (Nilsson 2009, 2013) (Fig. 1.1a).

1.2 From a Simple Light Sensor to a Sophisticated Eye

Considering a single light detecting structure, the simplest visual task is nondirectional light detection (referred to as Class I visual behavior by Nilsson (2013, 2021)) (Fig. 1.1b). Such simple light detection can provide information about the time of day and even guide simple behavioral responses by comparing light intensity from different time periods. If screening pigment is added to a photoreceptor, or the animal's body itself limits the directions from which light can be detected, then the light sensing structure is considered to be capable of directional photoreception, or Class II visual behaviors. Such animals (as our imaginary organism illustrates in Fig. 1.2a) thus have a light detector that can facilitate simple phototactic light

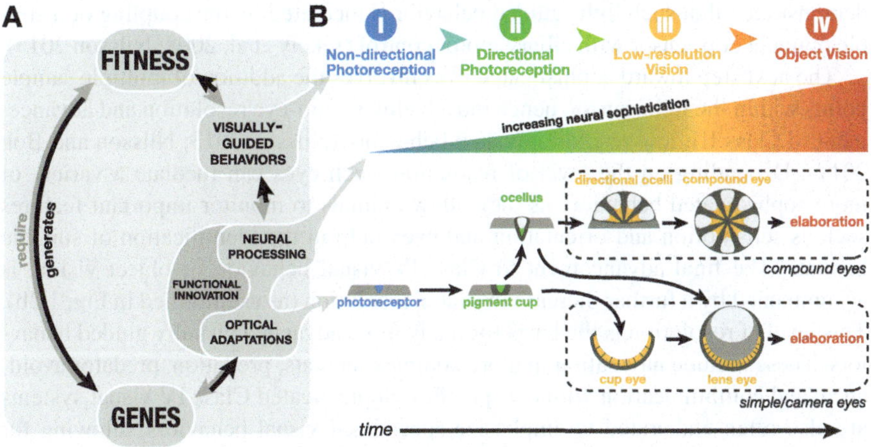

Fig. 1.1 The evolution of complexity in visual systems and visually guided behaviors. (**a**) Selection based on behavioral success drives the evolution of all sensory systems that acquire external information. Therefore, eye evolution must be driven by the fitness generated by visually aided behaviors, with existing visual systems being harnessed, integrated, and elaborated to support emergent visual tasks. Thus, there is necessarily an optical, neural, and behavioral continuity underpinning the advancement from simple to sophisticated visual behaviors (**b**). Classes I–IV of visually guided behaviors are described in the text and have been adapted from Nilsson (2013, 2021). Coloration refers to the required functional advancements (optical innovations and neural processing) that are necessary to support the classes of visual behaviors at the top

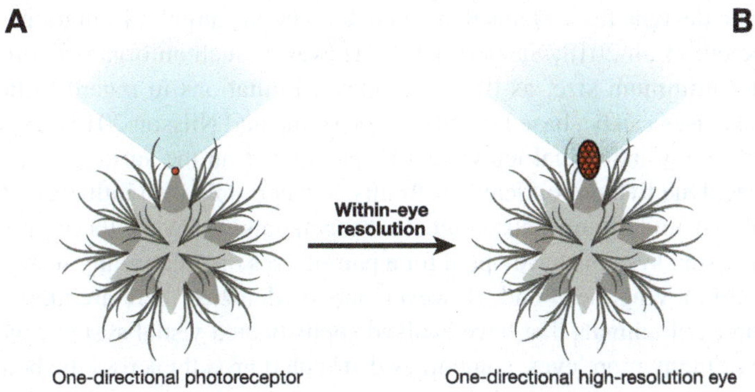

Fig. 1.2 One way to evolve an eye is through the acquisition of spatial resolution within a light sensor. (**a**) Hypothetical organism with a simple sensor that monitors light from a general angle. (**b**) By adding resolution, details can be resolved within that space; however, unless the animals were to be transparent, a single sensor could not facilitate all-around vision

responses, such as moving toward or moving away from a light source. If only one such receptor exists, however, it continues to remain necessary to compare light reception between different time points. A good example of this mechanism is found in marine zooplankton. For example, in *Platynereis dumerilii*, it has been demonstrated that such light-guided behavior is mediated by the coupling of a simple pigment cup sensor with ciliary motor control (Jékely et al. 2008; Nilsson 2013).

The next step toward sophisticated eyes involves the addition of multiple sample points within the light sensor, hence introducing within-eye resolution and advancement to Class III, low-resolution visual behaviors (Nilsson 2013; Nilsson and Bok 2017). Depending on the level of resolution, such eyes can mediate a variety of more sophisticated behaviors as they allow animals to monitor important features such as self-motion and orientation and even help in the identification of suitable habitats. The final advancement to Class IV visual behavior, or object vision, is accompanied by a further increase in spatial resolution (as exemplified in Fig. 1.2b). This level of resolution is the basis for many more advanced visually guided behaviors. These include navigation in more complex habitats, predation, predator avoidance, and communication with conspecifics. Sophisticated Class IV visual systems are also often elaborated to emphasize specialized visual behaviors, allowing for trade-offs between spatial resolution, temporal resolution, vision at low-light sensitivity, polarization vision, and color vision (Meece et al. 2021).

No matter how sophisticated the optics, there are limitations to vision with a single eye. In part, such limitations depend on the eye type. For example, single-chamber eyes are limited in their maximum field of view. The presence of a single lens typically restricts the visual field to a cone-shaped region of space in front of the eye. Photographers who work with wide-field lenses are well aware of ever-increasing distortions that result from increasing the visual field of a single image, with a general limitation of 180°. Compound eyes, on the other hand, are good at sampling many different directions, a remarkable attribute that has even inspired engineering designs for systems that can overcome the angular limitations of single lenses (Keum et al. 2018; Sanders 1997). However, such an organization needs to maintain a minimum size, as there are optical limitations in regard to how small individual lenses can be, based on diffraction (Land and Nilsson 2012). This, in turn, limits how many individual lenses can be positioned on the surface of an eye of a certain size. Due to these optical constraints, and certainly also influenced by developmental and underlying neurological arrangements, many of the more complex bilaterian visual systems have opted for a pair of eyes on the, usually forward facing and forward moving, head end. However, across Metazoa, there are many fascinating examples of animals that have evolved sophisticated visual systems incorporating three or many more eyes, sometimes distributed broadly across the body.

1.3 Sophisticated Vision Through a Distributed Visual System

In a distributed light detecting system, expanded directionality of light detection also may be accomplished by multiple sensors that are oriented toward different directions in space (Fig. 1.3). For multicellular organisms, it is usually the case that

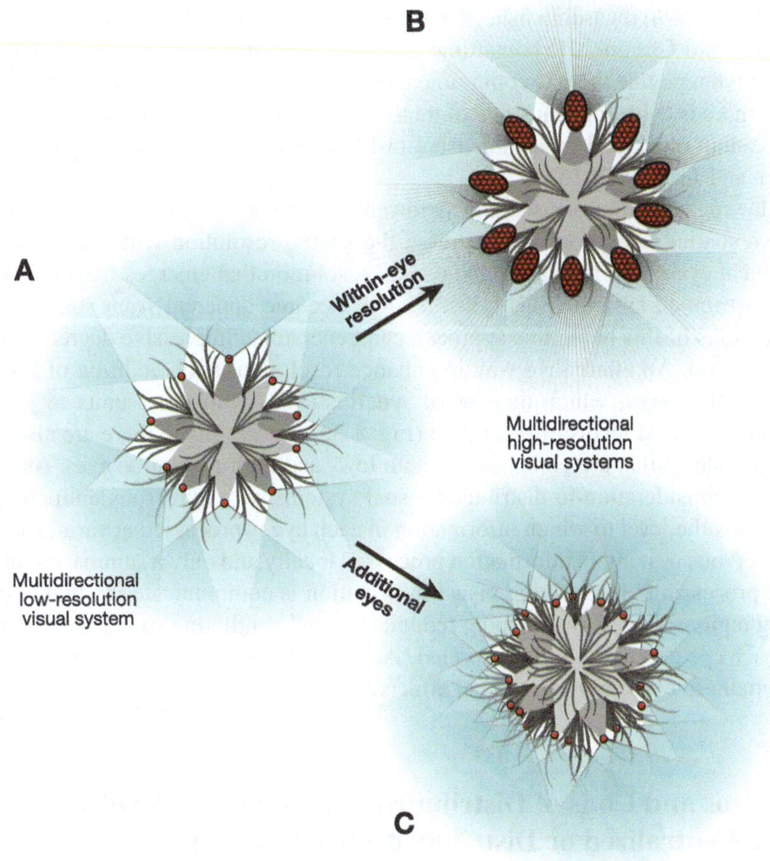

Fig. 1.3 In principle, a distributed visual system can gain resolution in two different ways. (**a**) Distributed system with relatively low resolution. (**b**) One way to improve the visual system is to increase the within-eye resolution so that the resulting total visual resolution is contributed by a combination of within unit and in-between unit inputs. (**c**) The second way is to increase the number of sampling units

their body occludes certain directions, and hence for comprehensive sampling, it becomes necessary for the organism to distribute sensors strategically to sample different directions. In the simplest case, the number of points that can be depicted would be equal to the number of individual Class II directional light sensing structures. Hence, even the simpler distributed systems functionally are more equivalent to Class III visual systems as defined by Nilsson (2013). At least this is the case if the necessary neural substrate exists to integrate information between those units or at a minimum to allow for comparisons between individual sample points. Simple distributed light sensing systems are relatively common, and even organisms with relatively sophisticated visual systems have been found to have additional extraocular photoreceptors in various regions of the body (for examples, see (Kasai and Oshima 2006) regarding fish, (Kingston et al. 2015) regarding cephalopods, and (Kingston and Cronin 2015) regarding crayfish). In many cases, it remains elusive if and to what degree there is communication between such light sensing structures. In fact, there is evidence that some of them are important for localized regulation of the circadian clock and other nonvisual tasks, rather than a contribution to vision (Cronin and Johnsen 2016).

Following an evolutionary pathway toward visual sophistication, another way to improve spatial resolution is to increase the spatial resolution within each of the component eyes, thus increasing the overall resolution that emerges from the combined distributed system (Fig. 1.3b). As will become apparent from some of the contributions of this book, this approach can generate an impressive degree of spatial resolution. An alternative way to enhance resolution is the addition of further low-resolution eyes, which then work synergistically with other units to jointly facilitate increased levels of resolution (Fig. 1.3c). Interestingly, there are also distributed systems that are composed of both low- and high-resolution eyes. Another important consideration in distributed visual systems with numerous multicomponent eyes is the level to which information in each eye is processed at various levels of the nervous system. Is information processed locally and only a summary sent for higher processing? If resolving visual information is communicated to higher processing regions, how is overlapping redundancy and conflicting spatial information from each eye processed and interpreted? Are there higher processing levels or does vision in these eyes only cause local effects?

1.4 Pros and Cons of Distributed Vision (Or "To Evolve a Centralized or Distributed Visual System")

Conceptually, there are pros and cons of distributing visual sampling throughout the organism. One of the advantages includes a high level of directional flexibility. For any multicellular organism, unless it were completely transparent, the only way to truly see into all directions is to break up from where vision is being sampled and to distribute sensors over the surface or at least to some strategic locations. The field of view of each sensor may then be adjusted to fit the spacing so that seamless surround vision can be achieved, and considerable computation and integration is necessary to combine the information. Alternatively, it may be acceptable to have some gaps in a surrounding visual field, especially if combined with eye, head, or whole-body movement. However, if such computational resources are in place, an animal that samples equally well from all directions can then monitor its surroundings without moving. This would be beneficial for predation, allowing ambush of the prey without having to give away position through eye, head, or body readjustment prior to striking. This strategy also can help to facilitate rapid escape responses, for example, if an organism can detect a threat equally well from all directions. Thus, many sessile animals utilize distributed eyes to govern alarm responses that see them withdrawing into a protective tube or shell (Chap. 5). For motile animals, detecting a predator is just the first step in a successful escape, and if vision is omni-directional, it also would be beneficial to have the necessary motor control to directly move away from the detected danger without first having to turn, a strategy that indeed is implemented in starfish (see Chap. 4).

Another advantage of such distributed systems is that there is some level of redundancy in these relatively delicate systems. Depending on the level of overlap, losing one of the units may lead to relatively minor deficiencies, as appears to be the case in chitons (Chap. 6), in which the eyes constantly decay, while new ones are formed and added (Sigwart and Sumner-Rooney 2021). For these strategies to work, the kinds of samples that are taken from different directions need to be similar in nature.

A further strategy for organisms with distributed systems is to use sensors that are placed at different locations to specialize for different visual functions. For example, some of the sensors may be specialized for orientation, while others are tuned to detect shadows, moving objects, specific colors, or even polarization, as observed among the most sophisticated distributed systems. As beautifully summarized in Chap. 10 on spiders and Chap. 2 on cnidarians, in such sophisticated visual systems, the field of view of individual eyes varies greatly according to the task at hand.

Given such advantages, why have distributed visual systems not evolved in more organisms? In reality, few organisms exist that do not have at least some level of distributed visual systems. In fact, having two eyes as most animals do (Fig. 1.4a) could be considered to constitute a distributed system. A rare example of an exceptionally centralized system is the minute larvae of ascidians, which have a single ocellus. If it is ablated, then these larvae lose the ability for phototaxis (Tsuda et al. 2003). Closer examination of the tunicate *Ciona* reveals that their visual repertoire includes the ability to escape from looming responses in addition to phototaxis and that these behaviors are mediated by distinct photoreceptors (Salas et al. 2018). These data suggest that remarkable complexity already exists in this very simple cyclopic eye that also has served as inspiration for biorobotics (Long et al. 2004).

There also are distinct benefits to having a centralized visual system. To achieve sophisticated eyes that are fully distributed, there needs to be a relatively sophisticated neural substrate that allows for the integration of the many different sources of visual information into one coherent system. Such computation might be easier where neural substrate accumulates, because there are shorter distances between inputs and processing units. Furthermore, vision often needs to be a high-speed sensory modality, so a short distance between photoreceptors and their processing substrate minimizes latency in behavioral responses. It has been argued that centralization also could have been driven by an iterative process (Martinez and Sprecher 2020). For example, if receptors were present in certain areas of an animal, then the processing of acquired information could have resulted in the addition of necessary neural substrate, the presence of which then could have favored the accumulation of additional receptors in that area.

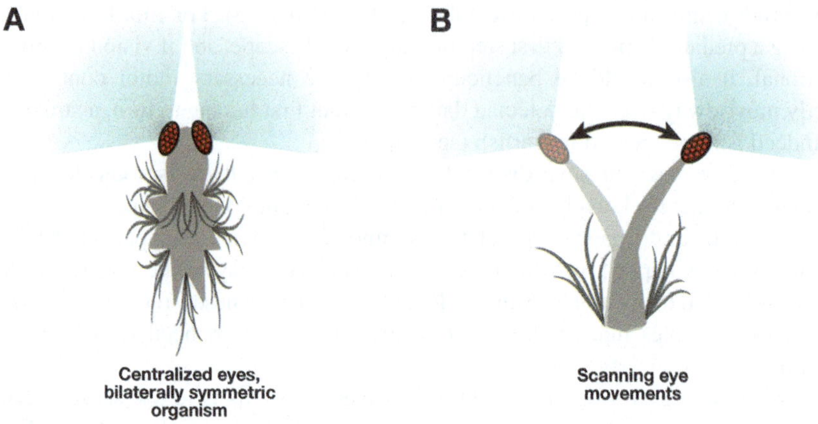

Fig. 1.4 (**a**) Cephalization and the presence of two relatively well-developed eyes are a common strategy among bilaterally symmetric organisms. (**b**) Another way to expand the visual field or obtain directional visual information is to perform scanning, saccadic, or other eye movements

Evidence for the benefit of centralization of neural computation comes from the evolution of centralization in metazoan brains, which is thought to have emerged independently at least four times, originating from distributed systems (Northcutt 2012). Here the transition involved the formation of ganglia, which are aggregations of neural cell bodies that already bring together neural substrates. There are likely economic reasons that favor the evolution of at least some level of aggregation for the process of vision. Well-ordered centralized eyes can cover a visual area at a high resolution with a minimum of photoreceptor and optical resources. Keeping input receptors together furthermore enables retinotopic mapping and processing in centralized visual systems which is less costly and requires less processing to compensate for ambiguities or overlapping fields of view.

Many organisms do not confine themselves to either a centralized or a distributed system, but their visual system contains components of both, or may transition between dominant centralized and distributed systems at different life stages. In part, this relates to light detecting systems that were likely very common relatively early during evolution, possibly even long before they served the function of vision, perhaps to help avoid light-induced stress (Swafford and Oakley 2019). The diversity of ways in which light is being sampled is also apparent from the presence of many different types of photopigment (such as the c- or r-type opsins) and related components (such as g-proteins) that are found in animals (Nilsson 2004; Porter et al. 2012). In addition to eyelike structures, opsins also may be found on the skin or in deeper parts of the nervous system (Porter et al. 2012; Ramirez et al. 2011) as described for echinoderms (Chap. 3). Among the simplest versions of eyelike structures is the combination of a photoreceptor cell and a pigment cell (Arendt 2003). Evidence exists for multiple independent origins of distributed visual systems, presumably from such substrates. As a corollary, there are interesting points of convergence in the types of cells used and the structures of the eyes in some cases. This suggests that there are some optimal configurations that evolution tends to move toward with its available toolkit, in each respective phylum. In fact, as we hope to capture in this volume, distributed visual systems are quite common in a diverse group of different animals.

Comparing the many different systems, it becomes clear that the distribution of visual sensors often relates to the body symmetry – the radial symmetry of echinoderms and cnidarians, for example, facilitates sampling of many different directions. However, the bilateral symmetry of many other organisms includes the evolution of a body axis that is well-defined by a series of homeobox genes (McGinnis and Krumlauf 1992). This leads to organisms with distinct heads and tails and visual systems that are relatively symmetrically distributed on the two sides of the body. Another important pattern that emerges relates to the animal's ability to locomote. Many of the sessile creatures, in the absence of the ability to adjust their body position, sample approximately equally from the different directions (see Chap. 5 and Bok et al. 2016). When locomotion is directed primarily in one direction, having the visual system primarily on the head end allows the animal to detect what lies ahead and plan appropriate action during locomotion (Fig. 1.4a). Regardless of the level to which a visual system is distributed, additional spatial resolution and/or an expanded visual field can also be gained through eye movements (Land (2019) and Fig. 1.4b).

1.5 Survey of Diverse Distributed Visual Systems

In this volume, we seek to gather information on a wide variety of unusual visual systems that, to some extent, emphasize the properties of distributed visual systems described above (Fig. 1.5). In organizing the volume, we endeavored to include examples from most major animal phyla with an emphasis on some of the most dramatic instances of distributed and alternative visual systems. We provide a brief introduction to the various creatures and visual systems discussed in this volume. Generally, these visual systems consist of three or more eyes, sometimes positioned on a head but in many cases more widely dispersed over the body, especially in organisms that lack a head. These visual systems arose separately across a rich tapestry of independent evolutionary trajectories, culminating in unique optical innovations and neuronal processing strategies that influence visually guided behaviors.

1.5.1 Cnidarians

Cnidarians are one of the earliest branching phyla of animals and are of great interest in exploring the early stages in the evolution of vision. A number of unique opsins and photoreceptors have been identified in the group, and the box jellies

Fig. 1.5 Distributed and multi-eyed visual systems in animals discussed in this volume. Arrowheads indicate the position of the eyes. In panels **a**, **b**, **d**, **e**, and **g**, magnified views of the eyes are inset. (**a**) Box jellyfish have four rhopalia, each containing multiple simple lens eyes and pigment cups, set around the circumference of their bell. (**b**) Starfish, such as *Marthasterias glacialis*, have compound eyes at the tips of their arms. (**c**) Brittle stars, like *Macrophiothrix nereidina*, have dispersed photoreceptors positioned along the plates that cover their arms. (**d, e**) Sabellid fan worms, *Acromegalomma vesiculosum* (**d**) and *Bispira* sp. (**e**), have compound eyes on their radioles. While *Bispira* has dozens of small eyes along each radiole, *Acromegalomma* has a pair of

(continued)

(Cubozoa) in particular possess a surprising diversity of strange but sophisticated eyes (Fig. 1.5a). In Chap. 2, by Sydney Birch, Natasha Picciani, Todd Oakley, and David Plachetzki, the diversity of eyes and photoreceptive systems in the cnidarians is reviewed, and their place in our understanding of eye evolution is considered. Their radially symmetric body plans and decentralized neural processing naturally lend themselves to developing a distributed visual system. However, even within this context, their eyes are quite unusual, some including multiple lens and pigment cup eyes clustered together into the same rhopalium structures (Nilsson et al. 2005). Cnidarians have a polyp stage that is most likely derived from the ancestral lifestyle of these creatures. As is the case for many sessile animals, distributed visual systems serve them well for surveying the environment for threats and initiating a startle response. Interestingly, however, the function of these eyes in cubomedusa appears to include higher-level orientation and navigation tasks, such as maneuvering around obstacles and for maintaining position in mangrove habitats through Snell's window. Here, we see a parallel to another group of radially symmetrical organisms, the echinoderms (described in the next section), in that both groups of animals can use their distributed photoreceptor networks for relatively complex navigation through their habitats. This contrasts with many distributed visual systems that have simpler roles related to alarm responses or posture control. Perhaps the lack of a cranial centralization of neural resources in these groups has led to the development of distributed eyes specifically for these more complicated tasks, rather than evolving a centralized visual system.

1.5.2 Echinoderms (Deuterostomes)

Besides in echinoderms, elaborate distributed visual systems are rare in deuterostomes, perhaps owing to the large size and active lifestyles of many craniates, which have developed sophisticated paired cephalic eyes to drive high-resolution visual tasks like predation and communication. Exceptions to this are the medial pineal and parietal photoreceptive organs (Dodt 1973; Eakin and Westfall 1960). These are usually simple luminance detectors buried in the chordate brain but can

Fig. 1.5 (continued) enlarged eyes on the tips of the dorsal-most radioles in addition to smaller eyes on the tips of several lateral radioles (small arrowheads). (**f**) Chitons, such as *Acanthopleura granulata*, have clusters of ocelli along their shell plates. (**g**, **h**) Bivalves have eyes distributed along their mantles. While scallops have simple eyes with mirror optics (**g**, *Argopecten irradians*), arc clams have compound eyes (**h**, *Barbatia cancellaria*). (**i**) Myriapods, like the centipede *Scolopendra heros*, have clusters of lateral ommatidia with various degrees of complexity. (**j**, **k**) Crustaceans and insects have diverse tripartite eyes. In copepods like *Calanus finmarchicus*, these are the primary visual system (**j**), while in insects like the damselfly *Argia apicalis* (**k**), the dorsal ocelli perform secondary visual tasks alongside the compound eyes. (**l**) Spiders have an array of eight eyes with various specializations. This is perhaps most dramatically demonstrated in jumping spiders (Salticidae). (Photo credits: **a** Jan Bielecki, Dan-Eric Nilsson (inset); **b** Camilla Elinor Kosvig-Nielsen; **c**–**e** Michael Bok; **f** Alexandra Nahm Kingston; **g** Sönke Johnsen, Dan-Eric Nilsson (inset); **h** Dan-Eric Nilsson; **i** Ted C. MacRae; **j**–**l** Michael Bok) be more elaborate in some taxa. Especially notable is the well-developed parietal eye of some lizards. However, likely due to their radial symmetry, echinoderms are a particularly interesting group of animals in the exploration of distributed systems. Accordingly, there are two chapters that are devoted to this peculiar group of animals in this book. Chapter 3, authored by Lauren Sumner-Rooney and Ullrich Lüter, provides a good overview of the relatively limited information that thus far exists on extraocular vision of some of its members, most notably sea urchins and brittle stars (Fig. 1.5c). These are characterized by a multitude of different visual organs and even have intrinsically photoreceptive nerves. Visual organs possess r-opsin and c-opsin containing visual structures, as demonstrated in a series of immunohistochemical studies that are summarized in the chapter. In some cases, known signal transduction genes are present, and pax6, a deeply conserved gene that in many species can act as a switch for eye development, has also been found in several echinoderms. In both sea urchins and brittle stars, c-opsins and r-opsins are found in part in relatively close proximity, and in part at different locations, which raises interesting questions on the synergy of these different types of visual senses.

A detailed analysis of the dispersed visual system of starfish is provided in Chap. 4 by Anders Garm, Ditte Sunberg, and Camilla Elinor Kosvig-Nielsen. While starfish vary in regard to how many arms they process, they are commonly characterized by small compound eyes on their terminal tube feet, some of which are capable of image formation, albeit at relatively low spatial frequencies, and enable low-pass filtering (Fig. 1.5b). These visual structures presumably function synergistically with extraocular structures, which are less well understood. One of the most peculiar features of starfish, and perhaps echinoderms more generally, is that the different regions appear to have relatively even contribution to gaining visual input and to using that input in their guidance of behavior. For example, as is explored further in Chap. 4, it has been observed that as starfish move, they can do so by using different arms to lead into specific directions (Pearse et al. 1987). Components of the visual system are connected to a nerve ring, which appears to play an important role in the integration of the animal's visual input and behavioral responses. If the nerve ring is bisected, the two halves appear to attempt to take independent paths. These are fascinating findings that also highlight the power of this relatively simple invertebrate system to inform our queries on how distributed information can contribute to a relatively complex decision-making process.

1.5.3 Polychaetes (Annelida)

The annelids have a number of distributed and many-eyed visual systems. These include the lateral cerebral eyes on the head, which often number four in total in errant species like *Platynereis dumerilii*, while others have only two. Beyond the cerebral eyes, many polychaetes also have simple segmental ocelli along the length of the body that may function in exposure avoidance (Backfisch et al. 2013), or pygidial ocelli on the terminal segment, with murkier functionality (Ermak and

Eakin 1976). However, perhaps the most fascinating distributed eyes are found on the feeding tentacles, called radioles, of the sedentary families, Serpulidae and Sabellidae reviewed in Bok et al. (2016, 2017). Commonly referred to as fan worms, these animals project a crown of radioles from their mouths up out of their protective tubes into the water column. Since this is the only part of the animal with a view to the outside world, fan worms have evolved a number of strange ocelli on the radioles that are used in order to govern a startle response that initiates a rapid withdrawal of the crown into their tube. The ocelli are formed by a ciliary photoreceptor cell and various complements of lens and pigment cup cells, depending on the genus, and can be scattered broadly across the lengths of the radioles or consolidated into dozens of compound eyes (Fig. 1.5d, e). Interestingly, these eyes demonstrate fine gradations in the structural evolution of compound eyes in extant species, making them a great target to explore the evolution of complexity in visual systems and their neural processing.

Perhaps uniquely among distributed visual systems, certain genera of fan worms (*Acromegalomma* and *Spirobranchus*) seem to have secondarily evolved consolidated visual systems out of their distributed radiolar eyes. In both cases, there is a single pair of greatly enlarged compound eyes, with over a thousand facets each, positioned prominently on the dorsal-most pair of radioles. It could be that these eyes simply represent a more economical consolidation of visual resources, with less redundant overlap, or they could confer additional visual capabilities due to their increased organization and higher spatial resolution potential.

1.5.4 Bivalvia (Mollusks)

Like the fan worms, many bivalves use distributed eyes as a burglar alarm to detect threats and initiate an alarm response. In Chap. 5, Daniel Speiser, Daniel Chappell, Jorge Audino, Alexandra Kingston, and Jeanne Serb present a thorough review of these eyes in bivalves, with a particular focus on pteriomorphs, including the spectacularly odd mirror eyes of scallops (Fig. 1.5g). Like fan worms, the eyes of bivalves are quite diverse in complexity, arrangement, and positioning. However, they exceed the fan worms in optical diversity, photoreceptor cell type, and neural integration. In this group, we see both compound (Fig. 1.5h) and simple eyes (Fig. 1.5g), including the aforementioned mirror optics of scallops. Moreover, the photoreceptors in the eyes are ciliary in some cases and rhabdomeric in others, and sometimes both photoreceptor origins are found in separate retinas within the same eyes! This lends the bivalves to explorations of parallel photoreceptor specializations in the same eyes, perhaps akin to vertebrate retinas where there are visual ciliary photoreceptors parallel to rhabdomeric-derived, melanopsin-expressing, photosensitive ganglion cells. Furthermore, in regard to function, some species may use their eyes for more than detecting threats and initiating a startle response. There is some evidence that scallops can detect particle density in the water and possibly even navigate to better habitats while swimming.

1.5.5 Chitons (Mollusks)

A diverse set of distributed visual systems, from simple eye spots to image-forming organs, are found in chitons as reviewed in Chap. 6 by Daniel Chappell, Daniel Speiser, Doug Eernisse, and Alexandra Kingston. With a bilaterally symmetric body plan, this group of animals has a well-defined front end, with a cerebral nervous system. These slow-moving mollusks stand out for their variety of different types of distributed systems, which range from a system of simple aesthetes with a cluster of photoreceptors and pigment cells to shell eyes, visual structures with lenses that are made of aragonitic material (Fig. 1.5f). Over time, these lenses deteriorate, but new lenses are formed at the base of the shell plates, albeit in a less regular fashion. This unorthodox organization poses a challenge of ever-changing contributors to this distributed visual system of chitons. As discussed in the chapter, some evidence points to their ability to detect shadows cast by a potential predator, even if they are unable to locate objects in relation to their own position. Such observations stress the need for neural integration. However, how exactly do these very different visual system components contribute to the overall level of integration? One possible answer may lie in their peculiar nervous system organization, where most of the neurons are part of a medullary center rather than the more typical ganglia that are found in other organisms.

1.5.6 Myriapoda (Arthropoda)

A group with particularly diverse eye organizations, and many examples of somewhat different distributed systems, are the Arthropoda. They are generally divided into the Chelicerata and Mandibulata (for a review, see Giribet and Edgecombe 2019), the latter including the insects and crustaceans (also referred to as Pancrustacea). At the base of the Mandibulata are the Myriapoda, which makes them an important group to understand regarding the ancestral state of the Pancrustacea, which have become particularly specious within the terrestrial (insects) and marine environments (crustacea). Myriapods themselves represent a remarkably diverse group with four major subgroups, one potential relationship of which has been suggested through molecular studies (Miyazawa et al. 2014) but remains subject to debate (Szucsich et al. 2020). While most myriapods have a pair of lateral eyes, the diversity in organization mirrors the diversity of the group as a whole, with substantial differences in their sophistication (Fig. 1.5i).

In Chap. 7 by Andy Sombke and Carsten Müller dives deeply into the structural organization of the different eye types that are found in myriapods. As is the case for insects, eyes can typically be found bilaterally, on the side of the head, situated medially. Structurally some of the myriapod eyes are reminiscent of insect ommatidia, with remarkable conservation regarding the number and position of some of the photoreceptor and support cells, as well as the presence of a crystallin cone and cuticular lens. In contrast, some chilopods have much more sophisticated eye units with over 1000 photoreceptor (retinula) cells. Particularly notable is a layering organization of photoreceptors, which in some of the more sophisticated eye units become elaborate. Millipede eyes also have interesting modes of eye growth, with additional eye units being added from a peripheral proliferation zone during molts, a pattern that is reminiscent of what has been found in some groups of insects (Buschbeck and Friedrich 2008).

1.5.7 Pancrustacea (Arthropoda)

The most dominant type of eyes for both insects and crustaceans are the laterally located compound eyes, which may contain up to thousands of small ommatidia, as exemplified in dragonflies (Sherk 1978). Much variation exists in the way they are organized (Cronin et al. 2014; Land and Nilsson 2012; Meece et al. 2021), and some represent a distributed system within an eye through dramatic regional specialization. Good examples of this are the many differently functioning regions of ommatidia in the eyes of mantis shrimp (Marshall et al. 2007; Cronin et al. 2022), the polarization sensitive dorsal rim area of insects (Labhart and Meyer 1999), or the female-spotting sex zones in several species of fly (Land 1997; Land and Eckert 1985). In addition, a variety of different types of lateral eyes, typically referred to as stemmata, exist in the larval form or holometabolous insects (Gilbert 1994). These are particularly interesting as they have evolved from a compound eye ancestor and can manifest as anything from a compound eyelike organization to distributed ommatidial-like structures or sets of highly complex image-forming camera eyes (Buschbeck 2014). For example, in Lepidoptera and Trichoptera, stemmata typically are relatively simple (Paulus 1979; Paulus and Schmidt 2009), following the cellular organization of insect ommatidia. However, in contrast to ommatidia that are part of a compound eye, these visual units have drifted apart from each other, each sampling a different area, and hence they have evolved into a distributed system of stemmata rather than manifesting one cohesive eye. In other groups, such as Coleoptera, some insects have evolved stemmata that are greatly enlarged and comprise extended retinas, as exemplified by tiger beetle larvae (Toh and Okamura 2007) and diving beetle larvae (Mandapaka et al. 2006). The latter include stemmata that are particularly well suited for underwater prey capture, with complex optics and multiple retinas (Stowasser and Buschbeck 2014).

Yet another type of eye found among Pancrustacea is medial, tripartite eye structures that are thought to be homologous to the anterior-median eyes of spiders (Paulus 1979; Friedrich 2006; Morehouse et al. 2017). These are the naupliar eyes in crustaceans (Fig. 1.5j) and dorsal ocelli in insects (Fig. 1.5k). In crustaceans, most larval forms have a naupliar eye (also called frontal eyes) located adjacent to the brain, which persists into adulthood in many cases (Elofsson 2006; Cronin et al. 2017), while the dorsal ocelli in insects are mainly associated with adult forms. The dorsal ocelli and naupliar eyes are both composed of three pigment cup photoreceptors oriented in a manner to potentially aid in stabilization or orientation in flight or in open water, respectively. In their most simplistic forms, these visual systems may not mediate any visual behavior more complex than this role as an optical statocyst. However, in both cases, there are examples where these eyes are expanded significantly in size and complexity, suggesting more advanced functional roles, and, in the case of the naupliar eyes in copepods, even function as the primary visual system. The dorsal ocelli of insects are discussed in detail here in Chap. 8 by Baird and Yilmaz, while the extraordinary elaboration of the naupliar eyes of copepods is discussed in Chap. 9 by Mireille Steck, Kristina Theam, and Megan Porter.

1.5.8 Arachnida (Arthropoda)

Some of the most sophisticated, high-functioning distributed systems are found among the Chelicerata. Ancient visual systems, as exhibited by horseshoe crabs (*Limulus*), have laterally situated compound eyes and relatively simple median eyes. This method of distributing visual organs is reminiscent of the Pancrustacea, despite major structural differences (Barlow 2009; Battelle 2006; Fahrenbach 1968). Divergence from this organization, however, is apparent in many of the other chelicerate groups, notably Arachnida. For example, in scorpions, the lateral eyes are composed of groups of single-chamber lens eyes that vary in number and position (Miether and Dunlop 2016). Arguably the most sophisticated distributed systems here are the eyes of spiders (Fig. 1.5l). Following the amazing diversity of spiders themselves, there is considerable diversity as well in the layout and function of specific eyes, as synthesized neatly in Chap. 10 by Alex Windsor, Nathan Morehouse, and Elizabeth Jakob. These findings are particularly interesting as most of the spider eyes have evolved from compound eyes whereas the anterior-median eyes share their developmental origin with those of insect ocelli (Morehouse et al. 2017).

In many cases, the details as to how the visual system functions relate to the many different hunting strategies that are being employed, but the general layout most likely follows a relatively conserved developmental plan (Morehouse et al. 2017). Particularly elaborate are the visual systems of salticids, jumping spiders, which have high-resolution anterior-median eyes that scan their prey with variously sophisticated color vision. As reviewed in the chapter, these spiders exhibit an elaborate division of labor among their different eyes. Specifically, it has been established with the help of a sophisticated eye tracker that their anterior-lateral eyes, which have a relatively large visual field, precisely direct the boomerang-shaped retina of the high-resolution anterior-median eyes (Land 1969, 1985), onto particularly important parts of the visual field. If a female spider faces a male, this includes specific aspects of his beautifully iridescently colored body parts, as he performs an elaborate courtship dance in front of her.

1.6 Summary/Conclusions

As we have begun to summarize in this chapter, examples of distributed and many-eyed visual systems are extremely diverse in nature. Distributing the acquisition of information to different organs has various advantages, and a diverse set of impressive solutions have emerged in the animal kingdom in response to a variety of environmental, developmental, and ecological pressures faced by the animals that use them. Upcoming chapters in this volume provide up-to-date reviews on some of the most prominent examples of such distributed systems. They differ from consolidated visual systems in significant ways and offer unique benefits and challenges.

Distributed visual systems may be useful in the rapidly growing fields of bioinspired distributed sensing and processing applications. For example, what can nature teach us about how resilient distributed systems deal with redundancy, overlap, and conflicting information in order to generate an accurate picture of the environment? How is that picture or internalized view of the outside world used to drive the behavior of an animal, or, for our uses, the environmental monitoring system, or perhaps a swarm of autonomous robots?

Despite the wealth of fascinating research detailed in this volume, distributed visual systems historically have been relatively poorly understood, and we hope that this book will illustrate to our readers that there is great potential in continuing to unravel their secrets.

References

Arendt D (2003) Evolution of eyes and photoreceptor cell types. Int J Dev Biol 47:563–571

Backfisch B, Veedin Rajan VB, Fischer RM et al (2013) Stable transgenesis in the marine annelid Platynereis dumerilii sheds new light on photoreceptor evolution. Proc Natl Acad Sci U S A 110:193–198

Barlow RB (2009) Vision in horseshoe crabs. In: Tanacredi JT, Botton ML, Smith D (eds) Biology and conservation of horseshoe crabs. Springer, Boston, pp 223–235

Battelle B-A (2006) The eyes of Limulus polyphemus (Xiphosura, Chelicerata) and their afferent and efferent projections. Arthropod Struct Dev 35:261–274

Bok MJ, Capa M, Nilsson D-E (2016) Here, there and everywhere: the radiolar eyes of fan worms (Annelida, Sabellidae). Integr Comp Biol 56:784–795

Bok MJ, Porter ML, Ten Hove HA et al (2017) Radiolar eyes of serpulid worms (Annelida, Serpulidae): structures, function, and phototransduction. Biol Bull 233:39–57

Buschbeck EK (2014) Escaping compound eye ancestry: the evolution of single-chamber eyes in holometabolous larvae. J Exp Biol 217:2818–2824

Buschbeck EK, Friedrich M (2008) Evolution of insect eyes: tales of ancient heritage, deconstruction, reconstruction, remodeling, and recycling. Evol: Educ Outreach 1:448–462

Cronin TW, Johnsen S (2016) Extraocular, non-visual, and simple photoreceptors: an introduction to the symposium. Integr Comp Biol 56:758–763

Cronin TW, Johnsen S, Justin Marshall N, Warrant EJ (2014) Visual ecology

Cronin TW, Bok MJ, Lin C (2017) Crustacean larvae – vision in the plankton. Integr Comp Biol 57:1139–1150

Cronin TW, Porter ML, Bok MJ, Caldwell R, Marshall J (2022) Color vision in stomatopod crustaceans. Phil Trans R Soc. In press

Dodt E (1973) The parietal eye (pineal and parietal organs) of lower vertebrates. In: Berlucchi G, Brindley GS, Brooks B et al (eds) Visual centers in the brain. Springer Berlin Heidelberg, Berlin/Heidelberg, pp 113–140

Eakin RM, Westfall JA (1960) Further observations on the fine structure of the parietal eye of lizards. J Biophys Biochem Cytol 8:483–499

Elofsson R (2006) The frontal eyes of crustaceans. Arthropod Struct Dev 35:275–291

Ermak TH, Eakin RM (1976) Fine structure of the cerebral and pygidial ocelli in Chone ecaudata (Polychaeta: Sabellidae). J Ultrastruct Res 54:243–260

Fahrenbach WH (1968) The morphology of the eyes of limulus. Z Zellforsch Mikrosk Anat 93:451–483

Friedrich M (2006) Ancient mechanisms of visual sense organ development based on comparison of the gene networks controlling larval eye, ocellus, and compound eye specification in Drosophila. Arthropod Struct Dev 35:357–378

Gilbert C (1994) Form and function of stemmata in larvae of holometabolous insects. Annu Rev Entomol 39:323–349

Giribet G, Edgecombe GD (2019) The phylogeny and evolutionary history of arthropods. Curr Biol 29:R592–R602

Horváth G (2014) Polarized light and polarization vision in animal sciences. Springer. https://doi.org/10.1007/978-3-642-54718-8

Jékely G, Colombelli J, Hausen H et al (2008) Mechanism of phototaxis in marine zooplankton. Nature 456:395–399

Kasai A, Oshima N (2006) Light-sensitive motile iridophores and visual pigments in the neon tetra, Paracheirodon innesi. Zool Sci 23:815–819

Keum D, Jang K-W, Jeon DS et al (2018) Xenos peckii vision inspires an ultrathin digital camera. Light Sci Appl 7:80

Kingston ACN, Cronin TW (2015) Short- and long-wavelength-sensitive opsins are involved in photoreception both in the retina and throughout the central nervous system of crayfish. J Comp Physiol A Neuroethol Sens Neural Behav Physiol 201:1137–1145

Kingston ACN, Kuzirian AM, Hanlon RT, Cronin TW (2015) Visual phototransduction components in cephalopod chromatophores suggest dermal photoreception. J Exp Biol 218:1596–1602

Labhart T, Meyer EP (1999) Detectors for polarized skylight in insects: a survey of ommatidial specializations in the dorsal rim area of the compound eye. Microsc Res Tech 47:368–379

Land MF (1969) Structure of the retinae of the principal eyes of jumping spiders (Salticidae: dendryphantinae) in relation to visual optics. J Exp Biol 51:443–470

Land MF (1981) Optics and vision in invertebrates. In: Handbook of sensory physiology, vol VII/6B, pp 471–592

Land MF (1985) The morphology and optics of spider eyes. In: Barth FG (ed) Neurobiology of arachnids. Springer Berlin Heidelberg, Berlin/Heidelberg, pp 53–78

Land MF (1997) Visual acuity in insects. Annu Rev Entomol 42:147–177

Land M (2019) Eye movements in man and other animals. Vis Res 162:1–7

Land MF, Eckert H (1985) Maps of the acute zones of fly eyes. J Comp Physiol A 156:525–538

Land MF, Nilsson D-E (2012) Animal eyes

Long JH, Lammert AC, Pell CA et al (2004) A navigational primitive: biorobotic implementation of cycloptic helical klinotaxis in planar motion. IEEE J Ocean Eng 29:795–806

Mandapaka K, Morgan RC, Buschbeck EK (2006) Twenty-eight retinas but only twelve eyes: an anatomical analysis of the larval visual system of the diving beetle Thermonectus marmoratus (Coleoptera: Dytiscidae). J Comp Neurol 497:166–181

Marshall J, Cronin TW (2011) Polarisation vision. Curr Biol 21:R101–R105

Marshall J, Cronin TW, Kleinlogel S (2007) Stomatopod eye structure and function: a review. Arthropod Struct Dev 36:420–448

Martinez P, Sprecher SG (2020) Of circuits and brains: the origin and diversification of neural architectures. Front Ecol Evol 8:82

McGinnis W, Krumlauf R (1992) Homeobox genes and axial patterning. Cell 68:283–302

Meece M, Rathore S, Buschbeck EK (2021) Stark trade-offs and elegant solutions in arthropod visual systems. J Exp Biol 224. https://doi.org/10.1242/jeb.215541
Miether ST, Dunlop JA (2016) Lateral eye evolution in the arachnids. Arachnology 17:103–119
Miyazawa H, Ueda C, Yahata K, Su Z-H (2014) Molecular phylogeny of Myriapoda provides insights into evolutionary patterns of the mode in post-embryonic development. Sci Rep 4:4127
Morehouse NI, Buschbeck EK, Zurek DB et al (2017) Molecular evolution of spider vision: new opportunities, familiar players. Biol Bull 233:21–38
Nilsson D-E (2004) Eye evolution: a question of genetic promiscuity. Curr Opin Neurobiol 14:407–414
Nilsson D-E (2009) The evolution of eyes and visually guided behaviour. Philos Trans R Soc Lond Ser B Biol Sci 364:2833–2847
Nilsson D-E (2013) Eye evolution and its functional basis. Vis Neurosci 30:5–20
Nilsson D-E (2021) The diversity of eyes and vision. Annu Rev Vis Sci 7:19–21
Nilsson D-E, Bok MJ (2017) Low-resolution vision – at the hub of eye evolution. Integr Comp Biol 57:1066–1070
Nilsson D-E, Gislén L, Coates MM et al (2005) Advanced optics in a jellyfish eye. Nature 435:201–205
Nityananda V, Tarawneh G, Rosner R et al (2016) Insect stereopsis demonstrated using a 3D insect cinema. Sci Rep 6:18718
Northcutt RG (2012) Evolution of centralized nervous systems: two schools of evolutionary thought. Proc Natl Acad Sci U S A 109(Suppl 1):10626–10633
Paulus HF (1979) Eye structure and the monophyly of the Arthropoda. In: Gupta AP (ed) Arthropod phylogeny. Van Nostrand Reinhold, New York, pp 299–383
Paulus VHF, Schmidt M (2009) Evolutionswege zum Larvalauge der Insekten: Die Stemmata der Trichoptera und Lepidoptera. J Zool Syst Evol Res 16:188–216
Pearse V, Pearse J, Buchsbaum M, Buchsbaum R (1987) Living invertebrates
Porter ML, Blasic JR, Bok MJ et al (2012) Shedding new light on opsin evolution. Proc Biol Sci 279:3–14
Ramirez MD, Speiser DI, Pankey MS, Oakley TH (2011) Understanding the dermal light sense in the context of integrative photoreceptor cell biology. Vis Neurosci 28:265–279
Salas P, Vinaithirthan V, Newman-Smith E et al (2018) Photoreceptor specialization and the visuomotor repertoire of the primitive chordate Ciona. J Exp Biol 221(7):jeb177972
Sanders JS (1997) Ultrawide-field-of-view imaging optical system designs based on arthropod vision systems. In: Acquisition, tracking, and pointing XI
Sherk TE (1978) Development of the compound eyes of dragonflies (Odonata). III. Adult compound eyes. J Exp Zool 203:61–80
Sigwart JD, Sumner-Rooney L (2021) Continuous and regular expansion of a distributed visual system in the eyed chiton *Tonicia lebruni*. Biol Bull 240:23–33
Stowasser A, Buschbeck EK (2014) How aquatic water-beetle larvae with small chambered eyes overcome challenges of hunting under water. J Comp Physiol A Neuroethol Sens Neural Behav Physiol 200:911–922
Swafford AJM, Oakley TH (2019) Light-induced stress as a primary evolutionary driver of eye origins. Integr Comp Biol 59:739–750
Szucsich NU, Bartel D, Blanke A et al (2020) Four myriapod relatives – but who are sisters? No end to debates on relationships among the four major myriapod subgroups. BMC Evol Biol 20:144
Toh Y, Okamura J-Y (2007) Morphological and optical properties of the corneal lens and retinal structure in the posterior large stemma of the tiger beetle larva. Vis Res 47:1756–1768
Tsuda M, Sakurai D, Goda M (2003) Direct evidence for the role of pigment cells in the brain of ascidian larvae by laser ablation. J Exp Biol 206:1409–1417
van der Kooi CJ, Stavenga DG, Arikawa K et al (2021) Evolution of insect color vision: from spectral sensitivity to visual ecology. Annu Rev Entomol 66:435–461
Warrant EJ (2017) The remarkable visual capacities of nocturnal insects: vision at the limits with small eyes and tiny brains. Philos Trans R Soc Lond Ser B Biol Sci 372:20160063

Chapter 2
Cnidarians: Diversity and Evolution of Cnidarian Visual Systems

Sydney Birch, Natasha Picciani, Todd Oakley, and David Plachetzki

Abstract Cnidarian photosensory systems exemplify distributed visual systems and are intriguing for a rich array of ecological and evolutionary questions. Here, we review what is known of photosensory systems in Cnidaria, in both larval and adult stages. We discuss the photobiology of cnidarians with attention to the phototransduction cascade, including cnidarian opsins, and summarize the visual organs known to be present in the phylum. Additionally, we summarize the diverse photobehaviors from medusozoan and anthozoan larvae and adults and discuss some ecological implications. We contextualize our discussion in light of distributed vision and highlight areas that warrant deeper investigation.

Keywords Vision · Evolution · Eyes · Cnidaria · Opsin · Photosensitivity · Medusozoa · Anthozoa

2.1 Introduction

Sensitivity to light is present across all domains of life, influencing critical organismal responses and behaviors. Photosensitivity in animals exists in a variety of modes, from extraocular photosensitivity to complex eyes capable of high spatial

S. Birch · D. Plachetzki (✉)
Department of Molecular, Cellular, and Biomedical Sciences, University of New Hampshire, Durham, NC, USA
e-mail: Sydney.Birch@unh.edu; David.plachetzki@unh.edu

N. Picciani
Department of Ecology and Evolutionary Biology, Yale University, New Haven, CT, USA

T. Oakley
Department of Ecology, Evolution, and Marine biology, University of California, Santa Barbara, CA, USA

E. Buschbeck, M. Bok (eds.), *Distributed Vision*, Springer Series in Vision Research, https://doi.org/10.1007/978-3-031-23216-9_2

resolution and vision. Although much is known about vision and photosensitivity in animals with bilateral symmetry (e.g., bilaterians), it is less clear how this extraordinary sense originated. Bilaterian visual systems commonly leverage both central nervous systems and brains to parse visual information, whereas non-bilaterians like cnidarians lack centralized nervous systems and brains and instead may possess a type of distributed visual system based on a diffuse nerve net. To gain a better understanding of the evolution of vision and eyes, we must examine non-bilaterian taxa that are capable of photosensitivity and, in some cases, vision that is based on a distributed visual system. In this chapter, we focus on the phylum Cnidaria.

Cnidarians are diploblastic, radially symmetrical organisms that mostly lack centralized nervous systems and instead possess diffuse nerve nets (Garm et al. 2006; Daly et al. 2007; Watanabe et al. 2009; Technau and Steele 2012; Bosch et al. 2017). Together, these features dictate that sensory information from the environment be processed in a distributed fashion, where the organismal behavioral response to a given stimulus emerges from the integration of numerous inputs from multiple sensory cell types and/or organs that are distributed around or across the surface of the organism. The distributed nature of sensory processing in cnidarians also suggests that some parsing of sensory information must take place locally, at the site of detection (e.g., photoreceptive neuron or visual organ), because there is little organization to the neuronal meshwork that connect and integrate distant visual sensors. These features of the organismal biology of cnidarians canalize many aspects of cnidarian photobiology.

The phylum Cnidaria is diverse with ~13,300 species encompassing primarily marine organisms that occur pan-globally (Daly et al. 2007; Kayal et al. 2018). Cnidarian body plans are usually polypoid or medusoid, each having distinct eco-morphological characteristics. Polyps are sessile and tube-shaped with tentacles extending from the oral end. Medusae are usually bell-, dish-, or umbrella-shaped, motile (usually free-swimming) life-history stages that often possess conspicuous sensory structures. Members of the cnidarian class Anthozoa exist as solitary or colonial polyps that may be sessile or motile, while medusozoans may possess both polyps and medusae. In addition, some medusozoan species possess only one of these body plans. Most cnidarian species exhibit complex life cycles involving larvae that undergo metamorphosis into a polyp; however, many variations on this theme are known. Cnidarians may also utilize distinct trophic strategies. Many cnidarian taxa are carnivores and capture prey using specialized stinging cells called cnidocytes. Nearly all cnidarian species possess cnidocytes, which is the defining characteristic of the phylum. In addition to carnivory, many cnidarian clades, from sea anemones and corals to some hydroids, harbor photosynthetic symbionts which produce energy for the organism (Falkowski et al. 1984; Muscatine et al. 1984; Steen 1988; Yellowlees et al. 2008; Burriesci et al. 2012; Kayal et al. 2018). Together, this morphological, life-history, and trophic diversity has allowed cnidarians to occupy a broad range of marine and aquatic ecological niches with different photic properties. Medusae can be found in open waters (shallow or deep) or in complex environments such as mangroves, kelp forests, and freshwater lakes and streams, while polyps may burrow into soft substrates or be attached to hard surfaces, in shallow waters, or in deep-sea habitats.

The objective of this chapter is to summarize the current understanding of photosensory systems in Cnidaria. We will examine cnidarian photosensory systems from the perspective of both larval and adult life-history stages and consider how their distributed nature impacts their organismal function.

2.2 Cnidarian Phylogenetic Relationships

Phylum Cnidaria can be divided into three major groups: Anthozoa, Endocnidozoa, and Medusozoa (Fig. 2.1) (Daly et al. 2007; Collins 2009; Kayal et al. 2018). Anthozoa has the greatest number of species and includes sea pens, sea fans, soft corals, stony corals, and sea anemones. Cerianthid tube anemones are also found within Anthozoa as a subclass due to major organismal differences with other anthozoans (Kayal et al. 2018). Endocnidozoa represents an enigmatic group of obligate parasites, which prey on both vertebrates and invertebrates. Finally, the Medusozoa is a highly diverse clade that contains two major groups, Hydrozoa and Acraspeda. Within Hydrozoa are the hydroids, siphonophores, and hydromedusae. Acraspeda

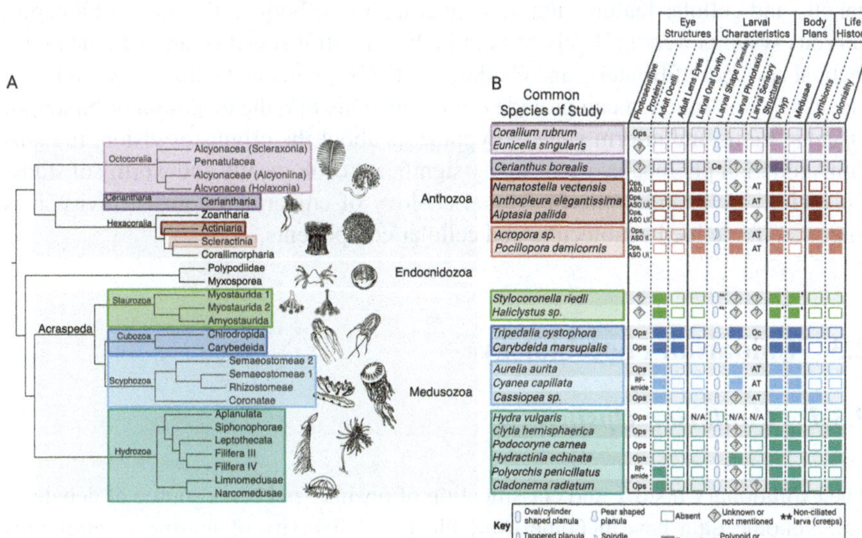

Fig. 2.1 Phylogeny of Cnidaria and characteristics of common species of study. (**a**) The phylogenetic relationships of the cnidarian classes. Cnidaria comprises Anthozoa, Endocnidozoa, and Medusozoa. Within Anthozoa are the Octocorallia (light purple), Ceriantharia (dark purple), and Hexacorallia (dark red = Actiniaria, pink = Scleractinia; both are common groups studied in Hexacorallia). Medusozoa includes Staurozoa (light green), Cubozoa (dark blue), Scyphozoa (light blue), and Hydrozoa (teal). Endocnidozoa includes Polypodiozoa and the enigmatic, parasitic Myxosporea. (**b**) Table of species that have been the focus of photobiological investigation. See key for symbol descriptions. Shaded boxes represent the presence of a trait. Empty boxes represent trait absence. Question marks represent missing information. *Ops* opsin, *ASO* anthozoan-specific opsin, *AT* apical tuft, *Oc* ocelli, *Ce* cerianthid larva. ((**a**) from Kayal et al. 2018)

contains three classes: Staurozoa, the stalked jellyfish; Cubozoa, the box jellyfish; and Scyphozoa, the true jellyfish (Collins 2002, 2009; Kayal et al. 2018). Given the branching order of the cnidarian classes (Kayal et al. 2018), the Endocnidozoa and the Medusozoa share a common ancestor to the exclusion of Anthozoa.

Photosensory and visual systems are well studied in bilaterian animals (e.g., protostomes and deuterostomes). Understanding the evolution of complex traits such as photosensitivity and vision is enhanced by studies of non-bilaterian animals that also possess these traits. Such non-bilaterian taxa include Ctenophora, Porifera, Placozoa, and Cnidaria. Each of these lineages possesses some sensory capacity, including the ability to detect light (Sleigh 1963; Srivastava et al. 2008; Collin et al. 2010; Rivera et al. 2012). However, of these taxa, only cnidarians and ctenophores possess nervous systems involving demonstrable synaptic transmission (Westfall and Kinnamon 1978; Galliot et al. 2009; Galliot and Quiquand 2011; Jager et al. 2011; Moroz et al. 2014; Bosch et al. 2017). While both cnidarians and ctenophores share this trait, cnidarians are the evolutionary sister lineage to bilaterians, splitting from bilaterians more than 600 million years ago (Dos Reis et al. 2015). This means that bilaterians and cnidarians likely inherited the same ancient neurogenetic toolkit. While evolutionary divergence in nervous systems has certainly occurred between the two lineages during the ensuing ~600 million years since their split, the genetic and cellular features that remain common to both cnidarian and bilaterian nervous systems were probably present in the ancestral eumetazoan (e.g., the hypothetical ancestor of Bilateria and Cnidaria). Therefore, understanding how cnidarian visual systems function can provide critical insights into the evolution of bilaterian visual systems and inform general hypotheses about the origins of vision. In addition, cnidarian photobehavior is highly significant ecologically and worthy of study. But before we examine the functional ecology of cnidarian photosensitivity, it is important to define its molecular and cellular components.

2.3 Cnidarian Photobiology

2.3.1 Cnidarian Opsins

The evolutionary history and classification of opsins has been a source of debate as new genome data have become available for a diversity of marine invertebrates including cnidarians (Plachetzki et al. 2007; Porter et al. 2011; Feuda et al. 2012, 2014; Schnitzler et al. 2012; Ramirez et al. 2016; Picciani et al. 2018; Fleming et al. 2020; Chari et al. 2021; Gornik et al. 2021). Prior to this data deluge, the opsin gene family was thought to contain three major gene clades: ciliary (c-opsins), rhabdomeric (r-opsin), and a group often referred to as G_o/RGR, which includes the G_o-coupled opsins (e.g., human neuropsin and peropsin) and the retinochrome and retinal G-protein-coupled receptors, which possess photoisomerase activity (Clay Smith and Goldsmith 1991; Shen et al. 1994; Del Pilar Gomez and Nasi 2000). More recently, a consensus has emerged of a fourth group of opsin termed cnidops, or cnidarian opsins, found only in cnidarians (Plachetzki et al. 2007; Porter et al. 2011; Feuda et al. 2012; Ramirez et al. 2016). In most analyses, cnidops is positioned close to either the ciliary clade (Suga et al. 2008; Plachetzki et al. 2010) or the G_o/RGR clade (Suga et al. 2008; Yau and Hardie 2009; Ramirez et al. 2016; Vöcking et al. 2017; Macias-Munõz et al. 2019). Xenopsin is another possible clade of opsins that contains cnidarian representatives (Ramirez et al. 2016; Vöcking et al. 2017), and two other opsin gene clades have been identified that are restricted to anthozoans: anthozoan-specific opsin 1 (ASO-1) and anthozoan-specific opsin 2 (ASO-2) (Vöcking et al. 2017; Gornik et al. 2021). Gornik et al. (2021) placed ASO-1 toward the base of the opsin tree, while ASO-2 had affinity with the ciliary opsins (Gornik et al. 2021).

Little consensus exists in the literature on cnidarian opsin phylogeny. This is not surprising because understanding where the cnidarian opsins fit into the larger framework of animal opsin phylogeny could be confounded by several factors that may be especially applicable to opsins. First, opsin genes encode relatively short proteins (~350 amino acids), meaning that the information content of the sequences themselves may be insufficient for accurate resolution of gene phylogenies that commonly include hundreds of sequences. In addition, there are many examples of opsin loci that are under strong selection, in some cases leading to rapid evolution and biased amino acid composition (Yokoyama 2000; Shichida and Matsuyama 2009; Owens and Rennison 2017; Fleming et al. 2020). Methods of phylogenetic reconstruction usually rely on modeling sequence evolution under neutral evolutionary processes and are vulnerable to compositional bias. Thus, the functional diversification of opsins by natural selection could obscure accurate resolution of metazoan opsin phylogeny. Moreover, due to dynamic and often lineage-restricted gene duplications and loss, opsin gene phylogenies may differ drastically from well-supported species trees, making it difficult to determine where to place the root of the opsin phylogeny. Because of this, outgroup genes, like the melatonin receptors, are commonly used to root the tree. However, studies of rhodopsin class GPCR phylogenies do not support melatonin receptors as the closest outgroup to the opsins (Fredriksson et al. 2003; Bjarnadóttir et al. 2006), suggesting that there may be other more closely related GPCR outgroups that could be used. Together, these observations may explain the current lack of consensus on metazoan opsin phylogeny and the position of cnidarian opsins within it.

While there are outstanding questions related to the structure of metazoan opsin phylogeny and the placement of cnidarian opsins, it is clear that cnidarians possess as many as four different paralogy classes of opsin: cnidops, xenopsins, ASO-1, and ASO-2. Cnidopsins are present across all cnidarian classes, including the morphologically reduced, parasitic Endocnidozoa (Gornik et al. 2021), and if present in medusozoan species, cnidops may be expressed in the eyes or visual organs. ASO-1 and ASO-2 are present only in Anthozoa, which lack visual organs. ASO-1 from *Acropora* (acropsin 3) activates a G-protein in a light-dependent manner (Mason et al. 2012), but how ASO-1 modulates behavior is unclear as is the function of ASO-2. We note that both rhabdomeric opsins and xenopsins have also been recovered from cnidarian datasets, but these inferences are dependent on the placement of the root of the tree (Feuda et al. 2012, 2014, 2022; Hering and Mayer 2014; Macias-Munõz et al. 2019; Gornik et al. 2021). Future work should systematically address the placement of the root of opsin phylogeny.

2.3.2 *Cnidarian Phototransduction and the Origins of Metazoan Visual Cascades*

At least two types of photosensitive proteins are present in cnidarians: cryptochromes and opsins. Cryptochromes (CRYs) are DNA photolyase-like photoreceptor proteins that are sensitive to blue light. CRYs are found among all cellular organisms except Archaea and are commonly involved in circadian rhythm entrainment (Levy et al. 2007; Garm and Ekström 2010; Kim et al. 2014; Porter 2016; Gornik et al. 2021). Opsins are G-protein-coupled receptors (GPCRs) with seven-transmembrane domains and are the common type of protein involved in vision across the animal kingdom (Terakita 2005; Plachetzki et al. 2010; Feuda et al. 2012; Oakley and Speiser 2015).

The composition of opsin-based phototransduction cascades in bilaterians (e.g., ciliary, G_o-coupled, and rhabdomeric) has been reviewed in-depth (Yau and Hardie 2009; Fain 2020). In general, phototransduction occurs when the chromophore, usually 11-*cis*-retinal (a vitamin A derivative), interacts with light resulting in a change in molecular configuration to all-*trans*-retinal (Yau and Hardie 2009). The chromophore binds to opsin at a conserved lysine residue (K296) that is present in all known functional visual opsins. The change in molecular configuration of the chromophore causes the opsin to change conformation, which triggers a G-protein-coupled signal transduction cascade. In ciliary phototransduction, a G_i G-protein, also known as transducin (G_t), activates phosphodiesterase (PDE) which reduces the standing cellular pool of cGMP, resulting in the closure of cyclic nucleotide-gated (CNG) ion channels (Fig. 2.2a). G_o-coupled phototransduction shares many components with ciliary phototransduction including a cyclic nucleotide as the secondary messenger, PDE, and the involvement of a CNG ion channel (Fig. 2.2b). Rhabdomeric phototransduction diverges significantly from both ciliary and G_o-coupled phototransduction cascades, utilizing a different G-protein (G_q), a different intermediary enzyme (phospholipase C), and different secondary messengers (PIP_2, $InsP_3$, and DAG), which ultimately modulate transient receptor potential C (TRPC) ion channels (Fig. 2.2d).

Evidence suggests that the cnidops-mediated phototransduction cascade involves the activation of a G_s G-protein, followed by the activation of adenylate cyclase (AC) to produce cAMP as the secondary messenger (Fig. 2.2c) (Koyanagi et al. 2008; Plachetzki et al. 2010, 2012; Liegertová et al. 2015). cAMP is proposed to bind and open a CNG ion channel causing depolarization of the cell (Nilsson 2009). The components of this pathway (G_s-AC-cAMP-CNG) have all been implicated in cnidarian phototransduction, but not all have been confirmed using functional tests.

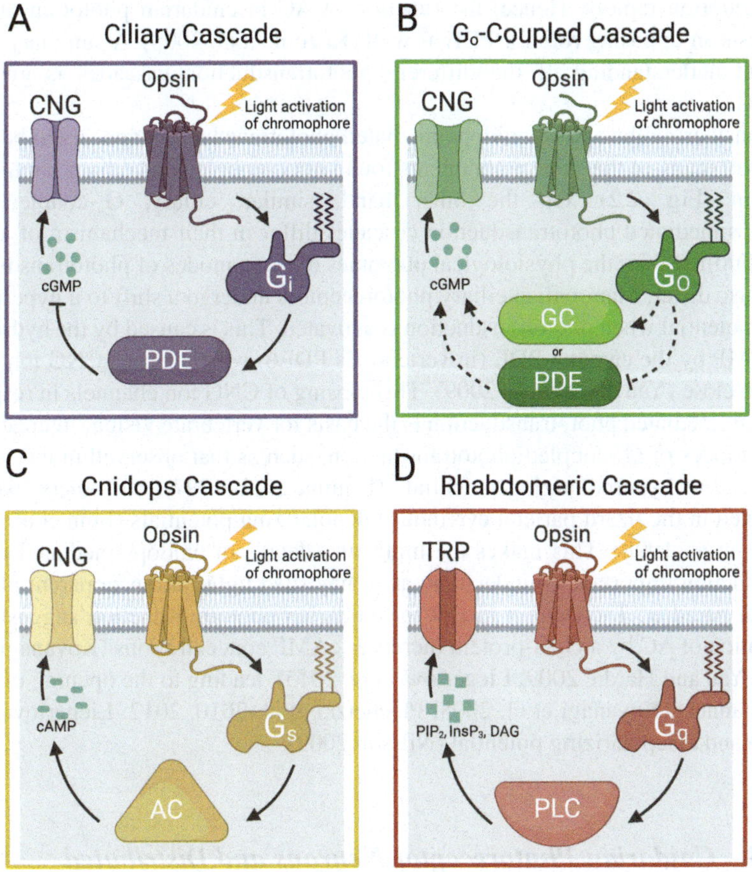

Fig. 2.2 Modes of phototransduction: (**a**) Key components in ciliary phototransduction. Light activates the chromophore which binds to c-opsin, activating G_i alpha. The G-protein activates phosphodiesterase (PDE), which removes cGMP causing a cyclic nucleotide-gated (CNG) ion channel to close and the hyperpolarization of the cell. (**b**) A depiction of the two types of G_O-coupled cascades. In scallops, the G_O alpha G-protein activates guanylate cyclase (GC), which increases cGMP, the opening of CNG ion channels, and the hyperpolarization of the cell. In a second pathway, found in the lizard parietal eye, G_O activation inhibits PDE leading to an increase in the concentration of cGMP, the opening of CNG ion channels, and depolarization of the cell. (**c**) The hypothesized cnidops cascade. A G_s G-protein activates adenylate cyclase (AC) which increases the concentration of cAMP leading to the opening of a CNG ion channel and depolarization of the cell. (**d**) Rhabdomeric cascade. A G_q alpha protein activates PLC which modulates the concentrations of PIP_2, $InsP_3$, and DAG causing a transient receptor potential (TRP) channel to open and the cell to depolarize

ACs and PDEs play important physiological roles in a wide range of GPCR-mediated cell signaling cascades, where ACs are the enzymes primarily responsible for cAMP in eukaryotes, and PDEs remove such cyclic nucleotides. For cell signaling cascades to be dynamic, signaling molecules like cAMP must change in concentration rapidly. Hence, the function of AC in cnidarian phototransduction suggests an opposing role for PDE as well (Kozmik et al. 2008). A summary of the current understanding of the different phototransduction cascades is given in Fig. 2.2.

Considering the above, cnidops-mediated phototransduction appears to be composed of many of the same components found in G_0-coupled and ciliary phototransduction (Fig. 2.2). On the other hand, similar, ciliary, G_0-coupled, and cnidops-mediated phototransduction cascades differ in their mechanism of activation. Furthermore, the physiological outcomes of these modes of phototransduction may also differ. Mammalian ciliary photoreceptors undergo a shift to a hyperpolarizing potential when phototransduction is activated. This is caused by the hydrolysis of cGMP by the enzyme PDE (in vertebrates PDE6) which causes CNG ion channels to close (Yau and Hardie 2009). This closing of CNG ion channels in response to opsin-mediated phototransduction is the basis for vertebrate vision. Interestingly, some modes of G_0-coupled phototransduction, such as that observed in the scallop eye, lead to hyperpolarizing potentials (Kojima et al. 1997), but others like that observed in the lizard parietal eye induce depolarizing potentials (Finn et al. 1997; Xiong et al. 1998). This makes the molecular details of cnidops-mediated phototransduction important to understand as it may shed light on this apparent physiological paradox. The current model for cnidarian phototransduction suggests that activation of AC by a G_s G-protein increases cAMP concentrations (Koyanagi et al. 2008; Yau and Hardie 2009; Liegertová et al. 2015), leading to the opening of CNG ion channels (Koyanagi et al. 2008; Plachetzki et al. 2010, 2012; Liegertová et al. 2015) and a depolarizing potential (Nilsson 2009).

2.3.3 *Cnidarian Photoreceptor Neurons and Distributed Sensory Systems*

Cnidarians demonstrate a wide range of light detection capabilities, from extraocular detection to vision with camera-type eyes. Extraocular photoreception occurs in cells outside of the eyes or photosensory organs and is observed across the phylum in adult and larval stages. Extraocular photoreception may occur by opsin-mediated phototransduction or by cryptochromes which are blue light sensitive proteins (Reitzel et al. 2013; Porter 2016). Cnidarian photoreceptors have been comprehensively reviewed by Martin (2002). In general, cnidarian photoreceptors are ciliated cells with bipolar morphology where one part forms the light-receptor process and the other forms an axon (Martin 2002). In cnidarians, extraocular photoreceptors have been identified by their expression of the peptidergic neurotransmitter RFamide and opsin where they are typically expressed in neural cells in the tentacles as demonstrated by in situ staining and single-cell sequencing (Martin 2002; Plickert and Schneider 2004; Plachetzki et al. 2007, 2012; Chari et al. 2021). Extraocular photoreception is the most common mode of photosensitivity in Cnidaria.

Cnidarian extraocular photoreceptors, especially among hydrozoans, are commonly connected in arrays by ganglion cells that function as interneurons (Fig. 2.3e). Together, ganglion cells function to stitch together a patchwork of extraocular photoreceptors distributed across the surface of the animal. In hydrozoans like the hydra, such extraocular photoreceptors are commonly associated with battery cell complexes, which house both sensory neurons and cnidocytes (Hyman 1940; Westfall and Kinnamon 1978; Hufnagel et al. 1985; Plachetzki et al. 2012). The discharge of cnidocytes is dependent on receiving a suitable signal from sensory neurons. The distributed nature of such battery cell complexes allows for the signal from one complex to propagate adjacent complexes. In this way, signals that permit cnidocyte discharge may spread to other complexes to serve the same role, allowing the coordination of feeding and defensive behaviors.

2.3.4 Cnidarian Eyes

In addition to extraocular sensitivity, some medusozoans also possess specialized light-detecting organs. Simple eye spots, termed ocelli, are composed of a cluster of photoreceptor cells interleaved with non-sensory pigment cells (Fig. 2.3a; Martin 2002; Picciani et al. 2018). This arrangement allows for the detection of the direction of light (Arendt and Wittbrodt 2001; Martin 2002). More complex ocelli also exist such as cup-shaped ocelli, where pigment cells produce a cup and photoreceptor cells project into the cup (Fig. 2.3b; Eakin and Westfall 1962; Martin 2002). Such ocelli are present across the Medusozoa, including in the stalked jellyfish (Staurozoa) (Fig. 2.3l; Blumer et al. 1995; Miranda and Collins 2019), although many variations of cup-shaped ocelli exist across the class, ranging in complexity. For example, the well-studied hydrozoan *Cladonema radiatum* possesses an everted pigment cup with lenslike bodies formed from cytoplasmic portions of pigment cells (Fig. 2.3c; Weber 1981a, b). Lastly, eyes composed of a retina, cornea, and a lens are also present in cnidarians. Such complex eyes are found in structures called rhopalia which are only present in the true jellyfish and box jellyfish (e.g., Scyphozoa and Cubozoa). The complex eyes of rhopalia can possess a retina made up of photoreceptor cells, a spherical lens that has a graded refractive index composed of crystallin proteins, a cornea, and, in the lower lens eye, a pupil (Fig. 2.3d; Pearse and Pearse 1978; Piatigorsky et al. 1989; Nilsson et al. 2005). These many types of cnidarian eyes originated independently at least nine times among medusozoans (Salvini-Plawen and Mayr 1977; Picciani et al. 2018; Miranda and Collins 2019) co-opting preexisting pigment and photoreceptor cells. No visual organs of any type have been identified from cnidarian species outside of the Medusozoa.

Rhopalia are club-shaped sensory structures found on bell margins of adult and juvenile medusae in scyphozoans and cubozoans (Fig. 2.3f–j; Garm et al. 2006; Helm 2018); however, they differ between the two groups. In adult scyphozoans, rhopalia are finger-shaped and are typically found in multiples of 4 (Hyman 1940) on the bell margin (Schafer 1878; Helm 2018) and can differ in the number and

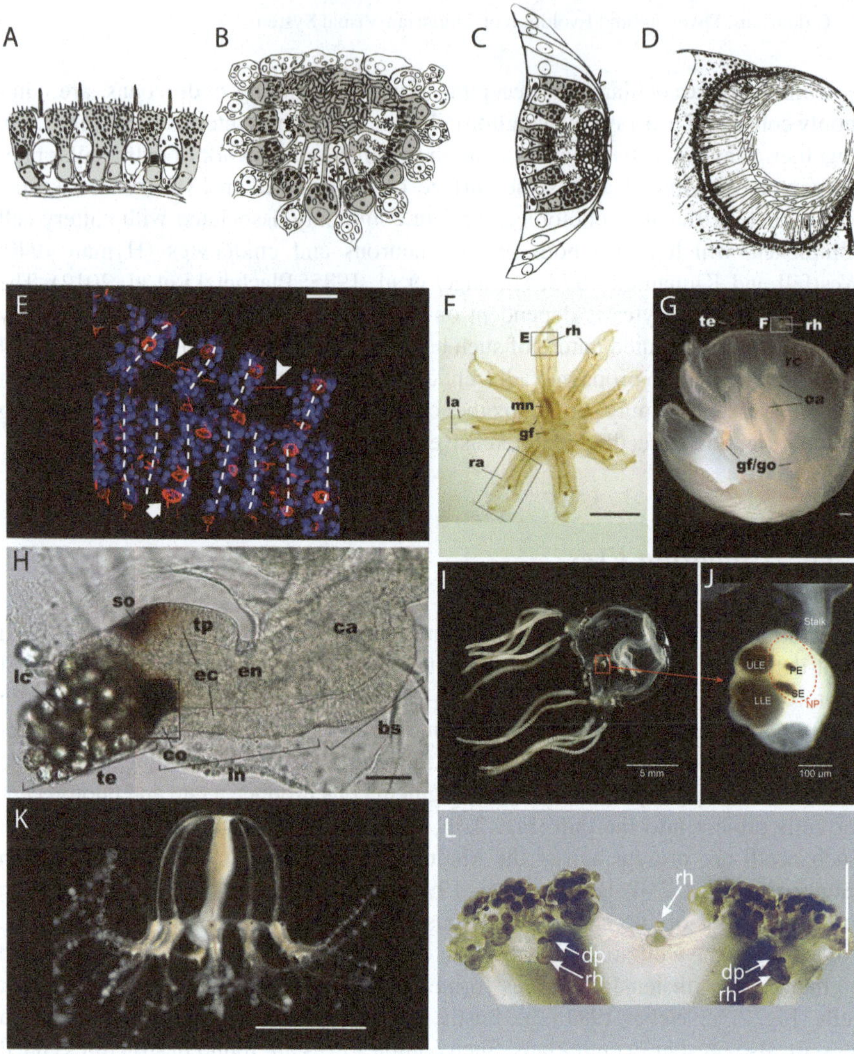

Fig. 2.3 Photoreceptors and eyes of Cnidaria. (**a**) An eyespot (ocellus) containing pigment cells interleaved with photosensory ciliated cells in a single layer. (**b**) An everted pigment cup where pigment cells form a cup and photosensory cells project into the cup. (**c**) An everted pigment cup with lenslike bodies from the hydrozoan *Cladonema radiatum*. (**d**) Lens eye of the cubozoan *Carybdea xaymacana*. The lens eye contains a crystallin lens, cornea, and retina. (**a–d**) Adapted from Picciani et al. (2018). (**e**) Confocal imaging of a *Hydra vulgaris* tentacle. Neurons including cnidocytes are stained in red (anti-acetylated alpha tubulin) and nuclei are stained in blue (DAPI). Adapted from (Plachetzki et al. 2012). (**f**) An ephyra of the scyphozoan *Aurelia aurita*, rh indicates rhopalia. (**g**) Juvenile *A. aurita* medusa. (**h**) Rhopalia of a juvenile *A. aurita* medusa. (**f–h**) Adapted from Nakanishi et al. (2009); *te* tentacle, *rh* rhopalium, *la* lappet, *mn* manubrium, *ra* rhopalar arm, *gf* gastric filaments, *gf/go* gastric filaments/gonads, *oa* oral arm, *rc* ring canal, *lc* lithocyst, *co* pigment cup ocellus, *so* pigment-spot ocellus, *tp* touch plate, *ca* rhopalar canal, *ec* ectoderm, *en* endoderm, *te* terminal segment, *in* intermediate segment, *bs* basal segment. (**i**) Cubomedusa *Tripedalia cystophora*. (**j**) *T. cystophora* rhopalia from I. Each rhopalium possesses six eyes, a lower lens eye (LLE), an upper lens eye (ULE), two pit eyes (PE), and two slit eyes (SE) and a neuropil (NP in red). (**i–j**) Adapted from (Bielecki et al. 2014). (**k**) Dark pigmented ocelli (Oc) of hydromedusa *Cladonema pacificum*. Adapted from (Fujita et al. 2019). (**l**) Rhopalioids (rh) and dark pigment spots (dp) of stauromedusae *Manania uchidai*. (Adapted from Miranda and Collins, 2019)

location across species (Fig. 2.3f–h; Russell 1970). Scyphozoan rhopalia typically possess a pigment-spot ocellus (Yamasu and Yoshida 1973; Yoshida and Yoshino 1980), a touch plate to sense tilt (Chapman and James 1973), and a statocyst that interacts with the touch plate to facilitate orientation in the water column (Fig. 2.3h; Hyman 1940; Helm 2018). The function of the pigment-spot ocellus is presumably photosensitivity, but rhopalia likely integrate chemosensory and mechanosensory information as well (Yamasu and Yoshida 1973; Nakanishi et al. 2009). Additionally, in scyphozoans, the rhopalia begin developing during the strobila stage where the statocyst and touch plate develop first and the pigment cup and pigment-spot ocelli are the last to develop during the late ephyra (juvenile medusa) stage (Nakanishi et al. 2009).

Rhopalia are more complex in cubozoans. Cubozoan rhopalia are club-shaped and, similar to scyphozoans, located at the margin of the bell, but cubomedusae have only four rhopalia (Berger 1898; Yamasu and Yoshida 1976; Coates 2003). Each cubozoan rhopalium has six eyes: a large complex eye (lower lens eye), a small complex eye (upper lens eye), a symmetrical pair of pit eyes, and a symmetrical pair of slit eyes (Fig. 2.3j). The complex eyes possess a hemispherical retina made up of photoreceptor cells, sometimes pigmented or with pigment cells, a lens composed of crystallin proteins, and a cornea (Pearse and Pearse 1978; Piatigorsky et al. 1989; Nilsson et al. 2005). Both lens eyes are capable of image formation and exhibit 10–20 degrees of spatial resolution; however, the retina is too close to the lens for sharp vision (Nilsson et al. 2005; O'Connor et al. 2009; Garm et al. 2013, 2016). The upper lens eye seems to be involved in navigation (Garm et al. 2011), while the lower lens eye is involved in obstacle avoidance (Garm et al. 2007b, 2013). Additionally, electrophysiological recordings and adaptation experiments have shown that cubozoan lens eyes are slow, use a single opsin (peak ~500 nm), and are probably color blind (Garm et al. 2007a; Bielecki et al. 2014). Both types of ocelli, pits and slits, are formed by invagination of the epithelium where the distal portion of the cells is heavily pigmented (Coates 2003). The rhopalial eyes of cubozoans are the most complex eyes found outside of Bilateria. While the complex camera eyes they possess are commonly of interest, rhopalia likely mediate other sensory modalities, including mechanosensitivity and possibly chemosensitivity (Garm et al. 2006; Skogh et al. 2006).

Insights on the evolutionary significance of cnidarian visual organs have also come from studies of their development. Development of bilaterian eyes may be determined by a developmental cascade of transcription factors known as the retinal determination gene network (RDGN) (Donner and Maas 2004; Silver and Rebay 2005). Much effort has gone into examining whether cnidarian genes that are orthologous to genes present in the bilaterian RDGN are also involved in cnidarian visual organ development. The bilaterian RDGN involves a number of genes including *Pax6*, *Six*, *Eya*, and *Dach* (Kumar 2009). In the hydrozoan *Cladonema*, all orthologs of *Pax* genes, except for *Pax6*, are present in the genome. Additionally, *Pax-B*, an ortholog to *Pax2/5/8*, is expressed in the larva, retina, lens, and statocyst of the cubozoan *Tripedalia cystophora* (Kozmik et al. 2003). *Pax-A*, which is an ortholog of the bilaterian *Pax* transcription factor *poxn*, is expressed in the developing eyes of *Cladonema* (Suga et al. 2010). In *Drosophila*, *poxn* is expressed in sensory cell progenitors and is an important determinant of sensory cell fate (Dambly-Chaudière et al. 1992). However, Nakanishi et al. (2015) did not find evidence of *Pax-A* expression in developing rhopalia eyes of the scyphozoan *Aurelia*, demonstrating that the requirement of *Pax-A* in cnidarian eye development is variable. Additionally, three *Six* genes were identified in *Cladonema* by Stierwald et al. (2004). Two of these, *sine oculus* (also known as *Six1/2*) and *optix* (also known as *Six3/6*), are expressed in the eye cup, a finding also recapitulated by Nakanishi et al. (2015) in studies of *Aurelia*; however, *Six3/6* did not change in expression with the onset of rhopalia development in *Aurelia*. The RDGN gene *eyes absent* (*Eya*) is also expressed in the developing rhopalia field in *Aurelia* (Nakanishi et al. 2015) and *Cladonema* (Graziussi et al. 2012). While the RDGN gene *Dach* (also known as *dachshund*) is missing from cnidarian genomes examined thus far, other bilaterian eye development genes (e.g., *Brn3* and *Pit1*) are expressed in the developing rhopalia of *Aurelia* (Nakanishi et al. 2010). Taken together, the current evidence suggests the involvement of *Sine oculus* and *Eya* in the development of both hydrozoan ocelli and scyphozoan rhopalia, while *Pax* genes are variable across the medusozoan classes. This suggests that some of the proposed components of the RDGN are bilaterian innovations.

2.3.5 Distributed Visual Systems in Cnidarians

Cnidarians generally lack centralized nervous systems. However, cubozoan, scyphozoan, and hydrozoan medusae possess neuronal regions of high organization that approximate centralization (Mackie 2004; Garm et al. 2006; Skogh et al. 2006). Such medusae nervous systems have been best studied in the cubomedusa *Tripedalia cystophora* and the hydromedusa *Aglantha digitale*. The marginal nervous system of cubozoans consists of a single ring nerve that links together the four rhopalia located around the margin of the bell (Fig. 2.3i–j). Within each rhopalium are highly structured, bilaterally symmetrical rhopalial nervous systems that are integrated with the ring nerve (Parkefelt et al. 2005; Garm et al. 2006; Skogh et al. 2006). In this way, cubozoan rhopalia and their integration with the ring nerve comprise a distributed visual system. Consistent with other distributed visual systems, the rhopalia are also sites of sensory integration, including visual processing (Garm et al. 2006; Skogh et al. 2006; Parkefelt and Ekström 2009).

Hydromedusae possess two (inner and outer) ring nerves that link the sensory organs that are distributed around the margin of the bell (Fig. 2.3k). The outer ring nerve of hydromedusae is morphologically similar to the cubozoan ring nerve (Satterlie 2002; Mackie 2004). Like cubozoan sensory systems, those of hydromedusae also share several features of a distributed visual system; however, hydromedusae sensory systems usually include a greater number of low complexity sensory organs like pigment cup ocelli compared to those of either cubozoans or scyphozoans. Interestingly, scyphozoans do not have a ring nerve, but they do possess complex rhopalial ganglia.

2.4 Photosensory Behaviors in Cnidarian Larvae

Given the diversity of Cnidaria, it is not surprising that a wide variety of photobehaviors are observed in both adult and larval life-history stages. In fact, the diversity in photobehaviors is often linked to specific life-history traits. Here we summarize photobehavior observations from different cnidarian clades and life-history stages.

2.4.1 Anthozoan Larvae

The central organizing element in anthozoan morphology is the polyp. Polyps may occur as solitary individuals (e.g., Actiniaria) or as colonies (e.g., Scleractinia) (Fig. 2.1b). Adult anthozoans are predominantly sessile as adults, and their motile larvae, usually a type of planula, are the primary dispersal life-history stage known from the class (Daly et al. 2007; Giribet and Edgecombe 2020). Additionally, many adult anthozoans (sea anemones and corals) possess photosynthetic symbionts, which requires the adult stages to reside in habitats that receive light (Pearse 1974a; Fredericks 1976; Yamashiro and Nishira 1995; Giribet and Edgecombe 2020). These features could explain the photobehaviors exhibited by anthozoan larvae, such as the preference for different wavelengths of light and preference for substrate color during settlement (Mason et al. 2011; Mason and Cohen 2012; Strader et al. 2015; Sakai et al. 2020). For example, scleractinian coral planulae use light in the identification of suitable settlement sites (Mundy and Babcock 1998; Mason et al. 2011, 2012; Mason and Cohen 2012; Strader et al. 2015; Mulla et al. 2020; Sakai et al. 2020). Here, due to the attenuation of long wavelengths of light by water, the wavelength of light detected by planulae may act as a type of depth meter, allowing settlement to occur in a light environment with suitable irradiance. Additionally, light cues have also been shown to inform coral larvae of settlement surface orientation (horizontal or vertical) (Strader et al. 2015). This photoresponse aligns with adult distribution of coral species found across different photic zones, indicating that light information may play an important role in niche differentiation and the structuring of benthic coral reef communities (Mundy and Babcock 1998; Mason and Cohen 2012; Strader et al. 2015; Sakai et al. 2020). Coral planulae express opsins and cryptochromes. Opsins have been linked to the perception of light wavelength (Mason and Cohen 2012; Strader et al. 2015; Sakai et al. 2020), while it has been suggested that cryptochromes are involved in diel migrations of planulae and possibly circadian entrainment (Levy et al. 2007, 2011; Brady et al. 2011). The planulae of sea anemones also display phototaxis; however, these behaviors have not been as thoroughly investigated as in corals (Müller and Leitz 2002). Opsins have been identified in *Nematostella* planula (Helm et al. 2013; Marlow et al. 2013), but their function has not been investigated.

2.4.2 Medusozoan Larvae

Like anthozoan larvae, medusozoan larvae are also tasked with identifying suitable settlement sites for polyp life-history stages; however, medusozoan larvae diverge in a few ways. Many medusozoan clades possess planulae that differ morphologically from anthozoans and between medusozoans themselves (Fig. 2.1b). Some medusozoan clades, such as the Aplanulata, lack planulae, but some species may possess a different type of larva called an actinula. To add to this, medusozoans typically have more complex life cycles than anthozoans, where there may exist a motile medusa stage in addition to the polyp stage. This life-history characteristic could explain many differences between anthozoan and medusozoan larvae (Fig. 2.1b). Interestingly, many cubozoan and scyphozoan planulae are negatively phototactic (Werner et al. 1971; Svane and Dolmer 1995; Holst and Jarms 2007), whereas most anthozoan planulae are positively phototactic. Negative phototaxis might be a way for larvae to avoid solar radiation, sedimentation, and predation (Svane and Dolmer 1995). Similar to anthozoan planulae, scyphozoan planulae possess apical tufts (Nakanishi et al. 2008; Yuan et al. 2008), sensory structures that may play a role in detecting environmental cues. While cubozoan planulae lack tufts, they do possess pigment spots which are hypothesized to be ocelli used for steering planula while swimming, but little is known about the photosensitive behavior or physiology of these larvae (Fig. 2.1b) (Nordström et al. 2003). Additionally, little is known about the apical tuft in scyphozoan larvae or what cues it can detect (Nakanishi et al. 2008; Yuan et al. 2008). The hydrozoan larvae seem to be the least complex across the phylum where there are no major sensory organs (ocelli or apical tufts) (Plickert and Schneider 2004). While some hydrozoan larvae display phototaxis (Birch et al., n.d.; Thorson 1964; Orlov 1997; Plickert and Schneider 2004), little is known about the physiology of photo-perception but opsin expression and RFamide positive neurons in hydrozoan larvae have been described (Birch & Plachetzki n.d.; Martin 2002; Katsukura et al. 2004; Plickert and Schneider 2004; Piraino et al. 2011).

2.5 Photosensory Behaviors in Adult Cnidarians

2.5.1 Anthozoan Adults

Like anthozoan larvae, adult anthozoans display a range of phototactic behaviors. Here, photosensitivity is observed in both symbiotic and nonsymbiotic species and includes behaviors such as tentacle expansion and retraction (Pearse 1974b;

Sweeney 1976; Gorbunov and Falkowski 2002; Levy et al. 2003; Hoadley et al. 2011; Reitzel et al. 2013), phototaxis (Pearse 1974a; Yamashiro and Nishira 1995), the regulation of circadian clocks (Reitzel et al. 2010; Leach et al. 2018; Leach and Reitzel 2019), and the synchronization of reproductive spawning (Levy et al. 2007; Brady et al. 2009; Sweeney et al. 2011; Rosenberg et al. 2017; Tarrant et al. 2019).

Light provides an important cue to all life-history stages, providing the opportunity for animals to tune their biology to annual, monthly, and daily rhythms. The synchronization of spawning in corals has been linked to both solar (Brady et al. 2009) and lunar cues (Jokiel et al. 1985; Sweeney et al. 2011; Grawunder et al. 2015; Kaniewska et al. 2015; Rosenberg et al. 2017). Rosenberg et al. (2017) performed an in-depth molecular study investigating the cellular pathways involved in the timing of gamete release in *Acropora digitifera*. They found that *A. digitifera* synchronizes gamete release with the lunar cycle using two light-sensing phototransduction pathways: one involved in the "setting" stage (rhodopsin-like photoreceptors and glutamate release) and the other in spawning (via masking effect of circadian genes) (Rosenberg et al. 2017).

Tentacle expansion and contraction behavior can also exhibit daily cycles in behavior and gene expression in both corals and sea anemones where tentacles expand at night for feeding but remain contracted during the day, while animals rely on symbionts for energy (Pearse 1974b; Sweeney 1976; Hoadley et al. 2011; Reitzel et al. 2013). Of course, this does vary across species, some do not contract tentacles while others show sensitivity to blue light, but it seems that the intensity of light is a large factor (Pearse 1974b; Sweeney 1976; Gorbunov and Falkowski 2002; Levy et al. 2003). Additionally, phototaxis also occurs in both corals and sea anemones. It is hypothesized that the positive phototaxis in corals is a way for corals to find an optimally illuminated habitat for their symbionts (Yamashiro and Nishira 1995). Sea anemones show both positive and negative phototaxis, depending on the intensity of light (Pearse 1974a; Fredericks 1976).

Interestingly, while anthozoans lack visual organs, they possess a higher diversity of opsins than medusozoans, where in addition to cnidopsin and possibly xenopsin, anthozoans also possess ASO-1 and ASO-2 (Gornik et al. 2021). Additionally, transcriptome evidence has revealed that opsin and cryptochrome genes exhibit rhythmic expression over a diel light cycle in sea anemones (*Nematostella*) and corals (*Acropora* sp.), but how such cycles of gene expression relate to photosensitive behaviors is poorly understood (Brady et al. 2011; Reitzel et al. 2013; Kaniewska et al. 2015; Leach et al. 2018; Leach and Reitzel 2020). Since many anthozoans possess dinoflagellate or algal symbionts, whose efficacy requires suitable conditions of irradiance, some have examined the role of the photosymbiont in anthozoan photosensitive behaviors. Corals with symbionts are capable of phototaxis, independent of their symbionts (Yamashiro and Nishira 1995; Gorbunov and Falkowski 2002). However, in symbiotic sea anemones, evidence suggests the involvement of dinoflagellate symbionts in phototactic behavior, but it is unclear how the symbiont contributes to photoreception or behavior (Pearse 1974a, b; Fredericks 1976; Foo et al. 2020).

2.5.2 Medusozoan Adults

The Medusozoa is highly diverse with four subclasses possessing different life histories and distributions. There are different behavioral requirements for photosensitivity in motile medusozoans than in sessile anthozoans which could explain why all complex photosensory organs within Cnidaria are associated with the medusa morphotype. Cubozoans are found in complex nearshore environments such as mangroves, kelp forests, and coral reefs, whereas scyphozoans and hydrozoan medusae are typically found in open waters (Coates 2003). This difference in habitat complexity has been hypothesized to underpin the rise of complex eyes and why we see an increase in complexity of rhopalia and visual acuity in cubozoans compared to scyphozoans and hydromedusae (Coates 2003; Garm et al. 2007b, 2011; O'Connor et al. 2009; Petie et al. 2013a; Seymour and O'hara 2020). Both scyphozoans and cubozoans use vision/light to inform the swim pacemaker, which is a neuronal central pattern generator that controls the rate of bell contractions (Passano 1965, 1973; Satterlie 1979, 2002; Mackie and Meech 1995; Schuyler and Sullivan 1997; Garm and Bielecki 2008; Garm and Mori 2009). The pacemaker is a major part of visually guided behaviors and has been implicated in horizontal directional swimming, vertical diel migrations, and shadow-avoidance behaviors (Passano 1965; Yoshida and Ohtsu 1973; Arkett 1985; Arkett and Spencer 1986; Schuyler and Sullivan 1997; Satterlie 2002; Garm and Mori 2009).

Cubozoans show additional visually guided behaviors such as navigation (O'Connor et al. 2009; Garm et al. 2011; Seymour and O'hara 2020), obstacle avoidance (Garm et al. 2007b; Petie et al. 2013a), and the identification of light shafts for feeding (Buskey 2003; Garm and Bielecki 2008). Electrophysiological changes in pacemaker activity, which cause alterations in swim behavior, can be driven by different light conditions (Satterlie 1979, 2002; Garm and Bielecki 2008; Garm and Mori 2009). For example, box jellyfish reduce contractions (sink) in high light intensity areas due to an inhibition of the pacemaker, while swimming contractions increase in lower light intensity areas. This behavior is linked to feeding. *Tripedalia cystophora* feeds on dense swarms of copepods (*Dioithona oculata*) which swarm in highly lit areas. *Tripedalia* positions itself within light shafts to maximize feeding (Buskey 2003; Garm and Bielecki 2008). Furthermore, electrophysiological studies on each of the four eye types of *Tripedalia* have shown that three of the four eye types modify the swim pacemaker frequency (Garm and Mori 2009).

Cubozoans, as well as scyphozoans and hydrozoans, also display a similar behavior called the shadow reflex where medusae increase swimming contractions in the presence of a shadow or rapid decrease in light intensity (Yoshida and Ohtsu 1973; Arkett 1985; Arkett and Spencer 1986; Hamner et al. 1995). In cubozoans, Garm and Mori (2009) showed that the lower lens eye was mainly involved in the shadow response and in reducing swim speed in light shafts. Meanwhile, the upper lens eye is permanently oriented to look up through the water column for navigation using terrestrial cues (cubomedusae identify the canopy of mangroves to stay along the edge of lagoons) (Garm et al. 2011). Cubozoans also present an obstacle avoidance behavior that requires spatial vision (Garm et al. 2007b, 2013). Furthermore, visual information can dictate what direction medusae swim and swimming speed by modulating the velarium, a ring of tissue around the bottom of the bell, (Petie et al. 2013b, a).

Pupillary responses (e.g., rate of pupil constriction) in the lower lens eye differ between species (O'Connor et al. 2009; Seymour and O'hara 2020). The pupillary response could provide different degrees of visual acuity which may correspond to differing complexity of habitats and ecologies, where more frequent pupillary responses are associated with more complex habitats (O'Connor et al. 2009; Seymour and O'hara 2020). Additionally, cubozoans have adapted visually to being active in either the day or night (Garm et al. 2012, 2016). The night-active cubomedusa *Copula sivickisi* has image-forming eyes and an opsin that is adapted to low light (peaks at 460 nm) (Garm et al. 2016). *C. sivickisi* uses vision to hunt at night by identifying areas where prey items generate bioluminescent flashes.

Much less is known about photosensitivity in staurozoans, the stalked polyps, which are members of Acraspeda and the sister to cubozoans and scyphozoans. All staurozoans are benthic, and like other cnidarians, staurozoans also have a complex life cycle. Staurozoans have planula larvae that metamorphose into juvenile stauropolyps, which undergo an apical metamorphosis into the adult form, a non-swimming Stauromedusae (Miranda et al. 2010; Miranda and Collins 2019). The planula larvae of staurozoans are quite different from other planulae. Staurozoans have a non-ciliated, creeping larva that moves by a series of extensions and retractions, instead of swimming (Otto 1976; Collins 2002; Miranda et al. 2010). No photosensitive structures have been identified on the staurozoan planula larva, nor have there been reports of photosensitivity in the larva.

Adults across the class possess pigment-spot ocelli; for a recent in-depth review of eyes in Staurozoa, see (Miranda and Collins 2019). Out of the 11 genera in the class, 4 possess pigment-spot ocelli: *Calvadosia*, *Stylocoronella*, *Manania*, and *Haliclystus* (Miranda and Collins 2019). The pigment-spot ocelli are typically found on the oral side of the calyx at the inner base of the tentacles (Blumer et al. 1995; Westlake and Page 2017). Additionally, pigment spots have been observed on rhopalioids in some genera, where rhopalioids (also called anchors) are structures found between each arm in some stauromedusae (Fig. 2.3l). Rhopalioids may be homologous to rhopalia (Westlake and Page 2017) and are best known for their role in adhesion rather than sensation (Miranda et al. 2013; Miranda and Collins 2019). Light may also play a role in synchronous spawning (Otto 1976) in staurozoans.

Hydromedusae do not possess rhopalia; however, they do have ocelli (Fig. 2.3k). Hydromedusae vary in the number of ocelli they possess, the type of ocelli they possess, and the locations of the ocelli (Yamasu and Yoshida 1973; Martin 2002). Like other medusae, hydromedusae display diel migrations and possess a shadow reflex (Arkett 1985; Arkett and Spencer 1986). Light is also involved in pacemaker activity and in oocyte maturation and release in hydrozoans. Studies in *Clytia hemisphaerica* have shown that light-induced spawning (Ikegami et al. 1978; Freeman 1987) is mediated by a gonad-expressed opsin (Takeda et al. 2006; Quiroga Artigas et al. 2018). Additionally, hydromedusae also coordinate gamete release with diel migrations to ensure fertilization (Mills 1983; Martin 2002).

Hydrozoan polyps also display photobehaviors that have been studied in *Hydra vulgaris*. Polyps do not possess photosensitive structures like ocelli but rather display extraocular photosensitivity, like anthozoan polyps. Light is known to mediate cnidocyte discharge in *Hydra* tentacles through an opsin-mediated phototransduction pathway (Plachetzki et al. 2012). Light-mediated discharge was first identified in *Hydra*, but it has recently been shown to pervade Cnidaria. Because of this, Picciani et al. (2021) concluded that this type of extraocular photosensitivity predated the origin of eyes in Cnidaria, as proposed by Plachetzki et al. (2012). *Hydra*, and other hydrozoan polyps, display body contractions in response to light pulses (Singer et al. 1891; Passano and McCullough 1964; Rushforth 1965; Taddei-Ferretti et al. 2004) and phototactic locomotor behaviors. Additionally, hydra show diel rhythms in their contraction behavior, which is supported by rhythmic gene expression patterns; however, *Hydra* do not possess core clock genes as reported in anthozoans (Kanaya et al. 2019). Other photobehaviors of hydrozoan polyps include light-induced spawning, as in *Hydractinia* where photic stimuli triggers meiosis in oocytes, the rupturing of gonadal walls, and release of gametes (Ballard 1942; Frank et al. 2001; Quiroga Artigas et al. 2018).

2.6 Future Directions

Studies in Cnidaria are critical to our understanding of the evolution of photosensory and visual systems in animals. Cnidaria is the sister group to bilaterians, which means that bilaterians and cnidarians inherited the same genetic toolkit from their common ancestor. Therefore, studying cnidarian photosensory systems can provide insight into the evolution of bilaterian visual systems and inform hypotheses about the origins of vision. In addition, cnidarians demonstrate a rich photobiology that is interwoven throughout the dynamic life histories and ecological niches that characterize the phylum. However, research is still needed to inform many aspects of cnidarian photobiology. A key area that requires resolution is the metazoan opsin phylogeny, including a diversity of cnidarian species, where important questions remain. Additionally, there is a critical need to identify the underlying components involved in the cnidarian phototransduction cascades. Some components have been identified for cnidops, but genetic manipulation and electrophysiology are needed to understand their physiological function. Finally, while there has been an interest in cnidarian photobiology for decades, our understanding still only extends to a handful of species and has focused primarily on adult stages. Only by advancing these critical areas of research will we be able to grapple with the ecological, organismal, and evolutionary implications of photobiology in Cnidaria.

References

Arendt D, Wittbrodt J (2001) Reconstructing the eyes of Urbilateria. Philos Trans R Soc B Biol Sci 356:1545–1563. https://doi.org/10.1098/rstb.2001.0971

Arkett S (1985) The shadow response of a hydromedusan (Polyorchis penicillatus): behavioral mechanisms controlling diel and ontogenic vertical migration. Biol Bull 169:297–312

Arkett SA, Spencer AN (1986) Neuronal mechanisms of a hydromedusan shadow reflex – I. Identified reflex components and sequence of events. J Comp Physiol A 159:201–213. https://doi.org/10.1007/BF00612303

Ballard WW (1942) The mechanism for synchronous spawning in Hydractinia and Pennaria. Biol Bull LXXXII 82:329

Berger EW (1898) The histological structure of the eyes of Cubomedusae. J Comp Neurol 8:223–230

Bielecki J, Zaharoff AK, Leung NY et al (2014) Ocular and extraocular expression of opsins in the rhopalium of Tripedalia cystophora (Cnidaria: Cubozoa). PLoS One 9. https://doi.org/10.1371/journal.pone.0098870

Birch S, Plachetzki D (n.d.) Multisensory integration (MSI) by polymodal sensory neurons dictates larval settlement in a brainless cnidarian larva. Prep

Birch S, Provencher C, DiMeo C, et al (n.d.) Phototactic preference and its genetic basis in the planulae of the colonial Hydrozoan Hydractinia symbiolongicarpus. Prep

Bjarnadóttir TK, Gloriam DE, Hellstrand SH et al (2006) Comprehensive repertoire and phylogenetic analysis of the G protein-coupled receptors in human and mouse. Genomics 88:263–273. https://doi.org/10.1016/j.ygeno.2006.04.001

Blumer MJF, Salvini-Plawen LV, Kikinger R, Büchinger T (1995) Ocelli in a Cnidaria polyp: the ultrastructure of the pigment spots in Stylocoronella riedli (Scyphozoa, Stauromedusae). Zoomorphology 115:221–227. https://doi.org/10.1007/BF00393802

Bosch TCG, Klimovich A, Domazet-Lošo T et al (2017) Back to the basics: cnidarians start to fire. Trends Neurosci 40:92–105. https://doi.org/10.1016/j.tins.2016.11.005

Brady AK, Hilton JD, Vize PD (2009) Coral spawn timing is a direct response to solar light cycles and is not an entrained circadian response. Coral Reefs 28:677–680. https://doi.org/10.1007/s00338-009-0498-4

Brady AK, Snyder KA, Vize PD (2011) Circadian cycles of gene expression in the coral, Acropora millepora. PLoS One 6. https://doi.org/10.1371/journal.pone.0025072

Burriesci MS, Raab TK, Pringle JR (2012) Evidence that glucose is the major transferred metabolite in dinoflagellate-cnidarian symbiosis. J Exp Biol 215:3467–3477. https://doi.org/10.1242/jeb.070946

Buskey EJ (2003) Behavioral adaptations of the cubozoan medusa Tripedalia cystophora for feeding on copepod (Dioithona oculata) swarms. Mar Biol 142:225–232. https://doi.org/10.1007/s00227-002-0938-y

Chapman DM, James R (1973) Intraepithelial flagella in the medusa of Aurelia Aurita (L.). Publ Seto Mar Biol Lab 20:731–743. https://doi.org/10.5134/175746

Chari T, Weissbourd B, Gehring J et al (2021) Whole-animal multiplexed single-cell RNA-seq reveals transcriptional shifts across Clytia medusa cell types. Sci Adv 7. https://doi.org/10.1126/sciadv.abh1683

Clay Smith W, Goldsmith TH (1991) The role of retinal photoisomerase in the visual cycle of the honeybee. J Gen Physiol 97:143–165. https://doi.org/10.1085/jgp.97.1.143

Coates MM (2003) Visual ecology and functional morphology of Cubozoa (Cnidaria). Integr Comp Biol 43:542–548. https://doi.org/10.1093/icb/43.4.542

Collin R, Mobley AS, Lopez LB et al (2010) Phototactic responses of larvae from the marine sponges Neopetrosia proxima and Xestospongia bocatorensis (Haplosclerida: Petrosiidae). Invertebr Biol 129:121–128

Collins AG (2002) Phylogeny of the Medusozoa and the evolution of cnidarian life cycles. J Evol Biol 15:418–432

Collins A (2009) Recent insights into cnidarian phylogeny. Smithson Contrib Mar Sci:139–149. https://doi.org/10.5479/si.1943667X.0
Daly M, Brugler MR, Cartwright P et al (2007) The phylum Cnidaria: a review of phylogenetic patterns and diversity 300 years after Linnaeus. Zootaxa 1668:127–182. https://doi.org/10.1016/j.biopsych.2005.09.016
Dambly-Chaudière C, Jamet E, Burri M et al (1992) The paired box gene pox neuro: a determiant of poly-innervated sense organs in Drosophila. Cell 69:159–172. https://doi.org/10.1016/0092-8674(92)90127-X
Del Pilar Gomez MDP, Nasi E (2000) Light transduction in invertebrate hyperpolarizing photoreceptors: possible involvement of a G(o)-regulated guanylate cyclase. J Neurosci 20:5254–5263. https://doi.org/10.1523/jneurosci.20-14-05254.2000
Donner AL, Maas RL (2004) Conservation and non-conservation of genetic pathways in eye specification. Int J Dev Biol 48:743–753. https://doi.org/10.1387/ijdb.041877ad
Dos Reis M, Thawornwattana Y, Angelis K et al (2015) Uncertainty in the timing of origin of animals and the limits of precision in molecular timescales. Curr Biol 25:2939–2950. https://doi.org/10.1016/j.cub.2015.09.066
Eakin R, Westfall J (1962) Fine structure of photoreceptors in the Hydromedusan, Polyorchis penicillatus. Proc Natl Acad Sci 48:826–833
Fain GL (2020) Photoreception. In: Sensory transduction, 2nd edn. Oxford University Press, pp 178–216
Falkowski PG, Dubinsky Z, Muscatine L, Porter JW (1984) Light and the bioenergetics of a symbiotic coral. Bioscience 34:705–709
Feuda R, Hamilton S, Mclnerney JO, Pisani D (2012) Metazoan opsin evolution reveals a simple route to animal vision. Proc Natl Acad Sci 109:18868–18872. https://doi.org/10.1073/pnas.1305990110
Feuda R, Rota-Stabelli O, Oakley TH, Pisani D (2014) The comb jelly opsins and the origins of animal phototransduction. Genome Biol Evol 6:1964–1971. https://doi.org/10.1093/gbe/evu154
Feuda R, Menon AK, Göpfert MC (2022) Rethinking opsins. Mol Biol Evol 39. https://doi.org/10.1093/molbev/msac033
Finn JT, Xiong WH, Solessio EC, Yau KW (1997) A cGMP-gated cation channel and phototransduction in depolarizing photoreceptors of the lizard parietal eye. Vis Res 38:1353–1357. https://doi.org/10.1016/S0042-6989(97)00365-9
Fleming JF, Feuda R, Roberts NW, Pisani D (2020) A novel approach to investigate the effect of tree reconstruction artifacts in single-gene analysis clarifies opsin evolution in nonbilaterian metazoans. Genome Biol Evol 12:3906–3916. https://doi.org/10.1093/gbe/evaa015
Foo SA, Liddell L, Grossman A, Caldeira K (2020) Photo-movement in the sea anemone Aiptasia influenced by light quality and symbiotic association. Coral Reefs 39:47–54. https://doi.org/10.1007/s00338-019-01866-w
Frank U, Leitz T, Müller WA (2001) The hydroid Hydractinia: a versatile, informative cnidarian representative. BioEssays 23:963–971. https://doi.org/10.1002/bies.1137
Fredericks CA (1976) Oxygen as a limiting factor in phototaxis and in intraclonal spacing of the sea anemone Anthopleura elegantissima. Mar Biol 38:25–28. https://doi.org/10.1007/BF00391482
Fredriksson R, Lagerström MC, Lundin LG, Schiöth HB (2003) The G-protein-coupled receptors in the human genome form five main families. Phylogenetic analysis, paralogon groups, and fingerprints. Mol Pharmacol 63:1256–1272. https://doi.org/10.1124/mol.63.6.1256
Freeman G (1987) The role of oocyte maturation in the ontogeny of the fertilization site in the hydrozoan Hydractinia echinata. Rouxs Arch Dev Biol 196:83–92. https://doi.org/10.1007/BF00402029
Fujita S, Kuranaga E, Nakajima YI (2019) Cell proliferation controls body size growth, tentacle morphogenesis, and regeneration in hydrozoan jellyfish Cladonema pacificum. PeerJ 2019:1–20. https://doi.org/10.7717/peerj.7579
Galliot B, Quiquand M (2011) A two-step process in the emergence of neurogenesis. Eur J Neurosci 34:847–862. https://doi.org/10.1111/j.1460-9568.2011.07829.x

Galliot B, Quiquand M, Ghila L et al (2009) Origins of neurogenesis, a cnidarian view. Dev Biol 332:2–24. https://doi.org/10.1016/j.ydbio.2009.05.563
Garm A, Bielecki J (2008) Swim pacemakers in box jellyfish are modulated by the visual input. J Comp Physiol A Neuroethol Sensory, Neural, Behav Physiol 194:641–651. https://doi.org/10.1007/s00359-008-0336-0
Garm A, Ekström P (2010) Evidence for multiple photosystems in jellyfish. Int Rev Cell Mol Biol 280:41–78. https://doi.org/10.1016/S1937-6448(10)80002-4
Garm A, Mori S (2009) Multiple photoreceptor systems control the swim pacemaker activity in box jellyfish. J Exp Biol 212:3951–3960. https://doi.org/10.1242/jeb.031559
Garm A, Ekström P, Boudes M, Nilsson DE (2006) Rhopalia are integrated parts of the central nervous system in box jellyfish. Cell Tissue Res 325:333–343. https://doi.org/10.1007/s00441-005-0134-8
Garm A, Coates MM, Gad R et al (2007a) The lens eyes of the box jellyfish Tripedalia cystophora and Chiropsalmus sp. are slow and color-blind. J Comp Physiol A Neuroethol Sensory, Neural Behav Physiol 193:547–557. https://doi.org/10.1007/s00359-007-0211-4
Garm A, O'Connor M, Parkefelt L, Nilsson DE (2007b) Visually guided obstacle avoidance in the box jellyfish Tripedalia cystophora and Chiropsella bronzie. J Exp Biol 210:3616–3623. https://doi.org/10.1242/jeb.004044
Garm A, Oskarsson M, Nilsson DE (2011) Box jellyfish use terrestrial visual cues for navigation. Curr Biol 21:798–803. https://doi.org/10.1016/j.cub.2011.03.054
Garm A, Bielecki J, Petie R, Nilsson DE (2012) Opposite patterns of diurnal activity in the box jellyfish Tripedalia cystophora and Copula sivickisi. Biol Bull 222:35–45. https://doi.org/10.1086/BBLv222n1p35
Garm A, Hedal I, Islin M, Gurska D (2013) Pattern-and contrast-dependent visual response in the box jellyfish Tripedalia cystophora. J Exp Biol 216:4520–4529. https://doi.org/10.1242/jeb.091934
Garm A, Bielecki J, Petie R, Nilsson DE (2016) Hunting in bioluminescent light: vision in the nocturnal box jellyfish Copula sivickisi. Front Physiol 7. https://doi.org/10.3389/fphys.2016.00099
Giribet G, Edgecombe GD (2020) 8. Cnidaria. In: The invertebrate tree of life. Princeton University Press, pp 55–79
Gorbunov MY, Falkowski PG (2002) Photoreceptors in the cnidarian hosts allow symbiotic corals to sense blue moonlight. Limnol Oceanogr 47:309–315. https://doi.org/10.4319/lo.2002.47.1.0309
Gornik SG, Bergheim BG, Morel B et al (2021) Photoreceptor diversification accompanies the evolution of Anthozoa. Mol Biol Evol 38:1744–1760. https://doi.org/10.1093/molbev/msaa304
Grawunder D, Hambleton EA, Bucher M et al (2015) Induction of gametogenesis in the cnidarian endosymbiosis model Aiptasia sp. Sci Rep 5:1–11. https://doi.org/10.1038/srep15677
Graziussi DF, Suga H, Schmid V, Gehring WJ (2012) The "eyes absent" (eya) gene in the eye-bearing hydrozoan jellyfish Cladonema radiatum: conservation of the retinal determination network. J Exp Zool Part B Mol Dev Evol 318:257–267. https://doi.org/10.1002/jez.b.22442
Hamner WM, Jones MS, Hamner PP (1995) Swimming, feeding, circulation and vision in the Australian box jellyfish, Chironex fleckeri (Cnidaria: Cubozoa). Mar Freshw Res 46:985–990. https://doi.org/10.1071/MF9950985
Helm RR (2018) Evolution and development of scyphozoan jellyfish. Biol Rev 93:1228–1250. https://doi.org/10.1111/brv.12393
Helm RR, Siebert S, Tulin S et al (2013) Characterization of differential transcript abundance through time during Nematostella vectensis development. BMC Genomics 14:266. https://doi.org/10.1186/1471-2164-14-266
Hering L, Mayer G (2014) Analysis of the opsin repertoire in the tardigrade Hypsibius dujardini provides insights into the evolution of opsin genes in Panarthropoda. Genome Biol Evol 6:2380–2391. https://doi.org/10.1093/gbe/evu193

Hoadley KD, Szmant AM, Pyott SJ (2011) Circadian clock gene expression in the coral favia fragum over diel and lunar reproductive cycles. PLoS One 6. https://doi.org/10.1371/journal.pone.0019755

Holst S, Jarms G (2007) Substrate choice and settlement preferences of planula larvae of five Scyphozoa (Cnidaria) from German Bight, North Sea. Mar Biol 151:863–871. https://doi.org/10.1007/s00227-006-0530-y

Hufnagel LA, Kass Simon G, Lyon MK (1985) Functional organization of battery cell complexes in tentacles of Hydra attenuata. J Morphol 184:323–341. https://doi.org/10.1002/jmor.1051840307

Hyman LH (1940) The invertebrates: protozoa through Ctenophora. Am Midl Nat 24:449–501

Ikegami S, Honji N, Yoshida M (1978) Light-controlled production of spawning-inducing substance in jellyfish ovary. Nature 272:611–612

Jager M, Chiori R, Alié A et al (2011) New insights on ctenophore neural anatomy: Immunofluorescence study in Pleurobrachia pileus (Müller, 1776). J Exp Zool Part B Mol Dev Evol 316(B):171–187. https://doi.org/10.1002/jez.b.21386

Jokiel PL, Ito RY, Liu PM (1985) Night irradiance and synchronization of lunar release of planula larvae in the reef coral Pocillopora damicornis. Mar Biol 88:167–174. https://doi.org/10.1007/BF00397164

Kanaya HJ, Kobayakawa Y, Itoh TQ (2019) Hydra vulgaris exhibits day-night variation in behavior and gene expression levels. Zool Lett 5:1–12. https://doi.org/10.1186/s40851-019-0127-1

Kaniewska P, Alon S, Karako-Lampert S et al (2015) Signaling cascades and the importance of moonlight in coral broadcast mass spawning. elife 4:1–14. https://doi.org/10.7554/elife.09991

Katsukura Y, Ando H, David CN et al (2004) Control of planula migration by LWamide and RFamide neuropeptides in Hydractinia echinata. J Exp Biol 207:1803–1810. https://doi.org/10.1242/jeb.00974

Kayal E, Bentlage B, Sabrina Pankey M et al (2018) Phylogenomics provides a robust topology of the major cnidarian lineages and insights on the origins of key organismal traits. BMC Evol Biol 18:1–18. https://doi.org/10.1186/s12862-018-1142-0

Kim YM, Choi J, Lee HY et al (2014) DbCRY: a Web-based comparative and evolutionary genomics platform for blue-light receptors. Database 2014:1–10. https://doi.org/10.1093/database/bau037

Kojima D, Terakita A, Ishikawa T et al (1997) A novel G(o)-mediated phototransduction cascade in scallop visual cells. J Biol Chem 272:22979–22982. https://doi.org/10.1074/jbc.272.37.22979

Koyanagi M, Takano K, Tsukamoto H et al (2008) Jellyfish vision starts with cAMP signaling mediated by opsin-Gs cascade. Proc Natl Acad Sci 105:15576–15580. https://doi.org/10.1073/pnas.0806215105

Kozmik Z, Daube M, Frei E et al (2003) Role of pax genes in eye evolution: a Cnidarian PaxB gene uniting Pax2 and Pax6 functions. Dev Cell 5:773–785. https://doi.org/10.1016/S1534-5807(03)00325-3

Kozmik Z, Ruzickova J, Jonasova K et al (2008) Assembly of the cnidarian camera-type eye from vertebrate-like components. Proc Natl Acad Sci 105:8989–8993. https://doi.org/10.1073/pnas.0800388105

Kumar J (2009) The molecular circuitry governing retinal determination. Biochim Biophys Acta 1789:306–314. https://doi.org/10.1016/j.bbagrm.2008.10.001.The

Leach WB, Reitzel AM (2019) Transcriptional remodelling upon light removal in a model cnidarian: losses and gains in gene expression. Mol Ecol 28:3413–3426. https://doi.org/10.1111/mec.15163

Leach WB, Reitzel AM (2020) Decoupling behavioral and transcriptional responses to color in an eyeless cnidarian. BMC Genomics 21:1–15. https://doi.org/10.1186/s12864-020-6766-y

Leach WB, Macrander J, Peres R, Reitzel AM (2018) Transcriptome-wide analysis of differential gene expression in response to light: dark cycles in a model cnidarian. Comp Biochem Physiol - Part D Genomics Proteomics 26:40–49. https://doi.org/10.1016/j.cbd.2018.03.004

Levy O, Dubinsky Z, Achituv Y (2003) Photobehavior of stony corals: responses to light spectra and intensity. J Exp Biol 206:4041–4049. https://doi.org/10.1242/jeb.00622

Levy O, Appelbaum L, Leggat W et al (2007) Light-responsive cryptochromes from a simple multicellular animal, the coral Acropora millepora. Science (80-) 318:467–470. https://doi.org/10.1126/science.1145432

Levy O, Kaniewska P, Alon S et al (2011) Complex diel cycles of gene expression in coral-algal symbiosis. Science (80-) 331:175. https://doi.org/10.1126/science.1196419

Liegertová M, Pergner J, Kozmiková I et al (2015) Cubozoan genome illuminates functional diversification of opsins and photoreceptor evolution. Sci Rep 5:1–19. https://doi.org/10.1038/srep11885

Macias-Munõz A, Murad R, Mortazavi A (2019) Molecular evolution and expression of opsin genes in Hydra vulgaris. BMC Genomics 20:1–19. https://doi.org/10.1186/s12864-019-6349-y

Mackie GO (2004) Central neural circuitry in the jellyfish *Aglantha*. Neurosignals 13:5–19. https://doi.org/10.1159/000076155

Mackie GO, Meech RW (1995) Central circuitry in the jellyfish Aglantha digitale. I. The relay system. J Exp Biol 198:2261–2270. https://doi.org/10.1242/jeb.198.11.2261

Marlow H, Matus DQ, Martindale MQ (2013) Ectopic activation of the canonical wnt signaling pathway affects ectodermal patterning along the primary axis during larval development in the anthozoan Nematostella vectensis. Dev Biol 380:324–334. https://doi.org/10.1016/j.ydbio.2013.05.022

Martin VJ (2002) Photoreceptors of cnidarians. Can J Zool 80:1703–1722. https://doi.org/10.1139/z02-136

Mason BM, Cohen JH (2012) Long-wavelength photosensitivity in coral planula larvae. Biol Bull 222:88–92. https://doi.org/10.1086/BBLv222n2p88

Mason B, Beard M, Miller MW (2011) Coral larvae settle at a higher frequency on red surfaces. Coral Reefs 30:667–676. https://doi.org/10.1007/s00338-011-0739-1

Mason B, Schmale M, Gibbs P et al (2012) Evidence for multiple Phototransduction pathways in a reef-building coral. PLoS One 7:1–9. https://doi.org/10.1371/journal.pone.0050371

Mills CE (1983) Verttical migration and diel activity patterns of hydromedusae: studies in a large ttank. J Plankton Res 5:619

Miranda LS, Collins AG (2019) Eyes in Staurozoa (Cnidaria): a review. PeerJ 7:e6693. https://doi.org/10.7717/peerj.6693

Miranda LS, Collins AG, Marques AC (2010) Molecules clarify a cnidarian life cycle – the "Hydrozoan" Microhydrula limopsicola is an early life stage of the staurozoan Haliclystus antarcticus. PLoS One 5:1–9. https://doi.org/10.1371/journal.pone.0010182

Miranda LS, Collins AG, Marques AC (2013) Internal anatomy of Haliclystus antarcticus (Cnidaria, Staurozoa) with a discussion on histological features used in staurozoan taxonomy. J Morphol 274:1365–1383. https://doi.org/10.1002/jmor.20185

Moroz LL, Kocot KM, Citarella MR et al (2014) The ctenophore genome and the evolutionary origins of neural systems. Nature 510:109–114. https://doi.org/10.1038/nature13400

Mulla A, Lin C-H, Takahashi S, Nozawa Y (2020) Species-specific phototaxis of coral larvae causes variation in vertical positioning during dispersal. bioRxiv 68:1–12

Müller WA, Leitz T (2002) Metamorphosis in the Cnidaria. Can J Zool 80:1755–1771. https://doi.org/10.1139/z02-130

Mundy CN, Babcock RC (1998) Role of light intensity and spectral quality in coral settlement: implications for depth-dependent settlement? J Exp Mar Bio Ecol 223:235–255. https://doi.org/10.1016/S0022-0981(97)00167-6

Muscatine L, Falkowski PG, Porter JW, Dubinsky Z (1984) Fate of photosynthetic fixed carbon in light- and shade-adapted colonies of the symbiotic coral Stylophora pistillata. Proc R Soc B Biol Sci 222:181–202

Nakanishi N, Yuan D, Jacobs DK, Hartenstein V (2008) Early development, pattern, and reorganization of the planula nervous system in Aurelia (Cnidaria, Scyphozoa). Dev Genes Evol 218:511–524. https://doi.org/10.1007/s00427-008-0239-7

Nakanishi N, Hartenstein V, Jacobs DK (2009) Development of the rhopalial nervous system in Aurelia sp.1 (Cnidaria, Scyphozoa). Dev Genes Evol 219:301–317. https://doi.org/10.1007/s00427-009-0291-y

Nakanishi N, Yuan D, Hartenstein V, Jacobs DK (2010) Evolutionary origin of rhopalia: insights from cellular-level analyses of Otx and POU expression patterns in the developing rhopalial nervous system. Evol Dev 12:404–415. https://doi.org/10.1111/j.1525-142X.2010.00427.x

Nakanishi N, Camara AC, Yuan DC et al (2015) Gene expression data from the moon jelly, Aurelia, provide insights into the evolution of the combinatorial code controlling animal sense organ development. PLoS One 10. https://doi.org/10.1371/journal.pone.0132544

Nilsson DE (2009) The evolution of eyes and visually guided behaviour. Philos Trans R Soc B Biol Sci 364:2833–2847. https://doi.org/10.1098/rstb.2009.0083

Nilsson DE, Gislen L, Coates MM et al (2005) Advanced optics in a jellyfish eye. Lett to Nat 31:201–205. https://doi.org/10.1029/2004GL019844

Nordström K, Wallén R, Seymour J, Nilsson D (2003) A simple visual system without neurons in jellyfish larvae. Proc R Soc B Biol Sci 270:2349–2354. https://doi.org/10.1098/rspb.2003.2504

O'Connor M, Garm A, Nilsson DE (2009) Structure and optics of the eyes of the box jellyfish Chiropsella bronzie. J Comp Physiol A Neuroethol Sensory, Neural, Behav Physiol 195:557–569. https://doi.org/10.1007/s00359-009-0431-x

Oakley TH, Speiser DI (2015) How complexity originates: the evolution of animal eyes. Annu Rev Ecol Evol Syst 46:237–260. https://doi.org/10.1146/annurev-ecolsys-110512-135907

Orlov DV (1997) The role of larval settling behaviour in determination of the specific habitat of the hydrozoan Dynamena pumila (L.) larval settlement in Dynamena pumila (L.). J Exp Mar Bio Ecol 208:73–85. https://doi.org/10.1016/S0022-0981(96)02673-1

Otto JJ (1976) Early development and planula movement in Haliclystus (Scyphozoa: Stauromedusae). Coelenterate Ecology and Behavior, pp 319–329

Owens GL, Rennison DJ (2017) Evolutionary ecology of opsin gene sequence, expression and repertoire. Mol Ecol 26:1207–1210. https://doi.org/10.1111/mec.14032

Parkefelt L, Ekström P (2009) Prominent system of RFamide immunoreactive neurons in the rhopalia of box jellyfish (Cnidaria: Cubozoa). J Comp Neurol 516:157–165. https://doi.org/10.1002/cne.22072

Parkefelt L, Skogh C, Nilsson DE, Ekstköm P (2005) Bilateral symmetric organization of neural elements in the visual system of a coelenterate, Tripedalia cystophora (Cubozoa). J Comp Neurol 492:251–262. https://doi.org/10.1002/cne.20658

Passano LM (1965) Pacemakers and activity patterns in medusae: homage to romanes. Integr Comp Biol 5:465–481. https://doi.org/10.1093/icb/5.3.465

Passano LM (1973) Behavioral control systems in medusae: a comparison between hydro- and Scyphomedusae. Publ Seto Mar Biol Lab 20:615–645

Passano LM, McCullough CB (1964) Co-ordinating systems and behaviour in hydra: I. Pacemaker System of the Periodic Contractions. J Exp Biol 41:643–664. https://doi.org/10.1242/jeb.41.3.643

Pearse VB (1974a) Modification of sea anemone behavior by symbiotic zooxanthellae: Phototaxis. Biol Bull 147:630–640

Pearse VB (1974b) Modification of sea anemone behavior by symbiotic zooxanthellae: expansion and contraction. Biol Bull 147:641–651

Pearse JS, Pearse VB (1978) Vision in cubomedusan jellyfishes. Science (80-) 199:458. https://doi.org/10.1126/science.22934

Petie R, Garm A, Nilsson DE (2013a) Contrast and rate of light intensity decrease control directional swimming in the box jellyfish Tripedalia cystophora (Cnidaria, Cubomedusae). Hydrobiologia 703:69–77. https://doi.org/10.1007/s10750-012-1345-0

Petie R, Garm A, Nilsson DE (2013b) Velarium control and visual steering in box jellyfish. J Comp Physiol A Neuroethol Sensory, Neural, Behav Physiol 199:315–324. https://doi.org/10.1007/s00359-013-0795-9

Piatigorsky J, Horwitz J, Kuwabara T, Cutress CE (1989) The cellular eye lens and crystallins of cubomedusan jellyfish. J Comp Physiol A 164:577–587. https://doi.org/10.1007/BF00614500
Picciani N, Roberts NG, Daly M et al (2018) Prolific origination of eyes in Cnidaria with co-option of non-visual opsins. Curr Biol 28:2413–2419.e4. https://doi.org/10.1016/j.cub.2018.05.055
Picciani N, Kerlin JR, Jindrich K et al (2021) Light modulated cnidocyte discharge predates the origins of eyes in Cnidaria. Ecol Evol 11:3933–3940. https://doi.org/10.1002/ece3.7280
Piraino S, Zega G, Di Benedetto C et al (2011) Complex neural architecture in the diploblastic larva of Clava multicornis (Hydrozoa, Cnidaria). J Comp Neurol 519:1931–1951. https://doi.org/10.1002/cne.22614
Plachetzki DC, Degnan BM, Oakley TH (2007) The origins of novel protein interactions during animal opsin evolution. PLoS One 2. https://doi.org/10.1371/journal.pone.0001054
Plachetzki DC, Fong CR, Oakley TH (2010) The evolution of phototransduction from an ancestral cyclic nucleotide gated pathway. Proc R Soc B Biol Sci 277:1963–1969. https://doi.org/10.1098/rspb.2009.1797
Plachetzki DC, Fong CR, Oakley TH (2012) Cnidocyte discharge is regulated by light and opsin-mediated phototransduction. BMC Biol:10. https://doi.org/10.1186/1741-7007-10-17
Plickert G, Schneider B (2004) Neuropeptides and photic behavior in Cnidaria. Hydrobiologia 530–531:49–57. https://doi.org/10.1007/s10750-004-2689-x
Porter ML (2016) Beyond the eye: molecular evolution of extraocular photoreception. Integr Comp Biol 56:842–852. https://doi.org/10.1093/icb/icw052
Porter ML, Blasic JR, Bok MJ et al (2011) Shedding new light on opsin evolution. Proc R Soc B Biol Sci 279:3–14. https://doi.org/10.1098/rspb.2011.1819
Quiroga Artigas G, Lapébie P, Leclère L et al (2018) A gonad-expressed opsin mediates light-induced spawning in the jellyfish Clytia. elife 7:1–22. https://doi.org/10.7554/elife.29555
Ramirez MD, Pairett AN, Pankey MS et al (2016) The last common ancestor of most bilaterian animals possessed at least nine opsins. Genome Biol Evol 8:3640–3652. https://doi.org/10.1093/gbe/evw248
Reitzel AM, Behrendt L, Tarrant AM (2010) Light entrained rhythmic gene expression in the sea anemone Nematostella vectensis: the evolution of the animal circadian clock. PLoS One 5:1–9. https://doi.org/10.1371/journal.pone.0012805
Reitzel AM, Tarrant AM, Levy O (2013) Circadian clocks in the cnidaria: environmental entrainment, molecular regulation, and organismal outputs. Integr Comp Biol 53:118–130. https://doi.org/10.1093/icb/ict024
Rivera AS, Ozturk N, Fahey B et al (2012) Blue-light-receptive cryptochrome is expressed in a sponge eye lacking neurons and opsin. J Exp Biol 215:1278–1286. https://doi.org/10.1242/jeb.067140
Rosenberg Y, Doniger T, Harii S et al (2017) Canonical and cellular pathways timing gamete release in Acropora digitifera, Okinawa, Japan. Mol Ecol 26:2698–2710. https://doi.org/10.1111/mec.14062
Rushforth NB (1965) Inhibition of contraction responses of hydra. Integr Comp Biol 5:505–513. https://doi.org/10.1093/icb/5.3.505
Russell FS (1970) The Medusae of the British Isles. Cambridge University Press, Cambridge
Sakai Y, Kato K, Koyama H et al (2020) A step-down photophobic response in coral larvae: implications for the light-dependent distribution of the common reef coral, Acropora tenuis. Sci Rep 10:1–12. https://doi.org/10.1038/s41598-020-74649-x
Satterlie RA (1979) Central control of swimming in the cubomedusan jellyfish Carybdea rastonii. J Comp Physiol A 133:357–367. https://doi.org/10.1007/BF00661138
Satterlie RA (2002) Neuronal control of swimming in jellyfish: a comparative story. Can J Zool 80:1654–1669. https://doi.org/10.1139/z02-132
Schafer E (1878) Observations on the nervous system of Aurelia Aurita. Philos Trans R Soc B Biol Sci 169:563–575

Schnitzler CE, Pang K, Powers ML et al (2012) Genomic organization, evolution, and expression of photoprotein and opsin genes in Mnemiopsis leidyi: a new view of ctenophore photocytes. BMC Biol 10. https://doi.org/10.1186/1741-7007-10-107
Schuyler Q, Sullivan BK (1997) Light responses and diel migration of the scyphomedusa Chrysaora quinquecirrha in mesocosms. J Plankton Res 19:1417–1428. https://doi.org/10.1093/plankt/19.10.1417
Seymour JE, O'hara EP (2020) Pupillary response to light in three species of cubozoa (Box jellyfish). Plankt Benthos Res 15:73–77. https://doi.org/10.3800/pbr.15.73
Shen D, Jiang M, Hao W et al (1994) A human opsin-related gene that encodes a retinaldehyde-binding protein. Biochemistry 33:13117–13125. https://doi.org/10.1021/bi00248a022
Shichida Y, Matsuyama T (2009) Evolution of opsins and phototransduction. Philos Trans R Soc B Biol Sci 364:2881–2895. https://doi.org/10.1098/rstb.2009.0051
Silver SJ, Rebay I (2005) Signaling circuitries in development: insights from the retinal determination gene network. Development 132:3–13. https://doi.org/10.1242/dev.01539
Singer RH, Rushforth NB, Burnett AL (1891) The photodynamic action of light on. Dev Biol:169–173
Skogh C, Garm A, Nilsson D-E, Ekstrom P (2006) Bilaterally symmetrical rhopalial nervous system of the box jellyfish Tripedalia cystophora. J Morphol 267:1391–1405. https://doi.org/10.1002/jmor
Sleigh MA (1963) Movements and co-ordination of the ciliary comb plates of the ctenophores Beroe and Pleurobrachia. Nature 199:620–621
Srivastava M, Begovic E, Chapman J et al (2008) The Trichoplax genome and the nature of placozoans. Nature 454:955–960. https://doi.org/10.1038/nature07191
Steen R (1988) The bioenergetics of symbiotic sea anemones (Anthozoa: Actiniaria). Symbiosis (Philadelphia, PA) 5:103–142
Stierwald M, Yanze N, Bamert RP et al (2004) The Sine oculis/six class family of homeobox genes in jellyfish with and without eyes: development and eye regeneration. Dev Biol 274:70–81. https://doi.org/10.1016/j.ydbio.2004.06.018
Strader ME, Davies SW, Matz MV (2015) Differential responses of coral larvae to the colour of ambient light guide them to suitable settlement microhabitat. R Soc Open Sci 2. https://doi.org/10.1098/rsos.150358
Suga H, Schmid V, Gehring WJ (2008) Evolution and functional diversity of jellyfish opsins. Curr Biol 18:51–55. https://doi.org/10.1016/j.cub.2007.11.059
Suga H, Tschopp P, Graziussi DF et al (2010) Flexibly deployed Pax genes in eye development at the early evolution of animals demonstrated by studies on a hydrozoan jellyfish. Proc Natl Acad Sci 107:14263–14268. https://doi.org/10.1073/pnas.1008389107
Svane I, Dolmer P (1995) Perception of light at settlement: a comparative study of two invertebrate larvae, a scyphozoan planula and a simple ascidian tadpole. J Exp Mar Bio Ecol 187:51–61. https://doi.org/10.1016/0022-0981(94)00171-9
Sweeney BM (1976) Circadian rhythms in corals, particularly Fungiidae. Biol Bull 151:236–246
Sweeney AM, Boch CA, Johnsen S, Morse DE (2011) Twilight spectral dynamics and the coral reef invertebrate spawning response. J Exp Biol 214:770–777. https://doi.org/10.1242/jeb.043406
Taddei-Ferretti C, Musio C, Santillo S, Cotugno A (2004) The photobiology of Hydra's periodic activity. Hydrobiologia 530–531:129–134. https://doi.org/10.1007/s10750-004-2680-6
Takeda N, Kyozuka K, Deguchi R (2006) Increase in intracellular cAMP is a prerequisite signal for initiation of physiological oocyte meiotic maturation in the hydrozoan Cytaeis uchidae. Dev Biol 298:248–258. https://doi.org/10.1016/j.ydbio.2006.06.034
Tarrant AM, Helm RR, Levy O, Rivera HE (2019) Environmental entrainment demonstrates natural circadian rhythmicity in the cnidarian Nematostella vectensis. J Exp Biol 222. https://doi.org/10.1242/jeb.205393
Technau U, Steele RE (2012) Evolutionary crossroads in developmental biology: Cnidaria. Development 139:4491–4491. https://doi.org/10.1242/dev.090472
Terakita A (2005) The opsins. Genome Biol 6:1–9. https://doi.org/10.1186/gb-2005-6-3-213

Thorson G (1964) Light as an ecological factor in the dispersal and settlement of larvae of marine bottom invertebrates. Ophelia 1:167–208. https://doi.org/10.1080/00785326.1964.10416277

Vöcking O, Kourtesis I, Tumu SC, Hausen H (2017) Co-expression of xenopsin and rhabdomeric opsin in photoreceptors bearing microvilli and cilia. elife 6:1–26. https://doi.org/10.7554/eLife.23435

Von Salvini-Plawen L, Mayr E (1977) On the evolution of photoreceptors and eyes. Evol Biol 10:207–263

Watanabe H, Fujisawa T, Holstein TW (2009) Cnidarians and the evolutionary origin of the nervous system. Develop Growth Differ 51:167–183. https://doi.org/10.1111/j.1440-169X.2009.01103.x

Weber C (1981a) Structure, histochemistry, ontogenetic development, and regeneration of the ocellus of Cladonema radiatum dujardin (cnidaria, hydrozoa, anthomedusae). J Morphol 167:313–331. https://doi.org/10.1002/jmor.1051670306

Weber C (1981b) Lens of the hydromedusan Cladonema studied by SDS gel electrophoresis and immunofluorescent technique. J Exp Zool 217:15–21. https://doi.org/10.1002/jez.1402170103

Werner B, Cutress CE, Studebaker JP (1971) Life cycle of Tripedalia cystophora Conant (Cubomedusae). Nature 232:582–583

Westfall JA, Kinnamon JC (1978) A second sensory-motor-interneuron with neurosecretory granules in Hydra. J Neurocytol 7:365–379. https://doi.org/10.1007/BF01176999

Westlake HE, Page LR (2017) Muscle and nerve net organization in stalked jellyfish (Medusozoa: Staurozoa). J Morphol 278:29–49. https://doi.org/10.1002/jmor.20617

Xiong WH, Solessio EC, Yau KW (1998) An unusual cGMP pathway underlying depolarizing light response of the vertebrate parietal-eye photoreceptor. Nat Neurosci 1:359–365. https://doi.org/10.1038/nn0998_359

Yamashiro H, Nishira M (1995) Phototaxis in Fungiidae corals (Scleractinia). Mar Biol 124:461–465. https://doi.org/10.1007/BF00363920

Yamasu T, Yoshida M (1973) Electron microscopy on the photoreceptors of an Anthomedusa and a Scyphomedusa. Publ Seto Mar Biol Lab 20:757–778. https://doi.org/10.5134/175744

Yamasu T, Yoshida M (1976) Fine structure of complex ocelli of a cubomedusan, Tamoya bursaria Haeckel. Cell Tissue Res 170:325–339. https://doi.org/10.1007/BF00219415

Yau KW, Hardie RC (2009) Phototransduction motifs and variations. Cell 139:246–264. https://doi.org/10.1016/j.cell.2009.09.029

Yellowlees D, Rees TAV, Leggat W (2008) Metabolic interactions between algal symbionts and invertebrate hosts. Plant Cell Environ 31:679–694. https://doi.org/10.1111/j.1365-3040.2008.01802.x

Yokoyama S (2000) Molecular evolution of vertebrate visual pigments. Prog Reti Eye Res 19:385–419

Yoshida M, Ohtsu K (1973) A preliminary note on the electrical response to shadows of the Anthomedusa, Spirocodon saltatrix. Publ Seto Mar Biol Lab 20:647–651. https://doi.org/10.5134/175753

Yoshida M, Yoshino Y (1980) Differentiation of ocelli in Ephyrae of Aurelia aurita. Developmental and cellular biology of coelenterates: Proceedings of the 4th International Coelenterates Conference, pp 343–346

Yuan D, Nakanishi N, Jacobs DK, Hartenstein V (2008) Embryonic development and metamorphosis of the scyphozoan Aurelia. Dev Genes Evol 218:525–539. https://doi.org/10.1007/s00427-008-0254-8

Chapter 3
Extraocular Vision in Echinoderms

Lauren Sumner-Rooney and Jack Ullrich-Lüter

Abstract Scientists have observed light sensitivity in a wide range of echinoderms over several centuries, despite the vast majority of the phylum lacking eyes. Opsin-expressing cells are found scattered across the echinoderm body surface and appear to mediate light responses in all five extant classes. Among the eyeless groups, some species nonetheless exhibit what appear to be visual abilities, such as orienting to distant stimuli or responding to the appearance of potential predators. This ability for "extraocular vision" has been the subject of decades of research and – despite substantial progress – remains enigmatic. Although only explicitly demonstrated in two species so far, there is evidence to support extraocular vision in a range of sea urchins and brittle stars, using photoreceptors spread across the body. Several mechanisms for light channeling, photoreceptor screening, and sensitivity adjustment have been proposed but, so far, the underlying workings of these strange visual systems are elusive. This chapter will synthesize existing work in these groups, including behavioral, morphological, and molecular studies, and evaluate some of the proposed mechanisms that could support extraocular vision. We also review the challenges posed by such unconventional visual systems and suggest future areas of study.

Keywords Echinoderms · Photoreception · Extraocular vision · Decentralized · Radial symmetry · Ophiuroids · Echinoids · Immunohistochemistry · Behavior

---

L. Sumner-Rooney (✉)
Oxford University Museum of Natural History, University of Oxford, Oxford, UK

Leibniz Institute for Biodiversity and Evolution, Museum für Naturkunde, Berlin, Germany
e-mail: lauren.sumner-rooney@mfn.berlin

J. Ullrich-Lüter
Leibniz Institute for Biodiversity and Evolution, Museum für Naturkunde, Berlin, Germany

E. Buschbeck, M. Bok (eds.), *Distributed Vision*, Springer Series in Vision Research, https://doi.org/10.1007/978-3-031-23216-9_3

3.1 Introduction

Echinoderms have proved a challenging group for neuroethologists. Their radial symmetry and highly unusual nervous system organization make comparative studies with other phyla, even within Ambulacraria, extremely difficult. Hyman (1955) characterized the echinoderm nervous system as "somewhat primitive" and summarized that "there is a poor development of sense organs in the phylum." However, biologists and natural historians have reported high sensitivity to light in echinoderms for more than a century, triggering long-standing searches for photoreceptors and eyes (Mangold 1909; Holmes 1912; Dubois 1913; Von Üxküll 1925; see Hyman 1955 for summary). Eyes are indeed found in echinoderms, but their distribution appears to be restricted to just one or two classes. In the sea stars (Chap. 4), a modified terminal tube foot forms the optic cushion, a derived compound eye that facilitates orientation to objects, gaze stabilization, and other relatively advanced visual behaviors (Garm and Nilsson 2014; Beer et al. 2016; Petie et al. 2016; Garm 2017). Paired patches of pigment found at the base of the tentacles in some synaptid holothurians (sea cucumbers) have been described as ocelli (Yamamoto and Yoshida 1978) and implicated in photoreception, but not in vision per se. This chapter will specifically address extraocular vision: the enigmatic ability to see without discrete, image-forming organs. Thus, the optic cushions of sea stars, and putative ocelli in synaptid sea cucumbers, will not be further discussed here. Instead, we will explore the use of photoreceptors distributed elsewhere in the body to facilitate vision. In echinoderms, these may include light-reactive cells in the spines, radial nerves, body wall, and tube feet (Fig. 3.1, Table 3.1) and have been studied primarily in sea urchins (Echinoidea) and brittle stars (Ophiuroidea).

3.2 A Brief History of Extraocular Photoreception and Vision in Echinoderms

The vast majority of echinoderms studied react to light, but the underlying mechanisms remain mysterious. More than a century of research has almost yielded more questions than it has answers. All groups exhibit extraocular photoreception, including sea stars with optic cushions. This has frequently been described as a "dermal light sense" or similar terms that imply scattered photoreceptors or intrinsically light-sensitive nerves (see Yoshida 1966 for a comprehensive review). Generally, extraocular photoreception mediates photonegative responses in echinoids, ophiuroids, holothurians, and comatulid crinoids (with some exceptions, e.g., the sea urchin *Temnopleurus toreumaticus*), whereas sea stars exhibit both photopositive and photonegative responses (see Table 18.2 in Yoshida 1966). Such responses include phototaxis (Perrier 1873; Pearse 1908; Mangold 1909; Cowles 1910; Holmes 1912; Crozier 1914, 1915; Olmsted 1917; Hendler 1984; Johnsen and Kier 1999; Delroisse et al. 2016; Sumner-Rooney et al. 2018), covering (Dubois 1913;

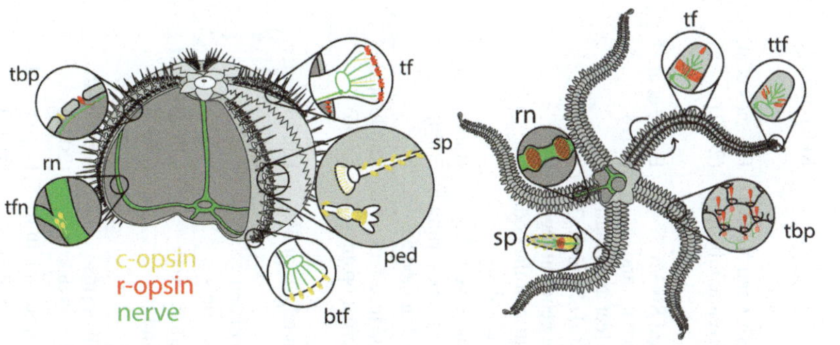

Fig. 3.1 Opsin localization in a generalized sea urchin and a generalized brittle star. A series of immunohistochemical surveys have demonstrated reactivity to antibodies raised against both c-opsin and r-opsin from *Strongylocentrotus purpuratus* (see Table 3.2). (**a**) In sea urchins, c-opsin reactivity is found in cells at the disc surface of the buccal tube feet, within the disc of non-buccal tube feet, in the base and surface of the spines, in the base and collar of the pedicellariae, in epithelial cells in the trabecular pores, and within the radial nerve cord at the base of the tube foot nerves. R-opsin reactivity is mostly associated with the tube feet, with strong reactivity in cells at the surface of the disc and lining the skeletal groove at the base of the tube foot, as well as reactivity in epithelial cells. (**b**) In brittle stars, c-opsin reactivity is mostly restricted to the spines, being reported in the interior base and the surfaces of the spines. R-opsin reactivity is somewhat species specific, but in various taxa, r-opsin-reactive cells are found within trabecular pores, at the tip and collar of the tube feet, within the spines, and in the swellings of the radial nerve cord. Btf, buccal tube foot; ped, pedicellariae; rn, radial nerve; sp., spine; tbp, trabecular pore; tf, tube foot; tfn, tube foot nerve; ttf, terminal tube foot

Mortensen 1948; Millott 1955; Lees and Carter 1972; Adams 2001), color change (Yoshida 1956; Hendler 1984; Byrne et al. 2004), defensive behaviors (Millott 1954; Millott and Yoshida 1960), and contraction of exposed body parts (Pearse 1908; Mangold 1909; Cowles 1910; MacCurdy 1912, 1913; Moore 1921; Yamanouchi 1929). Sensitivity to short (blue) and medium (green) wavelengths, and an apparent lack of sensitivity to long (red) wavelengths, has been highlighted by several authors (Stubbs 1983; Delroisse et al. 2014, 2016; Sumner-Rooney et al. 2018). In larvae, responses to illumination vary between species, as well as ontogenetic stages, with positive, negative, and a lack of taxes variously reported, as well as changes to swimming speed and tortuosity in response to ambient light (Pennington and Emlet 1986; Montgomery et al. 2018).

However, in some taxa, the sensitivity and/or sophistication of light responses is such that researchers began to investigate the possibility that they were mediated by vision (i.e., spatial resolution), and not "just" photoreception (the directional or nondirectional sensing of light), despite the apparent absence of discrete visual organs. Owing to the particularly unconventional nature of extraocular vision, it has been highly challenging to characterize and explain. Multiple hypotheses have been proposed, including several that have been discredited but still merit summarizing here.

Sea urchins have formed the focus of most vision research in echinoderms, likely thanks to their accessibility, easy husbandry, robustness, and often pronounced

Table 3.1 Summary of vision- and eye-related gene expression studies in echinoids and ophiuroids

Species	Gene/protein	Method	Investigated tissue(s)	Results	References
Echinoidea					
Paracentrotus lividus	Opsin4, Pax6, Retinal TFs, Six3, NeuroD	WMISH, IHC	Juveniles	Opsin4, Pax6, Six3 and NeuroD expression in juvenile tube feet	Paganos et al. (2022)
Heliocidaris erythrogramma	Pax6, Six1/2, Six3/6, Eya, and Dach1	WMISH, IHC	Juveniles	Pax6, Six1/2, and Six3/6 expressed in primary and secondary tube feet and putative sensory cells. Six1/2 and Six3/6 expressed in the neuropil region in the terminal disc of the podia. Dach localized to spines.	Byrne et al. (2018)
Strongylocentrotus purpuratus	pax6 Retinal TFs Sp-opsin 1, 5 and 4	Phylogenetic and gene expression analysis	Tube feet	Opsins, pax6, and several key mammalian retinal transcription factors expressed in tube fee	Burke et al. 2006)
S. purpuratus	Sp-Opsin1,2,3.1,3.2,4 and 5	Phylogenetic and gene expression analysis	Tube feet, pedicellaria	Pedicellariae and tube feet express six different opsin proteins	Raible et al. (2006)
S. purpuratus	Sp-opsin4 pax6	WMISH, IHC WMISH	Tube feet, epidermis Juveniles, tube feet	r-opsin+ cells in tube foot tips and bases pax6+ in tube feet and surrounding epidermis	Ullrich-Lüter et al. (2011)
S. purpuratus, Echinus esculentus, S. droebachiensis	r-opsins	IHC	Tube feet	r-opsin+ cell patterns vary across species and conspecifically	Ullrich-Lüter (2011)
S. purpuratus	Sp-opsin4, 1, 5	qRT-PCR	Tube feet, disc vs. stalk	Expression of all opsins severalfold higher in disc	Agca et al. (2011)
S. droebachiensis	opsin5	qPCR	Tube feet from different body regions	Highest opsin expression in tf exposed to lowest irradiance	Lesser et al. (2011)

S. purpuratus	Sp-opsin1	WMISH, IHC	Locomotory and buccal tube feet, spines, pedicellaria, and epidermis	c-opsin present in all investigated tissues	Ullrich-Lüter et al. (2013)
Brissopsis lyrifera	c-opsin, r-opsin	IHC	Diff. tube feet, spines	Both opsins only in spines	Blaue & Ullrich-Lüter, unpublished data
Diadema africanum	c-opsin, r-opsin	IHC	Tube feet, spines, epidermis	Diffuse r-opsin and c-opsin expression, no specific cells identified yet	Blaue & Ullrich-Lüter, unpublished data
Eucidaris tribuloides	c-opsin, r-opsin	IHC	Tube feet	r-opsin tube foot tip	Blaue & Ullrich-Lüter, unpublished data
Ophiuroidea					
Ophioderma brevispinum	Putative opsin	Western blot, immunolabeling against bovine rhodopsin	Arms, central disc	Stroma of spines immunopositive	Johnsen 1997
Amphiura filiformis, *Ophiocoma nigra*	c-opsin	IHC	Arm segments	c-opsin in spines	Ullrich-Lüter et al. (2013)
A. filiformis	Opsin candidates	Phylogenetic analysis	Draft genome, arm transcriptome	Six r-opsins One c-opsin	Delroisse et al. (2014)
A. filiformis	Transduction pathway genes	Paired-end Illumina HiSeq™ Phylogenetic gene expression analysis	Arm tissues	Key components of c- and r-opsin phototransduction pathways (e.g., TRP, cGMP)	Delroisse et al. (2016)

(continued)

Table 3.1 (continued)

Species	Gene/protein	Method	Investigated tissue(s)	Results	References
Ophiopsila aranea	Opsin candidates Transduction pathway genes	Paired-end Illumina HiSeq™ Phylogenetic gene expression analysis	Arm tissues	r-opsin, RGR opsin in *O. aranea*, key components of c- and r-opsin phototransduction pathways	Delroisse et al. (2016)
A. filiformis	r-opsin	IHC	Arm tissues	r-opsin in tube feet and radial nerve cord	(Delroisse et al. 2014)
Ophiocomina nigra, *Ophiothrix fragilis*	r-opsin c-opsin	IHC	Arm tissues	c-opsins in spines, r-opsins in tube feet	Blaue & Ullrich-Lüter, unpublished
Ophiura ophiura	r-opsin c-opsin	IHC	Arm tissues	r-opsin in spine tips c-opsin in spines	Blaue & Ullrich-Lüter, unpublished data
A. chiajei	c-opsin r-opsin	IHC	Arm tissues	c-opsin + r-opsin in spines, tube feet, and arm epidermis	Blaue & Ullrich-Lüter, unpublished data
A. filiformis	c-opsin	IHC	Arm tissues	c-opsin in spines	
Ophiomastix wendtii, *Ophiocoma echinata*, *Ophiocomella pumila*	r-opsin (c-opsin)	IHC	Arm tissues	r-opsin within trabecular pores, near surface of arm plates, (radial) nerves, spines c-opsin within trabecular pores, patchy?	Sumner-Rooney et al. (2018) Sumner-Rooney unpublished

Table 3.2 Proposed angular resolution of extraocular vision in echinoderms. These reflect the minimum stimulus size to elicit a behavioral response and therefore may be larger than the true limit of detection

Species	Proposed angular resolution	Method	References
Echinoidea			
Centrostephanus coronatus	22–27°	Orientation to a dark stimulus	Notar (2016)[a]
Diadema africanum	29–69°	Orientation to an isoreflectant stimulus	Kirwan et al. (2018)
Diadema africanum	~40°[b]	Orientation to a dark stimulus	Kirwan et al. (2018)
Diadema africanum	13–25°	Appearance of a dark stimulus	Kirwan et al. (2018)
Diadema antillarum	<72°	Orientation to a dark stimulus	Woodley (1982)[a]
Diadema setosum	9.35–11°	Orientation to a dark stimulus	Al-Wahaibi and Claereboudt (2017)
Echinometra lucunter	26–33°	Orientation to a dark stimulus	Blevins and Johnsen (2004)
Echinometra viridis	26–33°	Orientation to a dark stimulus	Blevins and Johnsen (2004)
Strongylocentrotus franciscanus	22–42°	Orientation to a dark stimulus	Notar (2016)[a]
Strongylocentrotus purpuratus	10° (Yerramilli and Johnsen 2010); 42° (Notar 2016)	Orientation to a dark stimulus	Yerramilli and Johnsen (2010), Jackson and Johnsen (2011), and Notar (2016)[a]
Ophiuroidea			
Ophiomastix wendtii	40–50°	Orientation to an isoreflectant stimulus	Sumner-Rooney et al. (2020) and Sumner-Rooney et al. (2021)
Ophiomastix wendtii	25–50°	Orientation to a dark stimulus	Sumner-Rooney et al. (2018) and Sumner-Rooney et al. (2020)
Ophiomastix wendtii	<70°	Appearance of a dark stimulus	Sumner-Rooney et al. (2021)

[a]These data are part of a thesis or conference abstract and have not yet been peer-reviewed
[b]Orientation to this stimulus was ambiguous

phototaxes and shadow responses. The particularly marked photosensitivity of diadematids was historically associated with striking blue iridiophores, highly reflective rounded or elongate structures on the genital plates (Fig. 3.2a). These were originally considered to be eyes (Doederlein 1885; Sarasin and Sarasin 1887), a characterization that persisted for many decades until Mortensen (1940) and Millott (1953) demonstrated histologically that they lacked pigmentation, sensory cells, or innervation and behaviorally that they were unresponsive to light.

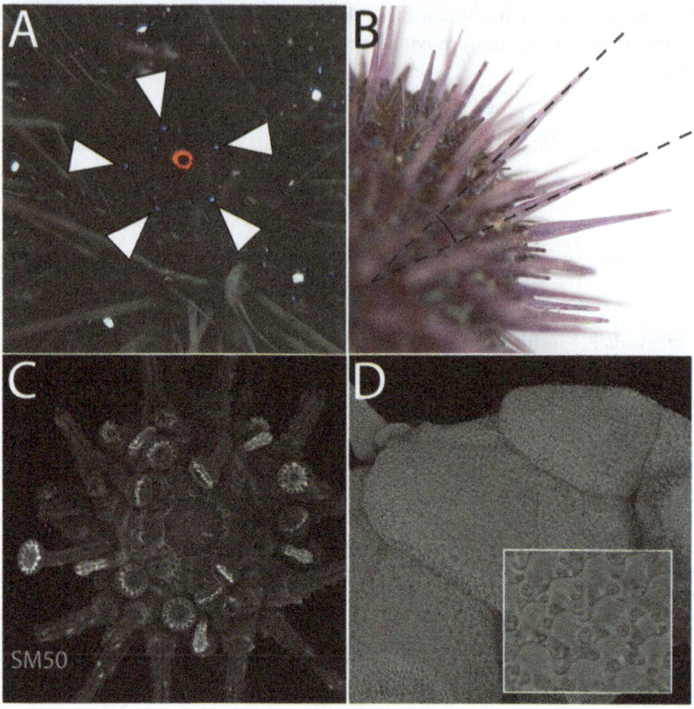

Fig. 3.2 Structures historically associated with photoreception in sea urchins and brittle stars. (**a**) Iridiophores (arrowheads) on the aboral side of the sea urchin *Diadema setosum*, erroneously reported to be eyes. Image: Philippe Bourjon, Creative Commons. (**b**) Sea urchin spines have been proposed to shade the test and photoreceptors therein, providing the screening required for resolution. The spacing of the spines would determine the acceptance angle (dotted lines). (Image: Adapted from A. Cutting, Creative Commons, after Blevins and Johnsen 2004). (**c**) Confocal image of juvenile *Paracentrotus lividus*, with immunoreactivity to skeletal marker SM50 highlighting the calcite ossicles in the tube feet, suggested to channel light. Image: Jeffrey Thompson (Thompson et al. 2021). (**d**) Expanded peripheral trabeculae (inset) on the dorsal arm plates of *Ophiomastix wendtii*, interpreted as microlenses. (Image: Sumner-Rooney et al. 2018)

Since this dismissal of the only proposed eyes in sea urchins, the idea that the body surface is analogous to a compound eye has become widespread. First proposed by Woodley (1982), the near-spherical nature of most echinoids lends itself to this analogy and provides a useful framework within which we can start to imagine how extraocular vision – and the challenges of integrating across wide regions of the body – might work. The "dermal" light sense had been well documented already in the context of photoreception, but the possibility that a diffuse sense could contribute to resolving vision was (and remains) novel and somewhat abstract.

The involvement of calcite skeletal components in sea urchin vision became a common theme across Echinoidea, having been proposed in various guises. Raup (1960) claimed that a certain amount of light transmits through the calcite skeleton depending in part on the refractive index of the contained soft tissue and, in part, by the crystallographic orientation of the plate; he later argued that this light may also become polarized (Raup 1965). Woodley (1982) proposed that the spines screened the light-sensitive test (the rounded body of the sea urchin) in such a way as to facilitate spatial resolution across the animal's body surface (Fig. 3.2b). Lesser et al. (2011) suggested that ossicles in the tube foot disc could function as a "light collector": they showed that illumination of a dissected ossicle at any given point scatters the light throughout the entire ossicle, not dissimilar to a fiber optic, and proposed an optical role (Fig. 3.2c).

Among brittle stars, ophiocomids have received the majority of research attention. Several species exhibit strong photonegativity and a striking change in body color, caused by the contraction and expansion of chromatophores in response to ambient light (Hendler 1984, 2005; O'Hara et al. 2004). Like sea urchins, an optical role for skeletal structures has also been proposed in brittle stars. Calcitic protuberances on the arm plates (termed expanded peripheral trabeculae, Fig. 3.2d) were proposed to act as "microlenses" in the ophiocomid *Ophiomastix wendtii*, focusing light onto an unknown photosensitive element below (Hendler 1984; Hendler and Byrne 1987; Aizenberg et al. 2001). It was suggested that these structures were shaped and positioned to reduce birefringence and spherical aberration (Aizenberg et al. 2001) and could contribute to a global visual system as ommatidia would to a compound eye. Light-responsive chromatophores were proposed to regulate the amount of light entering the "microlens," with their contraction into the stereom at night allowing more light to reach the photoreceptors and thus increase sensitivity in the dark.

In another brittle star, *Ophioderma brevispinum*, the arm plates and ossicles were demonstrated to polarize transmitted light (Johnsen 1994). It was suggested that this could confer polarization sensitivity through the perpendicular orientation of adjacent ossicles and the comparison of signals from underlying photosensitive elements. Behavioral experiments showed that animals oriented significantly and preferred shade under polarized, but not unpolarized, light (Johnsen 1994; Johnsen and Kier 1999). Proposed uses for such an ability included monitoring depth and avoiding potentially dangerous exposure to UV radiation and predation in shallow water, but this possibility has not been further investigated, to our knowledge.

Little is known to date about photosensitivity in holothurians and crinoids. Apart from Yamamoto and Yoshida's (1978) morphological description of ocelli in the synaptid holothurian *Opheodesoma spectabilis*, no further morphological or behavioral evidence for vision sensu stricto in sea cucumbers or crinoids has been obtained to date. Otherwise, sea cucumbers have been reported to exhibit light sensitivity in the tentacles, tube feet, spines, and trunk of the body, with the greatest sensitivity at the oral end (Crozier 1915; Hess 1915; Olmsted 1917; Lin et al. 2013; Liu et al. 2020). Responses to both illumination and shading include the contraction of the tentacles and body wall, and closure of the mouth, as well as negative phototaxis in some species (Crozier 1914, 1915). Among crinoids, *Antedon* is reported to exhibit "dermal" light sensitivity (Langeloh 1937) and initiate swimming upon illumination (Dimelow 1958). Strictly visual stimuli have not been applied to subjects in these two groups to our knowledge and will thus not be further discussed within this chapter.

3.3 Visual Behavior

Beyond a general sensitivity to light, some groups have been historically highlighted for their especially strong reactions to illumination or shadows, as outlined above. In several cases since Woodley's proposition, it has been suggested that these behaviors are mediated by extraocular vision – the ability to resolve a scene. Such responses *may* be facilitated by extraocular vision, but it is important to note that responses to stimuli that cause an overall change in light intensity do not strictly necessitate resolution. Animals that are capable of directional photoreception, for example, may be able to orient to dark objects. Responding to the sudden appearance of a shadow only requires nondirectional photoreception. However, as these behaviors could be mediated visually, and experiments explicitly testing the ability to resolve have only been conducted relatively recently, we have not restricted the below discussion to rigid tests of resolution.

3.3.1 Orientation to Static Stimuli

According to Nilsson (2013), orientation to and avoidance of inanimate objects are classical behaviors requiring coarse spatial resolution. Indeed, orientation to static stimuli is the most commonly explored visual behavior in echinoderms. As is evident from more than a century of research, most echinoderms avoid exposure to bright light, which can trigger covering, contraction, or shelter-seeking responses (Perrier 1873; Mangold 1909; Cowles 1910; Holmes 1912; Dubois 1913; Crozier 1914, 1915; Clark 1915; Olmsted 1917; May 1925). Many taxa, particularly among shallow-water ophiuroids and echinoids, remain hidden beneath rocks, sandy substrate, algae, or debris during the day to avoid predation or UV exposure (Holmes 1912; Yoshida 1966; Hendler 1984; Sides and Woodley 1985; Johnsen and Kier 1999). Thus, in the case of negatively phototactic species, it is expected that animals may be motivated to move toward dark stimuli that could represent potential shelters (scototaxis). Note that in order to focus this section on likely candidates for vision, we include below only experiments testing orientation to distant and discrete stimuli and not to proximal shade or along intensity gradients as the latter can be readily achieved without spatial resolution.

In echinoids (sea urchins), orientation to dark objects has been consistently demonstrated in a range of shallow-water species. Following an unpublished report that *Diadema antillarum* orients to solid black bars (Woodley 1982), Blevins and Johnsen (2004), Yerramilli and Johnsen (2010), and Jackson and Johnsen (2011) conducted a series of experiments demonstrating that *Echinometra lucunter, E. viridis*, and *Strongylocentrotus purpuratus* also move toward solid dark stimuli presented against a paler (more reflective) background. These authors reported animals responding to stimuli as small as 10° of the horizon (in *S. purpuratus* Yerramilli and Johnsen 2010). Four further species have since been tested in this way: *S. fraciscanus* and *Centrostephanus coronatus* by Notar (2016, unpublished thesis), *Diadema setosum* by Al-Wahaibi and Claereboudt (2017), and *Diadema africanum* by Kirwan et al. (2018). All were found to respond to dark stimuli of varying angular sizes, usually orienting toward them (see Table 3.1).

Several variations on this experiment have been conducted in sea urchins, testing responses to stimuli of different gray values (affecting the Weber contrast, Al-Wahaibi and Claereboudt 2017) and colors (reflected wavelengths, Al-Wahaibi and Claereboudt 2017) as well as varying ambient light intensity (Notar 2016) and absolute distance from the animal (Jackson and Johnsen 2011). Al-Wahaibi and Claereboudt (2017) found that *Diadema setosum* is able to orient to stimuli of 19° angular width with gray values of 50% and above, but not 37% or below; note that these figures refer to the percentage of black ink used to print stimuli, and not to reflectance measurements. The same authors found that *D. setosum* oriented to red, but not to blue or green stimuli of the same size, despite the "lightness" of the blue and green stimuli being lower than the red stimulus using the L*a*b color system. This broadly aligns with previous reports of animals being maximally sensitive to blue and green wavelengths, but insensitive to red. Notar (2016), in as yet unpublished data, found that *S. purpuratus* reacted to a 14° black stimulus with ambient illuminance of 10 lux and above but that *C. coronatus* only did so under 100 lux and above. Various authors have noted postural aspects of phototactic behavior: in *Arbacia* phototaxis occurs with spines lowered on the illuminated side and raised on the darker side (Mangold 1909; Holmes 1912), and Notar reported waving or scanning of the tube feet on the leading side.

In brittle stars, although negative phototaxis has been widely reported in a range of species (Mangold 1909; May 1925; Hendler 1984; Johnsen and Kier 1999; Delroisse et al. 2015; Sumner-Rooney et al. 2018), movement toward distant dark stimuli has only been reported in *Ophiomastix wendtii* (Cowles 1910; Sumner-Rooney et al. 2018, 2020, see Table 3.1). This species will respond to a black vertical bar occupying 50° of the horizon, but not 25° (Sumner-Rooney et al. 2020), although earlier work by Cowles (1910) suggested they were able to detect smaller shaded areas of unspecified size. This orientation response was absent in another ophiocomid, *Ophiocomella pumila*, despite exhibiting strong negative phototaxis and similarities in the photoreceptor system between the two species (Sumner-Rooney et al. 2018, 2020).

However, as described above, orientation to a stimulus that causes a local net decrease in light intensity does not strictly require vision. The results of these experiments could be scototactic, facilitated by directional photoreception, and the possibility of extraocular vision remains unconfirmed. In order to explicitly test for resolving vision, recent work has used stimuli that are, on average, isoreflectant with the background against which they are presented, thus requiring spatial resolution to be detected (Fig. 3.3). This approach was first used in echinoderms by Petie

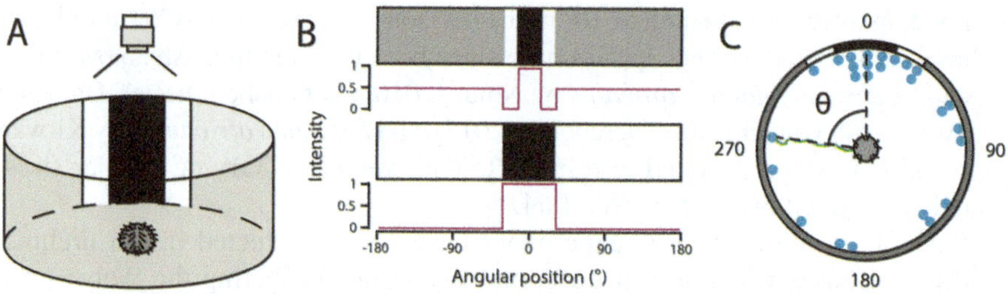

Fig. 3.3 Typical experimental setup for visual behavior assays. (**a**) The subject is placed at the center of a round arena, with a stimulus presented on the inner wall, and observed from above. (**b**) Typical stimuli include vertical bars or circles. Note that the presentation of a black stimulus against a white background causes mean light intensity to vary around the arena, but nested black and white bars against a gray background do not. The latter stimulus can therefore only be detected by spatial resolution. (**c**) Typical data from an orientation experiment. The subject's path of travel (in green) may be tracked, and its terminal bearing relative to the stimulus (θ) may be recorded, for statistical analysis

et al. (2016), who presented black and white stimuli against a gray background to confirm that the optic cushions of sea stars were capable of spatial resolution (see Chap. 4). Similar stimuli have since been applied to both diadematid sea urchins (Kirwan et al. 2018) and ophiocomid brittle stars (Sumner-Rooney et al. 2020). Kirwan et al. (2018) presented *Diadema africanum* with a difference of Gaussians (DoG) stimulus, in which the background transitions gradually from gray, to white, to black at the center. They found that animals oriented toward a stimulus with an arc width (distance between the two white maxima) of 69°, but not to one of 29° – the first unequivocal evidence for vision in any animal lacking eyes. The authors of this study noted that a proposed resolution of 30–69° is considerably worse than reports from other echinoderms, but it is important that many of these studies may have been testing phototaxis and not vision and that a small dark object may be detected using a lower resolution than its size (Yerramilli and Johnsen 2010). Similarly, experiments using the brittle star *Ophiomastix wendtii* found that they responded to both a DoG and a black bar centered on a white bar (Sumner-Rooney et al. 2020), confirming the capacity for extraocular vision in ophiuroids as well as sea urchins. Sumner-Rooney et al. (2021) also found that prospective resolution, based on behavioral thresholds, was very coarse, with animals orienting to stimuli of 50° but not 40° angular width.

3.3.2 *Shadows and Looms*

Vision may also facilitate the detection of, and defensive responses to, potential predators. While shadows cast overhead inevitably decrease the overall ambient light intensity, in some (non-echinoderm) taxa, it has been demonstrated that spatial resolution of a dark shape is required to elicit defensive behavior, as opposed to a uniform decrease in intensity (e.g., Speiser et al. 2011). Responses to shading the body surface have been reported in several sea urchins, most thoroughly studied in diadematids (Von Üxküll 1925; Millott 1954; Millott and Yoshida 1960; Kirwan et al. 2018) but also in some other species (e.g., *Arbacia* Mangold 1909; Holmes 1912). Typically, this entails the pointing and/or oscillation of spines in the direction of the shadow, thought to be a defensive behavior. This can be provoked by shading of either the surface of the test or the radial nerve directly, and a similar "on" response can be elicited by point illumination (Millott and Yoshida 1960). In a series of experiments on *Diadema* (reviewed in Yoshida 1966), Millott and Yoshida determined that the duration, amplitude, or potentially speed of the response was affected by the intensity and duration of the stimulus. Responses to shading at the surface were found to vary according to position, with the greatest sensitivity toward the margins of the ambulacrum (Millott 1954). Millott (1954) also identified dark adaptation as a key factor heightening the sensitivity of the response. In the brittle star *Ophiura ophiura*, shading provokes a "freezing" response in active animals (Moore and Cobb 1985).

Alternatively, animals may be presented with a dark shape, as opposed to a shadow being cast over their body surface by an object. Kirwan et al. (2018) presented *Diadema africanum* with black circular stimuli on an adjacent computer screen. Animals responded with defensive spine-pointing to stimuli subtending 25° and 44°, but not to those subtending 10°. These authors also experimented with isoluminant stimuli to test the need for spatial resolution, but these were apparently inefficient in provoking a response (Kirwan et al. 2018); this could indicate that the shadow response is not, in fact, visually mediated. Similarly, the presentation of a dark shape on an overhead screen elicits jerks of the arm and sometimes the raising of arm spines in the brittle star *Ophiomastix wendtii* (Sumner-Rooney et al. 2021).

The appearance of looming shapes – those that grow exponentially in the field of view – can hold particular biological significance as they may represent an approaching object or organism and, thus, a threat of collision or predation. Specific responses to looms have been found in a wide range of animals (e.g., Oliva et al. 2007; Yamawaki and Toh 2009; Yilmaz and Meister 2013; De Franceschi et al. 2016). *Diadema africanum* exhibits its defensive spine-pointing response when presented with both appearing and looming shadows (Kirwan et al. 2018), but does not appear to discriminate between these. *Ophiomastix wendtii* appears to respond to a looming shadow that reaches an angular width of 70° (Sumner-Rooney et al. 2021), but it is unclear whether this response differs from that to a shadow of the same size appearing suddenly or passing overhead.

3.4 Physiology

3.4.1 Sea Urchins

Several species of sea urchins and brittle stars have been subject to electrophysiological investigation of their photosensitivity. Given the considerable challenges of identifying candidate photoreceptors (see below), these have exclusively focused on recording responses from the radial nerve cords. In all cases, these experiments have used a point light source, with some variation in intensity, duration, position, or wavelength, and not visual stimuli.

Physiological investigations of sea urchins have taken advantage of the ability to dissect the test, remove the viscera, and access the radial nerve cord directly, but electrophysiological recording has not been extensive. Millott (1954) demonstrated intrinsic photosensitivity of the radial nerve in *Diadema antillarum* by illuminating the nerve directly, which provoked the expected spine response. This was confirmed by Millott and Yoshida (Millott and Yoshida 1960), and later electrophysiologically in *D. setosum* by Takahashi (1964). Takahashi reported a longer duration response to shading than illumination, as well as a slight delay between stimulation and electrical activity, despite the speed of the spine response. Millott and Okumura (1968) recorded from the radial nerve cords of *Diadema*, *Arbacia lixula*, *Echinus esculentus*, and *Paracentrotus lividus* but could not replicate the results of Takahashi (1964).

3.4.2 Brittle Stars

Among brittle stars, *Ophiura ophiura* has been used extensively for physiological study, including shadow responses. Moore and Cobb (1985) described a "off" reaction recorded from the radial nerve, with spiking initiated in response to shadowing the arms and finishing when the shadow was removed. They demonstrated that darker shadows (produced using additional layers of neutral density filter) provoked higher spike rates and that the arm tips were the most sensitive to stimulation. Stubbs (1983, unpublished thesis) reported that the oral and aboral surfaces of the arm were equally sensitive and that responses were only recorded when more than one arm segment was stimulated (all subsequent experiments by Moore and Cobb stimulated five segments, presumably for optimal responses). In contrast to findings in sea urchins, Stubbs was unable to record responses from isolated radial nerve cords and suggested that the dermal plates were required and thus contained the photoreceptors. Cobb and Moore (1989) found that shadow responses in *O. ophiura* typically lasted for 7–17 s and that preparations eventually became habituated to repeated stimulation, with responses diminishing after 75 exposures. Although illuminating dark-adapted arms sometimes produced a response, this was less pronounced and less consistent than responses to shading. Stubbs (1983, unpublished thesis) recorded similar, but less pronounced, responses to shading from *Ophiothrix fragilis* and *Ophiocomina nigra*. However, there is no evidence yet that these three species are capable of extraocular vision, and the recorded responses are thus likely to demonstrate general photosensitivity.

Cobb and Hendler (1990) took extracellular recordings from the radial nerve in the putatively visual brittle star *Ophiomastix wendtii* while exposing the arm to a range of light stimuli. Animals exhibited a strong response to sudden illumination and a weaker one to shading, with the greatest sensitivity at the tips and dorsal side of the arm. Comparisons of light-adapted (chromatophores expanded, pigment distributed across the arm plate surface) and dark-adapted (chromatophores contracted, pigment restricted to skeletal pores and some surface bands) preparations showed that sensitivity was greater in dark-adapted animals, in agreement with behavioral evidence from Hendler (1984).

Working on the hypothesis of Hendler (1984) and Hendler and Byrne (1987) that the photoreceptors or photosensitive nerves were located beneath skeletal "microlenses," Cobb and Hendler applied a series of chemical bleaching treatments to several light-adapted arms, sequentially disrupting cells at increasing depth from the surface of the arm plate. They observed an initial increase in sensitivity following 20–45 s of treatment, followed by a marked decrease after >45 s (the latter preparations still responded to mechanical stimulation). Transmission electron microscopy (TEM) of the treated arms showed that the shorter bleaching treatments had destroyed the epithelium and chromatophores, removing the pigment and reminiscent of the increased sensitivity of dark-adapted arms. Longer treatments, which extinguished spiking, also disrupted deeper tissue layers including the axons projecting through the skeleton, in line with the hypothesis that the photosensitive elements are located beneath the arm surface.

3.5 Photoreceptors

Extraocular photoreceptors in echinoderms aren't readily identified by their morphology alone. Typical photoreceptor characteristics, such as membrane stacking, seem to be elusive in this group. Several authors have described candidate receptor cells from electron microscopy of the tube feet, pedicellariae, spines, arms, and test (Millott and Coleman 1969; Cobb and Moore 1986) but have been unable to confirm their nature. Molecular and immunohistochemical methods for locating opsins have therefore been crucial to unpicking their distribution in the body.

3.5.1 Molecular Characteristics

With the exception of a G_0-opsin expressing ciliary photoreceptor in the mantle eye of scallop (Kojima et al. 1997), all investigated photoreceptors facilitating vision per se express either a ciliary (c-opsin) or a rhabdomeric (r-opsin) type opsin. Each of the two receptor types has a specific phototransduction cascade as well as a specific pathway for the regeneration of opsin (Land and Nilsson 2012). While c-opsins are thought to be responsible for vision in the majority of deuterostomes, most protostome eyes express r-opsins in their visual photoreceptors.

The first attempt to characterize and locate echinoderm opsins was a study of *Asterias rubens* and *Ophioderma brevispinum* by Johnsen (1997). Protein extracts of arm tissue were subjected to Western blot analysis and found to contain a membrane-associated protein that reacted with two monoclonal antibodies raised against bovine rhodopsin (a c-opsin). The arms and central disk of *O. brevispinum* were also examined immunohistochemically: reactivity to both antibodies was detected in the tips of the arm spines. At higher magnification, the immunoreactive material was localized to small regions within the stroma of the ossicles.

3.5.2 Sea Urchin Opsins

Burke et al. (2006) and Raible et al. (2006) were the first to investigate genes potentially involved in echinoderm photoreception in the newly sequenced *Strongylocentrotus purpuratus* genome (Sodergren et al. 2006), combining phylogenetic and gene expression analyses. Burke et al. identified genes encoding four rhodopsin orthologs. Raible et al. (2006) concurrently identified six opsin proteins: Sp-Opsin1, belonging to the c-opsin family; Sp-Opsin4, a clear member of the r-opsin family; and two G_0-opsins, Sp-Opsin3.1 and Sp-Opsin3.2. Two remaining sea urchin opsins, Sp-Opsin2 and Sp-Opsin5, showed no clear affiliation to any of the larger opsin subfamilies in phylogenetic and sequence analyses and were considered to be highly derived members of one of the known groups, a finding later corroborated by D'Aniello et al. (2015). Using expression analysis, Burke et al. (2006) located Sp-opsin1, Sp-opsin5, and Sp-opsin4 in the tube feet of *S. purpuratus*. Concurrently, Raible et al. (2006) found transcripts of Sp-opsin1, Sp-opsin2, Sp-opsin3.1, and Sp-opsin4 present at high levels in the tube feet and pedicellariae. These findings provided the basis for subsequent work characterizing echinoderm photoreceptor cells using in situ hybridization as well as immunohistological studies using specifically designed antibodies against Sp-opsin.

Working concurrently on the basis of these findings, several authors demonstrated the expression of various opsins in the tube feet of sea urchins, particularly r-opsins. Ullrich-Lüter et al. (2011) investigated candidate visual opsins in *Strongylocentrotus purpuratus*. Via in situ hybridization and subsequent immunohistological staining using antibodies specifically designed against *S. purpuratus* r-opsin (Sp-Op4), they identified cells expressing a rhabdomeric opsin lining the disc of the locomotory tube feet and clustered at the base of each tube foot, embedded in a depression of the calcite skeleton. Immunogold labeling at the ultrastructural level identified the r-opsin in an unspecialized epidermal primary sensory cell type bearing an unmodified cilium and microvilli. The authors described axons connecting these receptor cells to the intraepidermal nervous system. Furthermore,

Ullrich-Lüter (2011, unpublished thesis) found that the number of r-opsin-reactive PRCs not only varied substantially between species (*S. purpuratus*, *Echinus esculentus*, *S. droebachiensis*) but also showed variation between conspecific specimens. Agca et al. (2011) also determined expression of three opsin genes in *S. purpuratus* adult tube feet via qRT-PCR analysis, namely, opsin4 (melanopsin), opsin1 (ciliary opsin), and opsin5. They found opsin4 expression to be localized mainly to the disk, where it was 26-fold higher than in the stalk. Expression of all three opsins was several times higher in disks than in stalks, suggesting that opsins are generally highly enriched in this area. Lesser et al. (2011) reported a "rhabdomeric-like" opsin in the tube feet of the green sea urchin *Strongylocentrotus droebachiensis*. Using qPCR, they found that the tube feet on the oral side of the test, which are exposed to the lowest irradiances of visible light, expressed significantly more opsin transcripts than those from more exposed areas. Phylogenetic analysis showed that this opsin in fact clusters between rhabdomeric and ciliary opsins (Lesser et al. 2011) and varies from the *S. purpuratus* Sp-opsin5 by only 4%. Sp-opsin5 was already suggested to be a novel sea urchin specific opsin by Raible et al. (2006), which was confirmed by D'Aniello et al. (2015) and accordingly termed echinopsin.

As deuterostomes, one might expect echinoderms to use c-opsin in their visual system. Ullrich-Lüter et al. (2013) later investigated the expression of a c-opsin, which, according to their phylogenetic analysis, was closely related to an assemblage of c-opsins involved in chordate vision. An antibody raised against *S. purpuratus* c-opsin protein (Sp-Opsin1) allowed them to localize epitopes in a variety of tissues of different echinoderms, which they found in locomotory and buccal tube feet, spines, pedicellaria, and epidermis of *S. purpuratus*. Using Western blots and a commercial c-opsin antibody (against bovine c-opsin), Delroisse et al. (2013) also found a c-opsin expressed in the oral but not in the aboral integument (each including spines, tube feet, and pedicellariae) of *Paracentrotus lividus*.

Since these first experiments, the expression of r-opsin and c-opsin has been studied in 16 echinoid species, in order to overcome the limitations of a single model organism and help to establish a ground pattern of PRC localization in echinoids (Blaue & Ullrich-Lüter, unpublished data, Fig. 3.4, Table 3.1). Both r-opsin and c-opsin expression were detected in almost all investigated species. The distribution pattern observed in *S. purpuratus* (r-opsin-positive PRCs around the rim of the tube foot discs, with c-opsin present throughout larger portions of the epidermis, spines, tube feet, pedicellaria) seems to be the ground pattern except for irregular echinoids, where c-opsin and r-opsin expression in the heart urchin *Brissopsis lyrifera* is limited to the spines. Ironically, given their prominent role in behavioral experiments, the only "exception" seems to be the investigated *Diadema* species, where both opsin types are expressed in different tissues, but no specific immunoreactive cells have been identified, yet. It has to be noted that the tube feet in *Diadema* are not of a locomotory function (they move via their long spines), unlike *S. purpuratus*, and their tube foot morphology is modified as an adaptation to ecological needs. Cidaroid sea urchins comprise the sister group to all remaining sea urchins (Kroh and Smith 2010), including *Diadema*. As the investigated cidaroid species, *Eucidaris tribuloides*, shows a localization pattern closer to that of *S. purpuratus*,

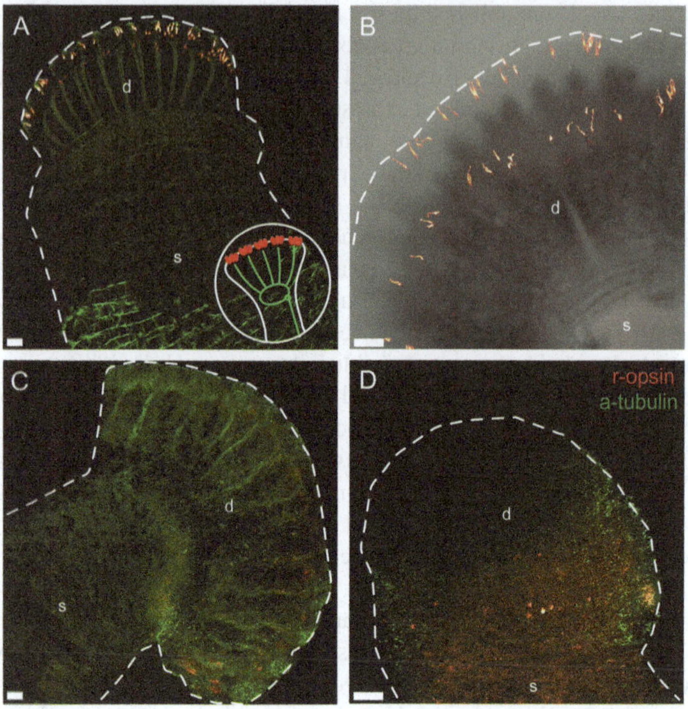

Fig. 3.4 R-opsin reactivity in representatives across echinoid phylogeny. Immunostainings: r-opsin (hot red, Sp-Op4-reactive), acetylated α-tubulin (green) in sea urchin tube feet. (**a**) *Strongylocentrotus purpuratus* (Strongylocentrotidae). Distinct r-opsin+ cells associated with neuronal tissue in tube foot disc. (**b**) *Loxechinus albus* (Parechinidae). R-opsin+ cells in tube foot disc. Note distance of cell bodies to skeletal disc rosette. (**c**) *Diadema africanum* (Diadematoida). R-opsin signal in tube foot disc. Signal resolution inconclusive. (**d**) *Eucidaris tribuloides* (Cidaroida). Few but clear r-opsin+ cells in tube foot tip and tube foot ganglion (lower right side of tube foot tip). **s**: tube foot stalk; scale bars: 50 microns

we can tentatively propose that r-opsin expression around the disc region in sea urchin locomotory tube feet, as well as c-opsin expression in the epidermis, comprises a ground pattern for potentially visual PRCs in sea urchins.

3.5.3 Brittle Star Opsins

Further molecular work began to explore opsin repertoires and expression patterns in ophiuroids. Delroisse et al. (2014) searched the *Amphiura filiformis* draft genome using opsin sequences identified from *Strongylocentrotus purpuratus* and found 14 candidates (9 bona fide opsin genes), as well as a further 3 from a new arm transcriptome. Their phylogenetic analysis found representatives of both primary types of opsins involved in vision: a surprising six r-opsin orthologs that form a clade with all other echinoderm r-opsins and a single c-opsin (Af-opsin1) that clusters with that of the sea urchins *S. purpuratus* (Sp-opsin1) and *Hemicentrotus pulcherrimus* (Hp-opsin1). The r-opsin clade was found to diverge at the base of the rhabdomeric opsin group, close to the vertebrate melanopsins, and the c-opsin clade basal to the chordate c-opsins, which includes vertebrate rhodopsins and pinopsins. Delroisse et al. (2016) using paired-end Illumina HiSeq™ technology analyzed transcriptomes from the arm tissues of two European brittle star species, *Amphiura filiformis* and *Ophiopsila aranea*. They detected an r-opsin and an RGR opsin in the arms of *Ophiopsila aranea* and found close sequence similarity of the former to one of the *Amphiura filiformis* rhabdomeric opsins (Af-op 4.2). They also recovered several key components of the ciliary and rhabdomeric phototransduction pathways in both *O. aranea* and *A. filiformis*.

Ullrich-Lüter et al. (2013) had already used an antibody raised against Sp-Opsin1 to locate c-opsin expression in the brittle stars *Amphiura filiformis* and *Ophiocoma nigra*. Contrary to their findings in *S. purpuratus,* they found the immunoreactivity to be located exclusively in cells within the spines (Fig. 3.5b, in agreement with Johnsen's original findings in Ophioderma brevispinum, 1997). These cells were in close vicinity/connection to nerve strands and associated with the calcite skeleton. Delroisse et al. (2014) confirmed the presence of c-opsin in the inner portion of the spines of *A. filiformis*, clearly contacting nearby nerve tracts and in close vicinity to dark pigmented areas of the spine bases. Sumner-Rooney et al. (2018) reported scattered c-opsin-reactive cells within the dorsal arm plates of *Ophiomastix wendtii* and *Ophiocomella pumila*. None of these authors reported c-opsin reactivity in the tube feet; however, recent investigations suggest that *Amphiura chiajei* is the only ophiuroid species so far showing c-opsin-reactive cells in their tube feet (Blaue & Ullrich-Lüter, unpublished data, Fig. 3.5).

Immunohistological labeling using anti-Sp-opsin4 antibodies in *Amphiura filiformis* revealed r-opsin-positive cells in the mid-region and tips of the tube feet and within the swellings of the radial nerve cord (Delroisse et al. 2014). Further investigations in *Ophiocomina nigra* and *Ophiothrix fragilis* corroborate the general pattern of c-opsin-reactive PRCs in the spines and r-opsin reactivity in both the tube feet and spines, but in *Ophiura ophiura*, r-opsin was located in the spines only (Ullrich-Lüter et al. 2013, Blaue & Ullrich-Lüter, unpublished data, Fig. 3.5).

Conversely, strong immunoreactivity to the same Sp-Op4 antibody was not found in the tube feet of the ophiocomids *Ophiomastix wendtii, Ophiocoma echinata*, or *Ophiocomella pumila* (Sumner-Rooney et al. 2018); some reactivity in the mid-region of the terminal tube foot was observed in *O. wendtii* but could not be strenuously supported (Sumner-Rooney unpublished data). These authors instead reported r-opsin-reactive cells located within trabecular pores, near the surface of the arm plates, in all three species. Using high-resolution microscopy, immunohistochemistry, and synchrotron tomography, they demonstrated that putative PRCs cover the animals' aboral, lateral, and oral surfaces, are associated with nerve bundles, and surround the EPTs ("microlenses"), their cell bodies being located just above the midline of those microlenses and projecting toward the surface of the arm. The presumed PRCs bear rounded terminal expansions that lack apparent

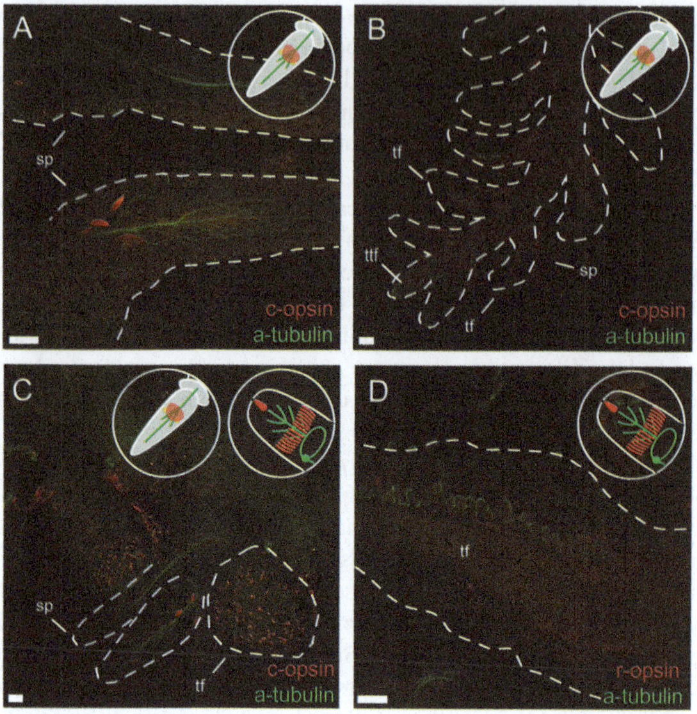

Fig. 3.5 C-opsin and r-opsin reactivity in *Ophiocomina nigra* and *Ophiothrix fragilis*. Immunostainings: opsin (hot red, either Sp-Op1- (c-opsin) or Sp-Op4- (r-opsin) reactive), acetylated α-tubulin (green) in brittle star arms. C-opsin+ cells are restricted to spines in (**a**) *O. nigra* and (**b**) *O. fragilis*. (**c**) R-opsin+ cells in *O. nigra* are most numerous in tube feet but also present in the spines. (**d**) In *O. fragilis*, r-opsin+ cells are found within tube feet only. **sp**: spines, **tf**: tube foot; scale bars: 50 microns

specialized membrane structures and react strongly and specifically to the sea urchin r-opsin antibody. These cells resemble descriptions of potential "photic receptors" described in *Ophiura ophiura* by Cobb and Moore (1986), but such cells were not detected using antibodies against r-opsin or c-opsin by Blaue and Ullrich-Lüter (unpublished data). Additional weak reactivity to Sp-Op4 was detected in the radial nerve cords and nerves within the arm plate, as well as the core of the spines (Sumner-Rooney et al. 2018). Observations of r-opsin reactivity did not differ substantially between the visual *Ophiomastix wendtii* and the apparently nonvisual *Ophiocomella pumila*; both possessed putative PRCs in trabecular pores. Sumner-Rooney et al. (2020) later sequenced transcriptomes from the arms of *O. wendtii* and *O. pumila*. They identified opsin candidates in both species, including multiple r-opsins and transduction pathway constituents, congruent with immunohistochemistry and studies of other echinoderms. In *O. wendtii*, three r-opsins, four G_o-opsins, two neuropsins, and a single c-opsin and RGR opsin were recovered, whereas in

O. pumila only two r-opsins and a single G_0-opsin and two neuropsins were detected; however, this may not reflect the full repertoire of these animals.

3.5.4 Opsins Across Echinodermata

D'Aniello et al. (2015) by surveying multiple genomic and transcriptomic resources corroborated the presence of all major ancestral bilaterian opsin groups (ciliary opsins, rhabdomeric opsins, neuropsins, peropsins, and RGR opsins) in echinoids and ophiuroids. They detected a total of 97 opsin sequences in echinoderms, 33 in echinoids, and 18 in ophiuroids. However, large gaps of knowledge remain here; for example, in the case of the sea urchin *Strongylocentrotus droebachiensis*, a close relative to *S. purpuratus*, only one opsin sequence was detected. It seems more likely that the apparent absence of additional sequences is the result of methodological limitations than a reflection of genuine biological differences. These results should therefore be interpreted as a known minimum for echinoderm opsin repertoires. In addition, these authors confirmed the presence of two novel echinoderm-specific opsin classes, found only in Echinoidea, Ophiuroidea, and Asteroidea. The previously identified Sp-opsin2 and Sp-opsin5 in *S. purpuratus* belong to echinopsins-A and echinopsins-B, respectively (Raible et al. 2006). The "rhabdomeric-like" opsin described in *S. droebachiensis* by Lesser et al. (2011) clusters with echinopsins-B and thus outside bona fide rhabdomeric opsins in this analysis.

3.5.5 Retinal Determinant Genes and Transcription Factors

Besides opsins, several authors have targeted other gene products associated with vision and eye development in other animal groups in the search for eyes and photoreceptors. This most commonly includes pax6, the so-called master regulator of animal eye development in many other species. However, the search for pax6 expression as an indicator for the presence of an "eye-organ" turned out to be somewhat difficult; in juveniles, it is found throughout the entire tube foot and the region of their origin (Czerny and Busslinger 1995; Agca et al. 2011; Ullrich-Lüter et al. 2011). Pax6 expression furthermore is shown to be transient in developing tube feet of the sea urchin *Heliocidaris erythrogramma* (Byrne et al. 2018) as well as in *Paracentrotus lividus* (Paganos et al. 2022). In adult tube feet, pax6 expression was identified in areas adjacent to r-opsin-positive cells in the tube foot disc, but not overlapping r-opsin expression (Ullrich-Lüter et al. 2011). Immunostaining of histological sections by Lesser et al. (2011) revealed the presence of pax6 protein in pigment cells within the perforated ossicles of the tube foot disc. Agca et al. (2011) concurrently examined the expression of seven retinal transcription factors, pax6, Pou4f, Six3/6, NeuroD, Rax, Isl1, and Ato6, in *S. purpuratus*. They detected transcripts of all seven pax6 sea urchin orthologs in both disk and stalk regions of the tube feet. While pax6 was generally expressed at higher levels in the tube foot stalk, its expression in the disk was restricted to the periphery, where they found small clusters of putative sensory neurons. All remaining retinal transcription factors were not significantly different expressed between stalk and disc region. Byrne et al. (2018) found Six1/2 and Six3/6 expression in the primary and/or secondary tube feet and putative sensory/neuronal cells of *Heliocidaris erythrogramma*. They also reported expression of Six1/2 and Six3/6 in the tube foot disc neuropil region and Dach expressed in spines. Paganos et al. (2022) found Six3 and NeuroD expressed in developing juvenile primary and secondary tube feet.

3.6 Achieving Spatial Resolution: Proposed Mechanisms

Thus far, we have emphasized the important difference between photoreception and vision but only outlined the nature and distribution of the photoreceptors. Precisely how these photoreceptors enable extraocular vision remains subject to investigation, but here we will outline and synthesize the existing evidence. Although they are extremely unusual, more than a century of vision research has taught us that there are certain requirements for spatial resolution that, we presume, also apply to extraocular visual systems. Only *Diadema africanum* and *Ophiomastix wendtii* have so far been proven to exhibit extraocular vision, using the strictest behavioral criteria. These species therefore make the best starting point to look for mechanistic explanations, even though their visual systems are likely to differ from one another.

3.6.1 Screening

To facilitate spatial resolution, photoreceptors must be screened from incoming light, reducing the angle of incident light and therefore the acceptance angle (Nilsson 2013). In conventional visual systems (reliant on eyes), this is universally achieved using pigment interspersing the photoreceptors. Both *Diadema africanum* and *Ophiomastix wendtii* are darkly pigmented, with light-responsive chromatophores that expand and contract to trigger an apparent change in color.

Hendler (1984) and Cobb and Hendler (1990) noted that *O. wendtii* was more sensitive to illumination in its dark-adapted (pale) form or with the pigment artificially removed and proposed that this was the result of more light reaching the photoreceptors within the skeleton. Sumner-Rooney et al. (2020) found that dark-adapted *O. wendtii* were still strongly negatively phototactic under natural darkness but observed that orientation to both visual (isoreflectant) and distant dark stimuli was extinguished. These orientation behaviors were not rescued by increased ambient light intensity, whether natural or artificial, if the animals were dark-adapted. The authors proposed that the significance of the chromatophores was in screening the photoreceptors (located in skeletal pores, r-opsin reactive) and demonstrated that the presence of pigment could considerably restrict their aperture, and therefore their acceptance angle, sufficiently to facilitate coarse spatial resolution. Presumably, the removal of screening pigment and increase in aperture would also increase photon capture, and thus sensitivity, in the dark as proposed by Hendler and colleagues. A screening role for the chromatophores would also explain the apparent absence of vision or orientation to distant dark stimuli in *Ophiocomella pumila*, despite having a similarly extensive photoreceptor network (Sumner-Rooney et al. 2018, 2020). Beyond *O. wendtii*, there are several additional ophiocomid species that have these light-dependent chromatophores (Hendler 1984; Byrne et al. 2004), of which at least two (*Ophiocoma echinata* and *O. scolopendrina*) may orient to distant dark stimuli (Michiels and Anthes pers. comm, Sumner-Rooney unpublished data).

The mechanism (or mechanisms) underlying photoreceptor screening in sea urchins is less clear. Ullrich-Lüter et al. (2011) did not observe pigment cells associated with photoreceptors but proposed that the photoreceptors located in the tube foot pore could be effectively screened by the surrounding opaque calcite skeleton. Using micro-CT, they calculated that in *S. purpuratus* this would confer an acceptance angle of up to 88°, depending on the position of the tube foot on the test. Kirwan et al. (2018) found that the estimated acceptance angle of pores screened in this way was also compatible with their behavioral results in *Diadema africanum*. In *Diadema,* amoeboid chromatophores similar to those in *O. wendtii* have been shown to expand under illumination and contract in the dark (Yoshida 1966). Kirwan et al. (2018) suggested that this pigment could also provide effective screening for the r-opsin-reactive photoreceptors located within tube foot pores in *D. africanum*, adding to the opacity of the surrounding skeleton. Whether the status of the chromatophores has any impact on specifically visual behavior in sea urchins remains to be investigated, but Millott (1954) reported greater sensitivity in the test of dark-adapted animals, as observed in *O. wendtii*, and it is noteworthy that the sea urchin species that have historically drawn attention for their photosensitivity are largely diadematids. Millott (1954) also commented that the "areas with most melanin… are most sensitive," a statement that could appear to be at odds with observations of *O. wendtii* (Hendler 1984; Cobb and Hendler 1990; Sumner-Rooney et al. 2020), except that these areas tend to be along the ambulacrum and within the tube feet, where photoreceptors have since been recognized to be located (Ullrich-Lüter et al. 2011; Lesser et al. 2011).

Alternatively, the possibility remains that sea urchin spines play a part in screening photoreceptors located on the test, as suggested by Woodley (1982). Close correlations between spine density and behaviorally derived estimates for resolution have been found in both *Strongylocentrotus* and *Echinometra* (Blevins and Johnsen 2004; Yerramilli and Johnsen 2010), suggesting that screening by the spines could confer and limit resolution. Kirwan et al. (2018) estimated that the potential resolution achievable by spine-screening in *D. africanum* was 60°, which is compatible with their behavioral results. There are a few potential hitches to the spine-screening model. Yerramilli and Johnsen (2010) noted that the distribution of spines was highly variable across the test, making resolution difficult to predict. Kirwan et al. (2018) also considered the erratic movement and responsiveness of the spines a potential disadvantage (though, presumably, the concentration of spines toward a threat would, under this model, temporarily provide greater resolution in this part of the "visual field"). A recent study by Notar et al. (2022) also found no correlation between visual environment (e.g. brightness, complexity) and resolution as predicted by spine density. Finally, the estimates of resolution for *Echinometra* and *Strongylocentrotus* that correlate with spine density were derived from orientation to dark stimuli and not isoreflectant ones and may therefore be overestimates. As the removal of spines often detrimentally affects animals' performance in behavioral experiments, it is challenging to test the impact of spine density experimentally. However, an unpublished undergraduate study by Frossard (2011) found that removing the spines in *Diadema savignyi* extinguished directional movement in response to a 5° but not a 9° dark stimulus; this result merits further investigation.

3.6.2 Limits of Resolution

In a contiguous retina, the angular spacing between screened photoreceptors approximates the spatial resolution. How resolution is limited in extraocular systems, where photoreceptors are scattered across the body, remains unclear. In both *D. africanum* and *O. wendtii*, the large angular aperture of each photoreceptor (92° and 68°, respectively) differs substantially from the inter-receptor angles (13° within a row of tube feet and 28° between parallel rows in *D. africanum*, around 7° on the dorsal arm plate of *O. wendtii* (Sumner-Rooney et al. 2021)). This implies dramatic oversampling in both systems, which could limit spatial resolution but improve sensitivity. Behavioral thresholds for these two species appear to align more closely to their angular apertures, from which acceptance angle can be estimated, than to inter-receptor angles. However, the limiting factor to resolution may not become clear until we have a better understanding of how information is compared and/or combined between individual photoreceptors in these systems.

3.6.3 Optics

In addition to photoreceptor screening, many visual systems employ optical structures to maximize light capture and improve resolution (Nilsson 2013). Although several authors have proposed other roles for the echinoderm skeleton, including photoreceptor screening (Woodley 1982), transmission (Raup 1960), and polarization (Raup 1965), as outlined at the beginning of the chapter, evidence for light-focusing or light-guiding structures contributing to extraocular vision is rarer.

The main mechanisms historically proposed are light scattering by the ossicles of the tube feet (Lesser et al. 2011) and light collecting or focusing by the EPT "microlenses" of ophiocomids (Hendler 1984; Hendler and Byrne 1987; Aizenberg et al. 2001). In the former, Lesser et al. (2011) found that illuminating any point on a dissected tube foot ossicle from *Strongylocentrotus droebachiensis* caused scattering across the ossicle. Combined with their concave shape and arrangement in the tube foot, these authors argued that this scattering could increase the chances of photon capture by putative photoreceptors within the tube foot, possibly located within the ossicle disc. It should be noted, however, that the tube feet in live animals are often darkly pigmented and that this likely obstructs incoming light before it reaches the ossicle. We therefore consider this possible functional role to be unlikely. The presumed general optical role of the skeleton more widely in echinoids has also been debated in the past, with several different authors demonstrating that dissected fractions of the skeleton conduct light (Raup 1960; Lesser et al. 2011). While they proposed that this functionality is persistent in live animals, others have found very little or no light penetrating through intact sea urchin tissues (Walker 2007, Ullrich-Lüter, personal observation in S. purpuratus). Additionally, the plausibility of this varies between species; Millott (1954) noted that the test of *Diadema* was "translucent," being much thinner and more fragile than species such as *S. purpuratus*.

The description of putative microlenses in *Ophiomastix wendtii* (Hendler and Byrne 1987; Aizenberg et al. 2001) paved the way for many subsequent publications that reinforced the interpretation of calcite structures as a general focusing apparatus in echinoderms (e.g. Gorzelak et al. 2014, 2017; Vinogradova et al. 2016; Márquez-Borrás et al. 2020). This has been particularly appealing in extinct taxa, where calcite structures are well preserved and behavioral and molecular investigations are impossible; for example, Gorzelak et al. (2014) proposed that similar structures in Late Cretaceous echinoderms demonstrated the existence of a complex visual system in sea stars and brittle stars. However, all such interpretations rely on these structures focusing light onto light-sensitive nerves or unspecialized PRCs beneath them. As the findings of Sumner-Rooney et al. (2018, 2020) have refuted the presence of r-opsin- or c-opsin-reactive cells in this location, this generalized interpretation should be treated with caution. The r-opsin-reactive cells detected by these authors instead sit alongside the EPTs, with distal expansions sitting level or slightly distal to the widest part of the EPTs, well out of the likely path of any light travelling through them. Although several other opsins were later recovered from transcriptomes of *O. wendtii* (Sumner-Rooney et al. 2020), the vast majority of visual systems are reliant on c- and r-opsins (Land and Nilsson 2012). From what we know of echinoderm vision, r-opsins are likely to be the key molecular component. Thus, although an unidentified photoreceptor could be present beneath the EPTs, we find no evidence to support their characterization as lenses. Although many low-resolution eyes have lenses or lenslike structures, these are not necessary to function and are absent in the pigment cup eyes of platyhelminthes, some bivalves and some gastropods, for example (Nilsson 2013). Given, also, that the photoreceptors are not concentrated into a retina, the potential optical benefits provided by many very small lenses could be relatively low.

From the existing evidence, it appears that neither of the confirmed extraocular visual systems use optical properties of the skeleton. The structure of the echinoderm stereom nonetheless provides the unique possibility to incorporate different cell types within and around the skeleton and its local modifications. We thus have no intention of generally rejecting the idea of the echinoderm skeleton being involved in photoreception of these animals. However, more evidence from integrative studies is important before reiterating this hypothesis in the future.

3.7 Nervous Systems and Processing

The integration of inputs from hundreds or thousands of individual photoreceptors remains the most enigmatic aspect of extraocular vision. The echinoderm nervous system comprises five radial nerve cords connected by a central nerve ring (Cobb 1987). It has been interpreted to be a fivefold duplication of a simple CNS (Burke 2011). This unusual nervous system architecture represents another obstacle investigating visual systems in the phylum.

We know relatively little regarding the integration mechanism of photic stimuli, besides that each of the five ambulacra and their respective radial nerve cords seem to operate as an independent unit (reviewed in Yoshida 1966). From the few existing behavioral and physiological studies, we know that dissection of the nerve ring diminishes the coordinated movement in sea urchins (Holmes 1912) upon photic stimulation. Yoshida and Ohtsuki (1968) found similar results when they cut the nerve ring of the sea star *Asterias amurensis*. In addition, these authors experimented with removing all but one of the ocelli to examine signal transmission around the nerve ring; they found that phototaxis was least impaired when the nerve ring was cut opposite the ocellus-bearing arm. They interpreted this to reflect the maintenance of symmetry in signal from the intact arm around the ring. Similarly, transecting the radial nerve cord near the central nerve ring impedes phototaxis in sea stars (Yoshida and Ohtsuki 1968) and sea urchins (Yoshida 1966). Of course, it is somewhat challenging to disentangle the impact of such dissections on sensation from their impact on locomotion, but it appears that the role of the nerve ring in phototaxis and visually guided orientation is primarily to communicate between the radial nerve cords. Cobb and Moore (1989) demonstrated that responses to a variety of stimuli applied to one arm could be recorded from the radial nerve cord of a different (even opposite) arm, indicating that information transmission does not terminate in the central nerve ring.

There may be further clues to gain from behavioral observation and additional psychophysical experiments on intact animals. Sumner-Rooney et al. (2020) observed that while the brittle star *Ophiomastix wendtii* most commonly approached a visual stimulus with two leading arms, *Ophiocomella pumila* typically moved with a single leading arm. They suggested that this could indicate the presence of signal comparison between adjacent arms in *O. wendtii*, but this hypothesis requires further investigation. It is unknown whether *Diadema africanum* demonstrates a typical body orientation toward similar stimuli; it could be interesting in the future to compare the proportions of animals leading with the ambulacral or interambulacral plates in analogy to *O. wendtii*.

If echinoderms are likely to be comparing signals between arms or ambulacrae via the nerve ring, the next question to arise is how signals are combined within the five radial nerve cords. If these nerve cords are indeed equivalent to duplicated CNS, "visual" processing could be expected to occur here. Indeed, the radial nerve cords and different epidermal tissues were found to contain many orthologs of vertebrate genes involved in vision (Burke et al. 2006; Raible et al. 2006) and in some cases exhibit opsin reactivity and intrinsic sensitivity (see above). The scale of this task seems staggering at first glance. Ullrich-Lüter et al. (2011) estimated the total number of r-opsin expressing PRCs to be around 200,000 per adult individual in the sea urchin *S. purpuratus*, and Sumner-Rooney et al. (2018) roughly estimated the presence of 300,000 dorsal EPTs, each surrounded by 6–7 PRCs, giving an estimate of 750,000 to a million dorsal PRCs for the brittle star *O. wendtii*. If these are integrated within whole radial nerve cords, that gives 40,000 and 150,000–200,000 PRCs per "visual unit."

Alternatively, integration could occur at the level of individual arm segments or ambulacral plates; in sea urchins, normal spine-pointing responses to shadows are intact in isolated pieces of test if there is a piece of radial nerve cord present (Millott and Yoshida 1959, 1960). Yoshida and Millott (1960) demonstrated that the shading response in *Diadema* could be prevented by illuminating an adjacent part of the radial nerve cord, a maximum of 6 mm from the first site of stimulation. Stubbs (1983, unpublished thesis) noted that electrical responses to photic stimulation in *Ophiura ophiura* changed in signature when recorded from the adjacent arm segment but then remained constant across the body; he proposed that responses could therefore be integrated at the closest "ganglionic" swelling of the radial nerve cord. However, he also reported that more than one arm segment had to be stimulated to detect any response. Besides the convergence of photoreceptor axons into nerve bundles in *O. wendtii* and *O. pumila* (Sumner-Rooney et al. 2018), there is no evidence for the further organization of photoreceptors.

Despite the potential breakdown of these vast photoreceptor systems into more digestible "visual units," the integration of information across such dispersed (and in the case of brittle stars, complex and mobile) body surfaces remains perplexing given their apparently modest nervous systems. It is quite probable that these animals are not generating and perceiving images in the same sense that our eyes do. This leads us to the very tangled question of how these systems could detect local contrast, i.e., resolve spatially, but potentially not form images.

3.8 Evolution of Extraocular Vision

Investigating the evolution of sea urchin and brittle star extraocular vision is a considerable challenge. As well as the unconventional nature of these visual systems, allowing for few assumptions and meaning almost all data must be collected from scratch, efforts are also hampered by patchy taxonomic sampling. This gives rise to two key problems.

The first problem stems from within-group comparisons. Robust behavioral reactions accounting for vision, in its strictest sense, have only been obtained from *Diadema africanum* and *Ophiomastix wendtii* so far. But molecular data and experimental tools for these species have been limited if not nonexistent. Characterizing potential photoreceptors via expression patterns of key genes and proteins is challenging in non-model systems, and it is often necessary to revert to more accessible species like the invertebrate model *S. purpuratus* and *Amphiura filiformis*. However, the wildly different ecologies of *S. purpuratus*, living in cold water kelp canopies, and tropical coral reef species such as *Diadema*, mean we must be cautious in extrapolating from one species' expressed photoproteins to another's behavior. The same holds true for comparisons between the cold water, mud digging, and filter feeding ophiuroid *Amphiura filiformis*, for which a genome and molecular tools are available, and the behavioral results obtained in *Ophiomastix wendtii*, an omnivorous tropical species living in shallow waters of coral reefs.

The second challenge lies with between-group comparisons. Within echinoderms, sea urchins and brittle stars are not even sister taxa (Reich et al. 2015). Instead, within the phylum, they are separated by more than half a billion years of divergence. While it is tempting to identify functional similarities between *Diadema africanum* and *Ophiomastix wendtii*, such as the potential role of chromatophores described above, it is crucial to remember that their visual systems (sensu stricto) are extremely unlikely to be homologous. Such similarities may be informative in unpicking the mechanisms by which extraocular vision is achieved, but this should not be mistaken for evidence for a single origin among echinoderms: half a billion years of evolution allows plenty of opportunity for convergence, too.

With these caveats in mind, evaluating our current knowledge of echinoderm photoreception and extraocular vision in a comparative setting does allow us to tentatively draw some initial conclusions. Key elements of these systems (opsin repertoires, PRC distributions, and pigmentation) can be examined individually to try and clarify their evolutionary history within echinoderms.

D'Aniello et al. (2015) conducted the most comprehensive survey to date of echinoderm opsins, including members of all five extant classes. The identification of eight bona fide opsin genes in the sea urchin *S. purpuratus* supports the presence of both visual candidate opsins, rhabdomeric and ciliary, in the common ancestor of extant echinoderms. Indeed, both are present in other sampled eleutherozoans (D'Aniello et al. 2015). The role of the enigmatic echinopsins remains something of a mystery, but these have also been recovered in all four eleutherozoan classes. Whether they were present in the ancestor of Eleutherozoa and Crinoidea is unclear. Among the crinoids, only a single r-opsin candidate was recovered from adult transcriptomes of two comatulids, *Antedon* and *Florometra*. Although this may not capture all retained opsins, it may reflect a dramatic reduction in opsin repertoires in crinoids, which are largely and ancestrally sessile. Other crinoid groups have yet to be sampled for this purpose, but only the motile comatulids have historically been reported to be photosensitive (Hyman 1955). Conversely, we cannot yet comment on the potential expansion of opsin repertoires in taxa exhibiting extraocular vision. *Ophiomastix wendtii* was found to express three r-opsins and four candidate

G_0-opsins, for example. The former does not appear to be a lineage-specific expansion according to phylogenetic analysis (Sumner-Rooney et al. 2020), and two r-opsins were also recovered in the closely related but apparently nonvisual *Ophiocomella pumila*. In contrast, *Amphiura filiformis* has a larger repertoire of r-opsins, but there is no evidence so far to suggest that this species is capable of vision. No survey of opsins has been performed in *Diadema* to date, but this will certainly be of interest in the future.

Tentative comparisons of the expression patterns of the putative visual opsins also offer some insight to the evolutionary history of photoreceptor distribution in echinoderms. With some caution regarding the inconclusive expression patterns in diadematids, we propose that the presence of r-opsin expressing PRCs in the tube feet is a common theme across both echinoids and ophiuroids. As several asteroids also display r-opsin-reactive PRCs in their tube feet (Ullrich-Lüter 2011 unpublished thesis, Blaue & Ullrich-Lüter, unpublished data), it is reasonable to hypothesize that this pattern was also present in the last common ancestor of these groups (Reich et al. 2015). C-opsins, on the other hand, seem to be expressed throughout large portions of the epidermis, including spines, tube feet, pedicellariae, body wall, and radial nerves in sea urchins. In most investigated brittle stars, c-opsin expression is restricted to cells within their spines, but in *Amphiura chiajei*, c-opsin expression was found also within tube feet and the arm epidermis, thus resembling more the pattern described in echinoids. Those findings amplify the need for a broader taxon sampling to clarify the ground pattern of c-opsin expression in the last common ancestor of echinoids and ophiuroids. With regard specifically to the strictly visual species, it is notable that both *D. africanum* and *O. wendtii* seem to differ in their expression patterns from other members of Echinoidea and Ophiuroidea, respectively. The similarities between r-opsin reactivity in *O. wendtii*, *O. echinata*, and *O. pumila* indicate that they likely share a widespread extraocular PRC network, which only mediates vision in selected species. This implies some additional modification in visual species beyond the photoreceptors themselves, underlining the important distinction between mapping photoreceptor distributions and drawing conclusions about vision and behavior.

While we have been increasingly broadening our view of opsin expression patterns over the past decade (see Table 3.2), general insight into the nature, and therefore evolution, of other essential components of extraocular vision remains scarce. Many echinoderms have dark pigment cells within their epidermis, particularly those inhabiting shallow waters or even the intertidal, often at low latitudes. Combined with the reaction of investigated pigment cells to UV radiation, it is reasonable to speculate that these pigment cells are likely to be an adaptive trait in order to reduce UV exposure and consequent radiation damage (Gras and Weber 1983). In some taxa the pigment cells have been implicated in visual processes (e.g. Millott 1976; Hendler 1984; Hendler and Byrne 1987; Cobb and Hendler 1990; Sumner-Rooney et al. 2020), but that does not necessitate that vision is their primary function. Pigmentation for protection from UV radiation may instead be an exaptation for vision, with pigments becoming functionally integrated into preexisting PRC networks in some echinoderm species. This aligns with evidence from

O. wendtii and *O. pumila* and would support a relatively recent origin(s) of extraocular vision within Ophiocomidae (Sumner-Rooney et al. 2020).

Indeed, seen from an ecological viewpoint, the need for a sophisticated (extraocular) visual system in echinoderms isn't immediately obvious. Interestingly, all behavioral evidence for true spatial vision in the phylum so far comes from animals being relatively agile compared to their fellow species. This is the case for the starfish *Acanthaster planci* (Petie et al. 2016), which finds its preferred coral food over relatively large distances by sight; the brittle star *Ophiomastix wendtii* (Sumner-Rooney et al. 2020), a cryptic omnivore inhabiting shallow reefs with high predation pressure; and the sea urchin *Diadema africanum* (Kirwan et al. 2018), which deters predatory fishes by vigorous spine jerks. It is obvious how each of these animals benefits from a sophisticated visual system that can convey quick and relatively high-resolution visual information.

At the opposite end of the spectrum, we have a plethora of echinoderm species that have a very slow-moving lifestyle but might still benefit from visual capabilities to a lesser degree. Many of those have been demonstrated to display photosensitivity (Yoshida 1966, Table 3.1) and to express different opsin proteins in putative PRCs (see Table 3.2). However, representatives of species in key phylogenetic positions have been painfully hard to investigate, in regard to either their availability in sufficient numbers or their ability to reproduce a robust behavioral response given a stimulus that can only be seen with true spatial resolution abilities (see Kirwan et al. 2018). Some of the species demonstrated to orient to dark stimuli may indeed be capable of vision as well (see Table 3.1). We therefore urgently need additional such experiments in "eyeless" echinoderms in order to derive a clearer picture regarding the presence and evolution of extraocular vision in this phylum. From what we know today, we can only speculate that true image-forming vision might have evolved convergently multiple times inside echinoderms.

3.9 Future Research and Challenges

This field of research, though borne out of more than a century of work on echinoderm photoreception, is still in its infancy. At the time of writing, the first explicit behavioral evidence of extraocular vision is less than 3 years published (Kirwan et al. 2018). Although visual behavior has strictly only been documented in two species, the very specific cases of diadematids and ophiocomids are extremely unlikely to be the only groups with this ability. Other species that have featured in this chapter are clearly photosensitive; might they be capable of vision? There is much to learn, and we suspect that this includes many "unknown unknowns." For example, it is unclear whether we can apply some of the fairly fundamental functional principles of vision (in an eye) to these systems. Acceptance angles and inter-receptor angles, for instance, are closely allied and a good indicator of maximal resolution in most eyes (Cronin et al. 2014). As we have seen in diadematids and ophiocomids, there appears to be a dramatic mismatch between these values, as a result of photoreceptors being sparsely distributed across the body surface. The functional implications of this are not immediately obvious, but we would urge caution when applying relationships extracted from eyes to extraocular visual systems.

Because we know relatively little about extraocular vision and its occurrence, both in echinoderms and beyond, it is difficult to determine which existing data – gene expression, morphological or behavioral – might be relevant in the search to identify and understand it. This uncertainty adds considerably to our workload. To move forward with the greatest efficiency, we have identified four key priorities for future work that we hope will help to shape the future of the field.

1. **Integration and processing.** The combination of information from individual photoreceptors remains one of the most enigmatic aspects of extraocular systems. We suggest a combined approach to this problem: first, carefully designed psychophysical experiments could provide invaluable insights to signal transmission and hierarchy across whole animals. In the context of the orientation experiments described above, this could include multiple presented stimuli, removal or manipulation of photoreceptors, or the movement of stimuli during trials. Second, a modeling approach could be highly informative. Characterizing the optic capabilities of individual and clusters of photoreceptors, as well as neuroinformatic processing models for the decentralized nervous system, could provide testable hypotheses to apply to whole-animal systems.
2. **The relationship between morphology and resolution.** There remain several candidates for attaining and limiting spatial resolution, and comparative data involving species with varied morphology and behavior will be necessary to identify the true mechanism(s). In particular, *Strongylocentrotus purpuratus* and *Paracentrotus lividus* may offer an opportunity to examine the impact of changing morphology. Both species have the fewest and smallest photoreceptors in the discs of the tube feet at the oral side of the test. Whereas *P. lividus* has the largest accumulation of photoreceptor cells in the discs of the tube feet located at the horizontal plane of the ambulacrum, *S. purpuratus* has the highest numbers in the aboral area (Ullrich-Lüter, unpublished data). Although this finding is preliminary, this difference offers a chance to find the behavioral – and potentially ecological – implications of photoreceptor distribution in the sea urchin visual system.
3. **Behavior beyond orientation to static objects.** Nilsson (2013) includes collision avoidance and monitoring self-motion in his series of Class III tasks mediated by low-resolution vision. Although shadow responses have been reported in several species, specific reactions to moving stimuli have not been thoroughly tested to date. This seems especially pertinent owing to the putatively defensive nature of these shadow responses and merits further work. We know little about the impacts of stimulus speed, approach (i.e., looming or translation), or position in the field of view – including whether these responses are, indeed, visually mediated.
4. **Extraocular vision in other echinoderm classes.** Almost the entirety of this chapter, the reader will have noticed, has been dedicated to sea urchins and brit-

tle stars. No candidates for extraocular vision have been identified in holothurians (sea cucumbers) or crinoids (such as sea lilies and feather stars), despite many holothurians and at least comatulid crinoids being photosensitive (Perrier 1873; Clark 1915; Hyman 1955). Both are known to express r-opsins in selected species (D'Aniello et al. 2015); localization of r-opsin-reactive cells and behavioral assays in these groups would be hugely informative and should be prioritized. Presumably sea stars have little need of a second visual system, but we stand to be proven wrong; from Ullrich-Lüter (2011) we know that juveniles of the extant asteroid *Asterias rubens* possess r-opsin-reactive PRCs along the aboral epidermis of their arm grooves in addition to those in the optic cushions and tube feet. We do not know what the function of those aboral PRCs might be and how the nervous system of the animals handles sensory input from those very different body regions and light receptive organs. Experiments on blinded sea stars have indicated that phototactic behavior is affected by ocellus removal in the same way as by arm removal (Yoshida and Ohtsuki 1968), suggesting that this does not rely on extraocular photoreceptors, but we recommend further such work to identify their possible role.

The field of extraocular vision is still in its early development. The invaluable groundwork laid by more than a century of careful observation and experimentation on echinoderms has sketched out the landscape of their photosensitivity and identified crucial taxa such as *Diadema* and *Ophiomastix* for further work. We have tried to strike a balance here between drawing on what we know about these two visual taxa, contextualizing them within echinoderm photoreception more widely, and presenting potential "leads" for future investigation. The constant improvement of molecular and psychophysical tools will steadily make these non-model systems more accessible, and we fully expect that additional lineages will be discovered or confirmed to be capable of vision. We encourage further exploration of this bizarre and fascinating topic and hope that the coming decades will bring some clarity to both mechanistic and evolutionary aspects of extraocular vision.

References

Adams NL (2001) UV radiation evokes negative phototaxis and covering behavior in the sea urchin *Strongylocentrotus droebachiensis*. Mar Ecol Prog Ser 213:87–95

Agca C, Elhajj MC, Klein WH, Venuti JM (2011) Neurosensory and neuromuscular organization in tube feet of the sea urchin *Strongylocentrotus purpuratus*. J Comp Neurol 519:3566–3579

Aizenberg J, Tkachenko A, Weiner S et al (2001) Calcitic microlenses as part of the photoreceptor system in brittlestars. Nature 412:819–822. https://doi.org/10.1038/35090573

Al-Wahaibi MK, Claereboudt MR (2017) Extraocular vision in the sea urchin *Diadema setosum*. Mar Freshw Behav Physiol 50:31–40. https://doi.org/10.1080/10236244.2016.1265206

Beer S, Wentzel C, Petie R, Garm A (2016) Active control of the visual field in the starfish Acanthaster planci. Vis Res 127:28–34. https://doi.org/10.1016/j.visres.2016.05.007

Blevins E, Johnsen S (2004) Spatial vision in the echinoid genus *Echinometra*. J Exp Biol 207:4249–4253. https://doi.org/10.1242/jeb.01286

Burke RD (2011) Deuterostome neuroanatomy and the body plan paradox. Evol Dev 13:110–115. https://doi.org/10.1111/j.1525-142X.2010.00460.x
Burke RD, Angerer LM, Elphick MR et al (2006) A genomic view of the sea urchin nervous system. Dev Biol 300:434–460
Byrne M, Cisternas PA, Hoggett A et al (2004) Diversity of echinoderms at Raine Island, Great Barrier Reef. In: Heinzeller T, Nebelsick JH (eds) Proceedings of the 11th international echinoderm conference. Echinoderms, Munchen, pp 159–164
Byrne M, Koop D, Morris VB, Chui J, Wray GA, Cisternas P (2018) Expression of genes and proteins of the pax-six-eya-dach network in the metamorphic sea urchin: Insights into development of the enigmatic echinoderm body plan and sensory structures. Dev. Dyn., 247: 239–249. https://doi.org/10.1002/dvdy.24584
Clark AH (1915) A monograph of the existing crinoids, vol. 1. The comatulids, pt. 1:[General]. Bull US Natl Mus 56: i-vi, 1–406
Cobb JLS (1987) Neurobiology of the Echinodermata. In: Ali MA (ed) Nervous systems in invertebrates, NATO ASI series a: life sciences, vol 141. Plenum Press, pp 483–525
Cobb JLS, Hendler G (1990) Neurophysiological characterisation of the photoreceptor system in a brittlestar, *Ophiocoma wendtii* (Echinodermata: Ophiuroidea). Comp Biochem Physiol A 97:329–333
Cobb J, Moore A (1986) Comparative studies on receptor structure in the brittlestar Ophiura ophiura. J Neurocytol 108:97–108
Cobb JLS, Moore A (1989) Studies on the integration of sensory information by the nervous system of the brittlestar Ophiura Ophiura. Mar Behav Physiol 14:211–222. https://doi.org/10.1080/10236248909378708
Cowles RP (1910) Stimuli produced by light and by contact with solid walls as factors in the behaviour of ophiuroids. J Exp Zool 9:387–416
Cronin TW, Johnsen S, Marshall NJ, Warrant EJ (2014) Visual Ecology. Princeton
Crozier WJ (1914) The orientation of a holothurian. Am J Psychol-Legacy Content 36:8–20
Crozier WJ (1915) The sensory reactions of Holothuria surinamensis Ludwig. Contrib Bermuda Biol Station 3:233–297
Czerny T, Busslinger M (1995) DNA-binding and transactivation properties of Pax-6: three amino acids in the paired domain are responsible for the different sequence recognition of Pax-6 and BSAP (Pax-5). Mol Cell Biol 15:2858–2871. https://doi.org/10.1128/mcb.15.5.2858
D'Aniello S, Delroisse J, Valero-Gracia A et al (2015) Opsin evolution in the Ambulacraria. Mar Genomics 24:177–183. https://doi.org/10.1016/j.margen.2015.10.001
De Franceschi G, Vivattanasarn T, Saleem AB, Solomon SG (2016) Vision guides selection of freeze or flight defense strategies in mice. Curr Biol 26:2150–2154. https://doi.org/10.1016/j.cub.2016.06.006
Delroisse J, Mallefet J, Flammang P (2013) Light perception in brittle stars: detection of the molecular actors. In: International conference on invertebrate vision. Frontiers in Physiology. Fjälkinge, Sweden
Delroisse J, Ullrich-Lüter E, Ortega-Martinez O et al (2014) High opsin diversity in a non-visual infaunal brittle star. BMC Genomics 15:1035. https://doi.org/10.1186/1471-2164-15-1035
Delroisse J, Ortega-Martinez O, Dupont S et al (2015) De novo transcriptome of the European brittle star *Amphiura filiformis* pluteus larvae. Mar Genomics 23:109–121. https://doi.org/10.1016/j.margen.2015.05.014
Delroisse J, Mallefet J, Flammang P (2016) De novo adult transcriptomes of two European brittle stars: spotlight on opsin-based photoreception. PLoS One 11:e0152988. https://doi.org/10.1371/journal.pone.0152988
Dimelow EJ (1958) Pigments present in arms and pinnules of the crinoid, *Antedon bifida* (Pennant). Nature 182:812
Doederlein LHP (1885) Seeigel von Japan und den Liu-Kiu-Inseln. W. Moeser Hofbuchdruckerei
Dubois R (1913) Note sur l'action de la lumière sur les échinodermes. Compte Rendu de l'Association Française pour l'Avancement des Sciences 41:138–139

Frossard BA (2011) The anti-predator behaviour of the black longspine urchin (*Diadema savignyi*): spatial vision and the role of light in emergence. University of California, Berkeley
Garm A (2017) Sensory biology of starfish — with emphasis on recent discoveries in their visual ecology. Integr Comp Biol 57:1082–1092. https://doi.org/10.1093/icb/icx086
Garm A, Nilsson D-E (2014) Visual navigation in starfish: first evidence for the use of vision and eyes in starfish. Proc R Soc Lond Ser B Biol Sci 281:1–8
Gorzelak P, Salamon MA, Lach R et al (2014) Microlens arrays in the complex visual system of cretaceous echinoderms. Nat Commun 5:3576, 6pp. https://doi.org/10.1038/ncomms4576Microlens
Gorzelak P, Rahman IA, Zamora S et al (2017) Towards a better understanding of the origins of microlens arrays in Mesozoic ophiuroids and asteroids. Evol Biol 44:339–346
Gras H, Weber W (1983) Spectral light sensitivity of isolated chromatophores of the sea urchin, *Centrostephanus longispinus*. Comp Biochem Physiol 76:279–281
Hendler G (1984) Brittlestar color-change and phototaxis (Echinodermata: Ophiuroidea: Ophiocomidae). Mar Ecol 5:379–401
Hendler G (2005) An echinoderm's eye view of photoreception and vision. In: Heinzeller T, Nebelsick J (eds) Echinoderms: München: proceedings of the 11th international echinoderm conference. Taylor & Francis, München, pp 339–349
Hendler G, Byrne M (1987) Fine structure of the dorsal arm plate of *Ophiocoma wendti*: evidence for a photoreceptor system (Echinodermata, Ophiuroidea). Zoomorphology 107:261–272
Hess C (1915) Untersuchungen über den Lichtsinn bei Echinodermen. Pflugers Arch Gesamte Physiol Menschen Tiere 160:1–26
Holmes SJ (1912) Phototaxis in the sea urchin, *Arbacia punctulata*. J Anim Behav 2:126–136. https://doi.org/10.1037/h0076037
Hyman LH (1955) The invertebrates, Vol IV: Echinodermata. McGraw-Hill Book Company, New York
Jackson E, Johnsen S (2011) Orientation to objects in the sea urchin *Strongylocentrotus purpuratus* depends on apparent and not actual object size. Biol Bull 220:86–88
Johnsen S (1994) Extraocular sensitivity to polarised light in an echinoderm. J Exp Biol 195:281–291
Johnsen S (1997) Identification and localization of a possible rhodopsin in the echinoderms *Asterias forbesi* (Asteroidea) and *Ophioderma brevispinum* (Ophiuroidea). Biol Bull 193:97–105
Johnsen S, Kier WM (1999) Shade-seeking behaviour under polarized light by the brittlestar *Ophioderma brevispinum* (Echinodermata: Ophiuroidea). J Mar Biol Assoc UK 79:761–763. https://doi.org/10.1017/S0025315498000940
Kirwan JD, Bok MJ, Smolka J et al (2018) The sea urchin *Diadema africanum* uses low resolution vision to find shelter and deter enemies. J Exp Biol 221:jeb176271. https://doi.org/10.1242/jeb.176271
Kojima D, Terakita A, Ishikawa T et al (1997) A novel Go-mediated phototransduction cascade in scallop visual cells. J Biol Chem 272:979–982
Kroh A, Smith AB (2010) The phylogeny and classification of post-Palaeozoic echinoids. J Syst Palaeontol 8:147–212. https://doi.org/10.1080/14772011003603556
Land MF, Nilsson D-E (2012) Animal Eyes, Second. Oxford University Press, Oxford
Langeloh HP (1937) Über die Bewegungen von Antedon rosaceus und ihre nervöse Regulierung. Zoologische Jahrbücher, Abteilung Zoologie und Physiologie 57:235–279
Lees DC, Carter GA (1972) The covering response to surge, sunlight, and ultraviolet light in *Lytechinus anamesus*. Ecology 53:1127–1133
Lesser MP, Carleton KL, Böttger SA et al (2011) Sea urchin tube feet are photosensory organs that express a rhabdomeric-like opsin and PAX6. Proc R Soc B Biol Sci 278:3371–3379. https://doi.org/10.1098/rspb.2011.0336
Lin C, Zhang L, Liu S et al (2013) A comparison of the effects of light intensity on movement and growth of albino and normal sea cucumbers (*Apostichopus japonicus* Selenka). Mar Freshw Behav Physiol 46:351–366. https://doi.org/10.1080/10236244.2013.841350

Liu X, Lin C, Sun L et al (2020) Transcriptome analysis of phototransduction-related genes in tentacles of the sea cucumber *Apostichopus japonicus*. Comp Biochem Physiol Part D Genomics Proteomics 34:100675. https://doi.org/10.1016/j.cbd.2020.100675

MacCurdy HM (1912) Observations on the reactions of *Asterias forbesii* to light. Science 35:192

MacCurdy H (1913) Some effects of sunlight on the starfish. Science 38:98–100. http://www.jstor.com/stable/1640466

Mangold E (1909) Studien zur Physiologie des Nervensystems der Echinodermen. Archiv für die gesamte Physiologie des Menschen und der Tiere 126:371–406. https://doi.org/10.1007/BF01677797

Márquez-Borrás F, Solís-Marín FA, Mejía-Ortiz LM (2020) Troglomorphism in the brittle star *Ophionereis commutabilis* Bribiesca-Contreras et al., 2019 (Echinodermata, Ophiuroidea, Ophionereididae). Subterranean Biol 33:87–108. https://doi.org/10.3897/subtbiol.33.48721

May RM (1925) Les reactions sensorielles d'une ophiure (*Ophionereis reticulata*, Say). Bermuda Biol Station Res 59:372

Millott N (1953) Light emission and light perception in species of *Diadema*. Nature 171:973–974

Millott N (1954) Sensitivity to light and the reactions to changes in light intensity of the echinoid *Diadema antillarum* Philippi. Philos Trans R Soc Lond Ser B Biol Sci 238:187–220

Millott N (1955) The covering reaction of sea urchins. Nature 175:561

Millott N (1976) The photosensitivity of echinoids. In: Russell FS, Yonge M (eds) Advances in marine biology. Academic, New York, pp 1–52

Millott N, Coleman R (1969) The podial pit—a new structure in the echinoid Diadema antillarum Philippi. Z Zellforsch Mikrosk Anat 95:187–197. https://doi.org/10.1007/BF00968451

Millott N, Okumura H (1968) The electrical activity of the radial nerve in *Diadema antillarum* Philippi and certain other echinoids. J Exp Biol 48:279–287

Millott N, Yoshida M (1959) The photosensitivity of the sea urchin *Diadema antillarum* Philippi: responses to increases in light intensity. Proc Zool Soc London 133:67–71

Millott N, Yoshida M (1960) The shadow reaction of Diadema antillarum Philippi I. The spine response and its relation to the stimulus. J Exp Biol 37:363–375

Montgomery EM, Hamel JF, Mercier A (2018) Ontogenetic variation in photosensitivity of developing echinoderm propagules. J Exp Mar Biol Ecol 500:63–72. https://doi.org/10.1016/j.jembe.2017.12.003

Moore AR (1921) Stereotropic orientation of the tube feet of the starfish (*Asterias*) and its inhibition by light. J Gen Physiol 4:163

Moore A, Cobb JLS (1985) Neurophysiological studies on photic responses in Ophiura ophiura. Comp Biochem Physiol A Physiol 80:11–16. https://doi.org/10.1016/0300-9629(85)90669-3

Mortensen T (1940) A monograph of the Echinoidea: Vol. III. Aulodonta. C A. Reitzel Denmark, Copenhagen

Mortensen T (1948) Report on the Echinoidea Collected by the United States Fisheries Steamer "Albatross" During the Philippine Expedition, 1907-1910: Part 3: The Echinoneidae, Echinolampadidae, Clypeastridae, Arachnoididae, Laganidae, Fibulariidae, Urechinidae, Echinocoryth. US Government Printing Office

Nilsson D-E (2013) Eye evolution and its functional basis. Vis Neurosci 30:5–20. https://doi.org/10.1017/S0952523813000035

Notar JC (2016) A comparative study of Sea Urchin Visual ecology. Masters thesis, University of California, Los Angeles

Notar JC, Meja B, Johnsen S (2022) Testing mechanisms of vision: Sea urchin spine density does not correlate with vision-related environmental characteristics. Int Comp Biol 62(3):509–520. https://doi.org/10.1093/icb/icac119

O'Hara TD, Byrne M, Cisternas PA (2004) The *Ophiocoma erinaceus* complex: another case of cryptic speciation in echinoderms. In: Heinzeller T, Nebelsick JH (eds) Echinoderms: Munchen: Proceedings of the 11th international echinoderm conference. Taylor & Francis, London, pp 537–542

Oliva D, Medan V, Tomsic D (2007) Escape behavior and neuronal responses to looming stimuli in the crab *Chasmagnathus granulatus* (Decapoda: Grapsidae). J Exp Biol 210:865–880. https://doi.org/10.1242/jeb.02707

Olmsted JMD (1917) The comparative physiology of *Synaptula hydriformis* (Lesueur). Contributions of the Bermuda Biological Station, pp 333–379

Paganos P, Ullrich-Lüter E, Caccavale F, Zakrzewski A, Voronov D, Fournon-Berodia I, Cocurullo M, Lüter C, Arnone MI (2022) A new model organism to investigate extraocular photoreception: opsin and retinal gene expression in the sea urchin paracentrotus lividus. Cells 11(17):2636. https://doi.org/10.3390/cells11172636

Pearse AS (1908) Observations on the behaviour of the holothurian, *Thyone briareus*. Biol Bull 15:259–288

Pennington JT, Emlet RB (1986) Ontogenetic and diel vertical migration of a planktonic echinoid larva, Dendraster excentricus (Eschscholtz): occurrence, causes, and probable consequences. J Exp Mar Biol Ecol 104:69–95. https://doi.org/10.1016/0022-0981(86)90098-5

Perrier E (1873) L'anatomie et la régénération des bras de Comatula. Archives de Zoologie Expérimentale et Générale, Deuxième Série 2:29–86

Petie R, Garm A, Hall MR (2016) Crown-of-thorns starfish have true image forming vision. Front Zool 13:41. https://doi.org/10.1186/s12983-016-0174-9

Raible F, Tessmar-Raible K, Arboleda E et al (2006) Opsins and clusters of sensory G-protein-coupled receptors in the sea urchin genome. Dev Biol 300:461–475. https://doi.org/10.1016/j.ydbio.2006.08.070

Raup DM (1960) Ontogenetic variation in the crystallography of echinoid calcite. J Palaeontol 34:1041–1050

Raup DM (1965) Crystal orientations in the echinoid apical system. J Palaeontol 39:934–951

Reich A, Dunn C, Akasaka K, Wessel G (2015) Phylogenomic analyses of echinodermata support the sister groups of Asterozoa and Echinozoa. PLoS One 10:1–11. https://doi.org/10.1371/journal.pone.0119627

Sarasin PB, Sarasin CF (1887) Die Augen and das Integument der Diadematiden. Ergeb Naturw Forsch auf Ceylon Wiesbaden 1:1–18

Sides EM, Woodley JD (1985) Niche separation in three species of *Ophiocoma* (Echinodermata: Ophiuroidea) in Jamaica, West Indies. Bull Mar Sci 36:701–715

Sodergren E, Weinstock GM, Davidson EH et al (2006) The genome of the sea urchin *Strongylocentrotus purpuratus*. Science 314:941–952

Speiser DI, Eernisse DJ, Johnsen S (2011) A chiton uses aragonite lenses to form images. Curr Biol 21:665–670. https://doi.org/10.1016/j.cub.2011.03.033

Stubbs TR (1983) Some aspects of the neurobiology of ophiuroids: with special reference to *Ophiura texturata* (L.) (Echinodermata, Ophiuroidea. University of St Andrews

Sumner-Rooney L, Rahman IA, Sigwart JD, Ullrich-Lüter EM (2018) Whole-body photoreceptor networks are independent of 'lenses' in brittle stars. Proc R Soc B 285:20172590. https://doi.org/10.1098/rspb.2017.2590

Sumner-Rooney L, Kirwan JD, Lowe EK, Ullrich-Lüter E (2020) Extraocular vision in a brittle star is mediated by chromatophore movement in response to ambient light. Curr Biol 30:1–9

Sumner-Rooney L, Kirwan JD, Lüter C, Ullrich-Lüter E (2021) Run and hide: visual performance in a brittle star. J Exp Biol. https://doi.org/10.1242/jeb.236653

Takahashi K (1964) Electrical responses to light stimuli in the isolated radial nerve of the sea urchin, Diadema setosum (Leske). Nature 1343:1343–1344

Thompson JR, Paganos P, Benvenuto G et al (2021) Post-metamorphic skeletal growth in the sea urchin *Paracentrotus lividus* and implications for body plan evolution. EvoDevo 12:1–14. https://doi.org/10.1186/s13227-021-00174-1

Ullrich-Lüter E (2011) Morphological and molecular studies on photoreceptors in Echinodermata

Ullrich-Lüter EM, Dupont S, Arboleda E et al (2011) Unique system of photoreceptors in sea urchin tube feet. Proc Natl Acad Sci U S A 108:8367–8372. https://doi.org/10.1073/pnas.1018495108

Ullrich-Lüter EM, D'Aniello S, Arnone MI (2013) C-opsin expressing photoreceptors in echinoderms. Integr Comp Biol 53:27–38

Vinogradova E, Ruíz-Zepeda F, Plascencia-Villa G, José-Yacamán M (2016) Calcitic microlens arrays in Archaster typicus: microstructural evidence for an advanced photoreception system in modern starfish. Zoomorphology 135:83–87. https://doi.org/10.1007/s00435-015-0276-5

Von Üxküll J (1925) A stroll through the worlds of animals and men. In: Schiller CH (ed) Instinctive behaviour: the development of a modern concept. International Universities Press, Inc., New York, pp 5–80

Walker JW (2007) Effects of fine sediments on settlement and survival of the sea urchin Evechinus chloroticus in northeastern New Zealand. Mar Ecol Prog Ser 331:109–118

Woodley JD (1982) Photosensitivity in *Diadema antillarum*: does it show scototaxis? In: Lawrence JM (ed) Echinoderms: Tampa Bay: proceedings of the international echinoderm conference. Balkema and NH Salem, Rotterdam, p 61

Yamamoto M, Yoshida M (1978) Fine structure of the ocelli of a synaptid holothurian, Opheodesoma spectabilis, and the effects of light and darkness. Zoomorphology 90:1–17. https://doi.org/10.1007/BF00993740

Yamanouchi T (1929) Notes on the behaviour of the holothurian *Caudina chilensis*. Sci Rep Tôhoku Univ–1929–Ser 4:75–116

Yamawaki Y, Toh Y (2009) Responses of descending neurons to looming stimuli in the praying mantis *Tenodera aridifolia*. J Comp Physiol A 195:253–264. https://doi.org/10.1007/s00359-008-0403-6

Yerramilli D, Johnsen S (2010) Spatial vision in the purple sea urchin *Strongylocentrotus purpuratus* (Echinoidea). J Exp Biol 213:249–255. https://doi.org/10.1242/jeb.033159

Yilmaz M, Meister M (2013) Rapid innate defensive responses of mice to looming visual stimuli. Curr Biol 23:2011–2015. https://doi.org/10.1016/j.cub.2013.08.015

Yoshida M (1956) On the light response of the chromatophore of the sea-urchin, *Diadema setosum* (Leske). J Exp Biol 33:119–123

Yoshida M (1966) Photosensitivity. In: Boolootian RA (ed) Physiology of Echinodermata. Wiley, New York, pp 435–464

Yoshida M, Ohtsuki H (1968) The phototactic behavior of the starfish, *Asterias amurensis* Lütken. Biol Bull 134:516–532

Chapter 4
Dispersed Vision in Starfish: A Collection of Semi-independent Arms

Anders Garm, Ditte Sundberg, and Camilla Elinor Korsvig-Nielsen

Abstract The radially symmetric body of starfish has major implications on their nervous system including eyes and vision. All the up to 50 arms are structurally identical, and most examined species have a small compound eye basally on the terminal tube foot of each arm. The 20–300 ommatidia of the compound eyes are lens-less but hold approximately 100 photoreceptors with outer segments made of a combination of microvilli and a modified cilium. The eyes support image forming vision but of low spatial resolution and extremely low temporal resolution with flicker fusion frequencies $\leq$1 Hz. Starfish are color-blind, and vision seems to be based on a single rhabdomeric opsin although many other types of opsins are expressed in their eyes. Starfish also possess extraocular photoreceptors, but little is known about their identity and function. Not many visually guided behaviors are known from starfish so far, but habitat recognition is well documented in a couple of tropical species. More behavioral data are urgently needed, but interestingly, recent data suggest that at least in some situations vision is integrated with olfaction and rheotaxis forming a sensory hierarchy, where olfaction is dominating. Such processing and integration putatively take place in the central nervous system. The eyes are direct extensions of the radial nerve, which constitute the major part of the CNS of starfish and other echinoderms. In general, the echinoderm CNS is enigmatic and the functionality is at best speculative. Here we present new data showing differentiations of the radial nerve along the length of the arms and differences in radial nerve structure between eye-possessing and eyeless species.

Keywords Compound eyes · Ommatidia · Low resolution vision · Radial nerve · Sea star · Radial symmetry · Echinoderm · Temporal resolution · Habitat recognition

---

A. Garm · D. Sundberg · C. E. Korsvig-Nielsen
Marine Biological Section University of Copenhagen, Copenhagen, Denmark
e-mail: algarm@bio.ku.dk

E. Buschbeck, M. Bok (eds.), *Distributed Vision*, Springer Series in Vision Research, https://doi.org/10.1007/978-3-031-23216-9_4

4.1 Introduction

Echinoderms are remarkable in many ways. Despite being nested within Deuterostomia, most of the adult animals display penta-radial symmetry, which comes about through a truly astonishing metamorphosis of their larvae rivaling in complexity the metamorphosis of holometabolic insects (Lascalli 2000). This has major impact on the organization of the organs in the body, including the nervous system. Echinoderms are in general large animals, but many are semi-sessile with a rather simple body structure where, for example, the excretion and the blood-vascular systems are strongly reduced (Chia and Koss 1994). Most species also lack strong abilities to osmoregulate, and they are thus often restricted to areas of high salinity oceanic water. In these areas, though, they are often dominating the benthos, and echinoderms are found in all seas from the intertidal zone down to the deepest parts of the abyssal plains below 8000 m.

Asteroidea, starfish, is one of the most species-rich classes of echinoderms counting close to 2000 extent species (www.marinespecies.org/asteroidea). Most are specialized predators and display the most active behavior of all echinoderms even though the measured maximum walking speed is only approx. 80 cm/min (Mueller et al. 2011). In coastal areas, many species like *Asterias rubens* feed on the abundance of mussels often found in the area, whereas deep-sea starfish often prey on gorgonian corals or tiny planktonic or benthic crustaceans. The latter are prey for members of the specialized brisingid family, which are sit-and-wait predators putatively catching their crustacean prey in similar ways as cnidarian polyps using their 1000s of pedicellaria, instead of cnidocytes, to capture their planktonic prey (Zhang et al. 2019). Starfish are not exclusively predators, though; some are scavengers, whereas others are suspension feeders and a large number of species are detritivorous like the iconic blue starfish *Linckia laevigata* (Mueller et al. 2011). There are no doubts that starfish play major ecological roles in most marine habitats sometimes even at the level of single species. This is the case for the corallivorous species complex *Acanthaster planci*. Found all over the tropics, they specialize in feeding on scleractinian corals (Fig. 4.1). They occasionally occur in major outbreaks causing great damage to the coral reefs they inhabit and devour, and these outbreaks have increased in frequency over the last 50 years. It is estimated that approx. 20% of the coral decline globally is a result of these outbreaks (De'ath and Moran 1998; Fabricius et al. 2000; Moran 1986). For this reason, they have been intensively studied and become a model species for understanding starfish ecology and reproductive biology and lately also starfish neurobiology and sensory ecology including vision (Hall et al. 2017; Lowe et al. 2018; Lucas 2013; Motti et al. 2018; Petie et al. 2016b).

Starfish are clearly radially symmetric with a central disk from which a number of evenly distributed arms project to the sides (Fig. 4.1). Most species of starfish have 5 arms, but in some species, like the Antarctic *Labidiaster annulatus*, there can be as many as 50 arms. There is no known differentiation between the arms in a specimen, the arms are identical containing the same elements, and several species can reproduce asexually by autotomizing an arm, which will afterward regenerate

Fig. 4.1 The model starfish *Acanthaster planci*. (**a**) *A. planci* is a many-armed species with up to 24 arms each carrying a compound eye at the tip. The specimen here is an 8 cm juvenile seen from its aboral side clearly demonstrating the characteristic radial symmetry of starfish. The many spines have given it the trivial name crown-of-thorns starfish. (**b**) As late juveniles and adults, *A. planci* feeds almost exclusively on scleractinian corals, and here it is eating a staghorn coral. The white areas of the coral are areas already eaten by the starfish. It is estimated that 20% of all coral decline on the Great Barrier Reef in Australia is caused by outbreaks of *A. planci*

the rest of the body (Clements et al. 2019). Mainly due to this unorthodox organization of the body, starfish and other echinoderms are often considered to lack a central nervous system (CNS) as a central brain is not present (Clark et al. 2019). However, this is a far too simplistic view on the echinoderm nervous system. Detailed morphological studies have clearly demonstrated that they do possess a CNS but without a central brain. It consists of a ring nerve encircling the mouth opening and branching off a number of radial nerve cords, in starfish one projecting into each of the arms (Mashanov et al. 2006, 2009). Running in parallel with the CNS are the major parts of the water vascular system, unique to echinoderms, where the functional parts are the tube feet running in rows on the outside of the animals (Ullrich-Lüter et al. 2011). The tube feet serve a multitude of functions including locomotion, prey capture, excretion, respiration, and sensing. Almost all sensing in tube feet takes place in their distal end, normally forming an attachment disc, and here a high number of putative mechano- and chemoreceptors are found (Moore and Thorndyke 1993; Ullrich-Lüter et al. 2011). Accordingly, several behavioral studies have indicated that starfish, for a large part, are guided by rheotaxis and olfaction (Castilla and Crisp 1973; Valentincic 1975). Uniquely within echinoderms, most starfish possess an additional sensory system on the terminal tube foot, which is the first tube foot to develop during the metamorphosis and the only unpaired tube foot. This tube foot carries a compound eye basally on the oral side (Fig. 4.2) (Jourdain 1865). The eye constitutes the distal-most part of the radial nerve cord and is thus embedded in the CNS similar to the vertebrate eye, but since there is an eye on each arm tip, starfish eyes are also clear examples of a distributed visual system with between 5 and 50 eyes evenly distributed along the periphery of the body (Garm 2017).

Fig. 4.2 Diversity of starfish eyes. Most studied starfish have a compound eye on the terminal tube foot (TTF) of each arm. Here the TTF is seen for *Marthasterias glacialis* (**a**), *Pentaceraster mammillatus* (**b**), *Asterina* sp. (**c**), *Acanthaster planci* (**d**), and *Culcita novaeguineae* (**e**). The red screening pigment of the ommatidia is clearly seen on all species (arrow in **c**). Note the very large eyes on *P. mammillatus* and *C. novaeguineae*. Following the TTF are a number of smaller tube feet lacking the disk (asterisk), and they are believed to be the center of olfaction in starfish often called sensory tube feet

4.2 The Starfish Eyes

The starfish compound eye is sometimes called the optic cushion referring to the cushion shape with the ommatidia evenly distributed along the surface. Depending on species, there may be from fewer than 20 to more than 300 ommatidia with each of them clearly distinguishable as a ring of bright red screening pigment (Fig. 4.3). The chemical nature of this screening pigment is still unknown. It has been proven behaviorally that the eyes form true images and that the animal uses this image information (see later for details) (Petie et al. 2016a). Each ommatidium contains a large number of photoreceptors, but since the outer segments of the receptors are intermingled, it suggests that they sample light from the same area in space, thus collectively forming one separate part of the image (like a pixel in a digital image). In most cases, the ommatidia are round with a round pupil, but in some deep-sea species, the ommatidia are strongly elongated along the oral-aboral axis putatively gaining sensitivity without losing resolution along the horizon (Fig. 4.3) (Birk et al. 2018). So far, *Hippasteria phrygiana* is the examined species with the most ommatidia per eye (up to 320) (Birk et al. 2018), but preliminary data suggests that other

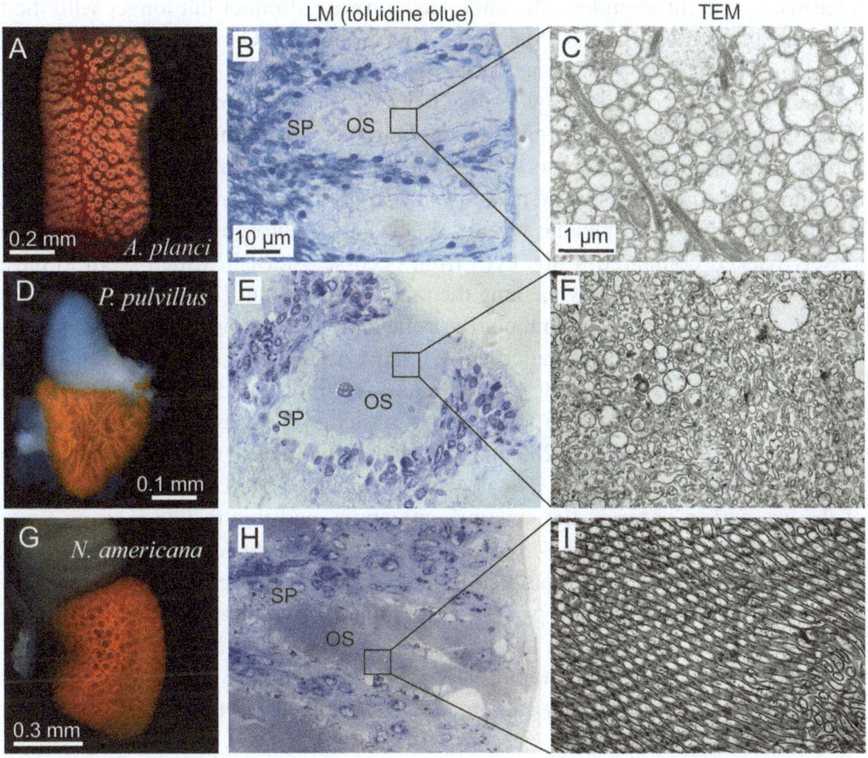

Fig. 4.3 Morphology of the starfish eye. (**a**, **d**, **g**) Eyes of *Acanthaster planci*, *Pteraster pulvillus*, and *Novodinia americana*, respectively. Note the round ommatidia in *A. planci* and *N. americana* and the vertically elongated ommatidia in the semi deep-sea species *P. pulvillus*. (**b**, **e**, **h**) Longitudinal sections of an ommatidium from each species. There are no lenses, and the inner part is filled with the outer segments (OS) of the photoreceptors surrounded by screening pigment (SP) mainly from pigment cells. (**c**, **f**, **i**) TEM micrographs of framed areas in (**b**, **e**, **h**) respectively. Note the rather loose arrangement of the photosensitive membranes in *A. planci* (**c**) and the dense packing for the deep-sea species *N. americana* (**i**). Scale bar in (**b**) also fits (**e**) and (**h**); scale bar in (**c**) also fits (**f**) and (**i**). (Modified after Birk et al., 2018)

species like *Culcita novaeguineae* and *Pentaceraster mammillatus* may have even more (Fig. 4.2).

The structure of the ommatidia is similar across all examined species. They are built by the same two cell types: pigmented photoreceptors and pigment cells. There seem to be no other cells involved in the eyes even though they are covered by an epithelium, which has been suggested to have an optical function (Penn and Alexander 1980). In most examined species, the epithelium is a monolayer of flat electron lucent cells, which rules out optical functions (Birk et al. 2018; Garm and Nilsson 2014). A fully developed ommatidium from an eye of *L. laevigata* or *A. planci* has approximately 100 pigment cells and a little fewer photoreceptors. The pigment cells are about 10 μm wide and 15–20 μm long with the apical part filled with pigment granules. The photoreceptors are slimmer but longer with their photosensitive outer segment alone being 10–30 μm long (Garm and Nilsson 2014). Interestingly, the outer segments of starfish eyes are made of a combination of microvilli and a modified cilium (Petie et al. 2016b), which might result in starfish photoreceptors having a different transduction mechanism than the normal ciliary and rhabdomeric photoreceptors. This is supported by structurally similar receptors in chiton larvae expressing xenopsins, a recently discovered family of opsins (Vöcking et al. 2017). In shallow water species, the photosensitive membranes are loosely packed compared to the eyes from deep-sea starfish, which seem to have increased their sensitivity by packing the membranes tighter (Fig. 4.3).

The starfish eye grows with the size of the animal. Contrary to many other animals where the eyes grow allometrically relative to the rest of the body, there is a close to linear growth in eyes of *A. planci*. As a few-month-old juveniles (3–5 cm in diameter), they have only 20–30 ommatidia in each eye, but as adults (40–50 cm in diameter), they have between 280 and 300 ommatidia in each eye (Korsvig-Nielsen et al. 2019; Petie et al. 2016b). The new ommatidia are added both in the periphery of the eye and in between the existing ommatidia, and this results in more acute vision as the animals grow, while the visual field of the eyes stays the same. Interestingly, the size of each ommatidium is the same in juveniles and adults indicating that they compromise spatial resolution as juveniles but not sensitivity (Korsvig-Nielsen et al. 2019). The terminal tube foot including the compound eye is formed during the larval metamorphosis, but here there are as few as four to five ommatidia. The visual capacity and functional significance of these tiny eyes are yet to be tested.

4.2.1 *Low Pass Filtering in Starfish Eyes*

As mentioned above, the highest number of ommatidia found in a starfish eye is just above 300, and this was for the North Atlantic species *H. phygiana* (Birk et al. 2018). Since the number of ommatidia putatively equals the number of resolved areas (pixels) in the formed image, it is obvious that the obtained spatial information is low. The maximum spatial resolution has been estimated in a few species through measurements of the interommatidial angles, and they are in the range of 7–17° (Birk et al. 2018; Garm and Nilsson 2014), which is comparable to cubomedusan and small insect eyes (Nilsson et al. 2005). Interestingly, the deep-sea species, *Novodinia americana*, is one of those with the highest resolution, which is putatively because they need to resolve bioluminescent patterns (Birk et al. 2018). Still, the relative low spatial resolution in all examined starfish results in low pass filtering – low spatial frequencies (large objects) are seen and high spatial frequencies (small objects) are filtered away. This again means that small and distant objects will not be seen whereas large relatively nearby objects will. In the only two species, *L. laevigata* and *A. planci*, where there is behavioral evidence for which part of the visual environment they see, it turns out to be the major structure in their habitat, the coral boulders (see below for details on visually guided behaviors) (Petie et al. 2016b; Sigl et al. 2016).

Low pass filtering is a central theme of starfish vision perhaps best seen in the temporal resolution. Again the initial data came from the model species *A. planci*, and it was found that this species have the lowest temporal resolution of all animal eyes examined so far (Petie et al. 2016b). The temporal resolution is measured as the flicker fusion frequency (fff) – the frequency where a sinusoidal change in the light intensity is no longer registered by the photoreceptors but seen as a constant stimulation with the average intensity. Depending on the absolute intensity, this typically varies between 20 and 55 Hz for the human eye, whereas the fff of *A. planci* is as low as 0.5–0.6 Hz (Fig. 4.4a). This means that changes happening faster than every other second are not seen! Such extreme low pass temporal filtering results in even slow moving objects causing severe image blur, and only stationary large objects – like the coral boulders – are seen by the animals. Importantly, there is also

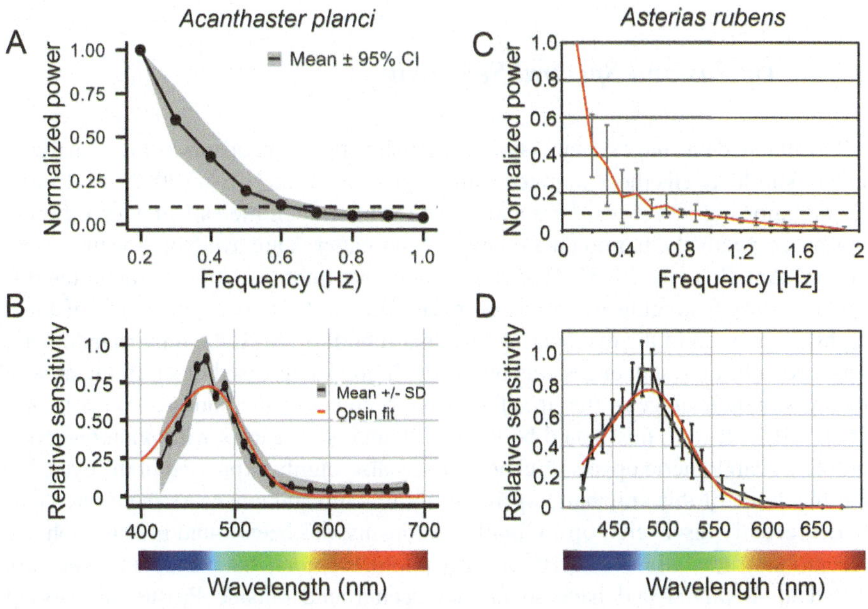

Fig. 4.4 Temporal resolution and spectral sensitivity. (**a**) Flicker fusion frequency (fff) curve for *Acanthaster planci*. There is a steep decline in the power of the response with increasing frequency, and the fff at 0.1 (broken line) is reached at approx. 0.6 Hz. (**b**) Spectral sensitivity curve for *A. planci*. The best fit using the least squares method is for an opsin peaking at 472 nm (red curve). Note that the curve is narrower than the opsin absorption curve indicating spectral filtering is taking place. (**c**) Flicker fusion frequency (fff) curve for *Asterias rubens*. Here there is also a steep decline in the power of the response with increasing frequency, and the fff at 0.1 (broken line) is reached at approx. 0.9 Hz. (**d**) Spectral sensitivity curve for *A. rubens*. The best fit is for an opsin peaking at 484 nm (red curve). A and B are modified after Petie et al., 2016b. Error bars in (**c**) and (**d**) indicate SD, $n = 8$ in (**c**) and 9 in (**d**)

a match with the self-motion of the starfish. With such a low temporal resolution, any fast movement of the eye bearing structures will also result in problems with image blur, but with maximum speeds of 60 cm/min, this is no problem for *A. planci* (Mueller et al. 2011). Importantly, the fff of a given eye is influenced not only by the intensity of light changes but also by the temperature of the photoreceptors. The warmer the eyes, the faster they are (Fritsches et al. 2005), and since *A. planci* is a tropical animal, this could mean that cold water starfish have even lower temporal resolution. This is not the case, though, since our previously unpublished data generated with the same protocol as for *A. planci* shows that *A. rubens* kept at 10 °C have a fff of approx. 1 Hz (Fig. 4.4c). This strongly indicates that it is not physiological constrains that set the fff in starfish photoreceptors but rather selective pressure on having low pass temporal filtering. The putative explanation is that in combination with the low pass spatial filter, it works as match filtering. Such filtering ensures that only large stationary objects are seen, whereas small moving objects, which are normally not important for the starfish, are removed from the visual input minimizing the need for processing power in the CNS.

4.2.2 Opsins and Spectral Sensitivity

All available data suggest that vision in starfish is based on opsins as the visual pigment – similar to all other examined animal eyes (Land and Nilsson 2012). Molecular examinations of *Asterias rubens* and *Patiria miniata* identified six and ten different opsins, respectively, but no data on expression patterns are available for these species (D'Aniello et al. 2015). Tissue specific transcriptomic data also found ten different opsins belonging to seven opsin families in *A. planci* with many of them highly expressed in the eye bearing terminal tube foot. Still, the only rhabdomeric opsin found in the sequences was by far the highest expressed opsin in the area of the eyes, which suggests that it is the visual pigment of the photoreceptors (Lowe et al. 2018). If true, this would be the first deuterostome eyes with photoreceptors utilizing rhabdomeric opsins. A chaopsin was also highly expressed in the eyes, but the functions of this enigmatic opsin family remains unknown. As mentioned earlier, a recently discovered opsin family, xenopsins, has been found in chiton photoreceptors structurally similar to starfish photoreceptors, but xenopsins were not found in *A. planci* and have so far not been found outside Prostomia (Döring et al. 2020).

Electrophysiological work performing electroretinograms (ERGs) supports the presence of a single opsin in the photoreceptors (Garm and Nilsson 2014; Petie et al. 2016b). In *A. planci*, *L. laevigata*, and *A. rubens*, the spectral sensitivity has a single peak in the blue part of the visual spectrum with λ-max = 472nm, 452nm, and 484nm, respectively. In the two first species, the obtained spectral sensitivity curves are narrower than the modeled absorption curve of an opsin (Garm and Nilsson 2014; Govardovskii et al. 2000; Petie et al. 2016b) (Fig. 4.4b). The suggested explanation here is that there is an external filtering happening putatively removing the damaging UV light present at high intensities in the shallow water tropical habitat of these two species (Garm and Nilsson 2014; Petie et al. 2016b). The spectral sensitivity peak in the deep blue part of the spectrum for these two species is matching the most abundant wavelengths in the clear ocean water they live in (McFarland and Munz 1975), thereby optimizing the contrast to objects in the water. The spectral curve of *A. rubens* has a much better match with a modeled opsin curve indicating that spectral filtering does not take place in this species (Fig, 4d). The peak sensitivity is also shifted a bit to about 485 nm, and both these differences match *A. rubens* living in temperate waters with a higher organic content, which removes most of the UV light and green-shifts the color of the water (Lythgoe 1979).

Behavioral experiments with blinded animals (eye-ablated) and specimens of eyeless species show that they still respond to light stimuli, which must then be controlled by yet unidentified extraocular photoreceptors (see later this chapter). This is well in line with data from members of other echinoderm classes such as sea urchins and brittle stars (Lesser et al. 2011; Sumner-Rooney et al. 2020; Ullrich-Lüter et al. 2013) (**see also Chapter 3 this volume**). Highly interesting, in both sea urchins and brittle stars there is recent evidence that these extraocular photoreceptors provide the animals with spatial vision although of very low resolution (Kirwan et al. 2018; Sumner-Rooney et al. 2020). It is currently unknown which of the opsins are involved in starfish extraocular photoreceptors, but in both sea urchins and brittle stars, it is a rhabdomeric opsin. What all the other opsins are doing in starfish is still unknown. There are no experimental or expression data to resolve this, but from their structure and through comparison with other systems, it has been suggested that some might be photoisomerases while others serve different physiological functions (Lowe et al. 2018).

4.3 Behavioral Repertoire of Starfish

Even though most echinoderms appear to be semi-sessile at first glance, they turn out to have a rather sophisticated behavioral repertoire, which is especially true for starfish. For over 100 years, starfish behavior has been examined with a focus on their foraging (Fenchel 1965; Kalmus 1929; Scheibling 1981) and reproduction (Boivin et al. 1986; Hamel and Mercier 1994). One of the functionally significant results from these studies is that there is not a leading arm per say – all arms take turn leading the animal pinpointing the uniqueness of their radial symmetric organization (Pearse et al. 1987). Most starfish are predators, many with a preference for bivalve prey, and their foraging has been shown to be at least partly olfactory guided but often in combination with negative rheotaxis (walking against the current) (Dale 1997). When it comes to reproduction, they are typically broadcast spawners, but little is known about the sensory cues behind the gamete release (Hamel and Mercier 1994). As all other animals, starfish are most likely multimodal in their behavioral control, and so far, chemoreception, mechanoreception, and photoreception have been documented (see [20], for review). The main sensory organs are the tube feet, and especially the young distal-most tube feet on each arm seem to be at play here. They lack the attachment disc and have no apparent mechanical function but are instead stretched out in front and above the arm (Sloan 1980). As mentioned earlier, this is also where the compound eyes are found: at the base of the unpaired terminal tube foot. Even though many behavioral studies have shown that starfish are olfactory guided, it is highly likely that control of most (if not all) behaviors involves multiple sensory input. Here we present some of the first evidence of multimodal behavior and reveal the included sensory hierarchy (see Sect. 4.5 on "Multimodal Control of Behavior" below).

4.4 Light Guided Behaviors

4.4.1 Shadow Response and Extraocular Photoreception

Extraocular photoreception seems to be common in echinoderms and has been studied in detail in the sea urchins *Strongylocentrotus purpuratus* and *Diadema africanum* (Kirwan et al. 2018; Ullrich-Lüter et al. 2011) and in brittle stars (**see Chapter 3 this volume**). Opsin expressing cells were found several places in *S. purpuratus* including at the base of the tube feet where they were suggested to support directional vision using the crust for directional screening (Ullrich-Lüter et al. 2011). Behavioral evidence from *D. africanum* supports such a system in sea urchins including proper image formation even though the spatial resolution is one of the lowest ever measured (30–70°) (Kirwan et al. 2018). Extraocular photoreception has also been suggested from starfish including from molecular data (Lowe et al. 2018), but little is known about the functional significance it may have in these animals.

In a previously unpublished behavioral study, we examined the shadow response of the sensory tube feet in adult specimens of four species of North Atlantic starfish: three with prominent eyes, *L. laevigata* ($n = 3$), *Marthasterias glacialis* ($n = 5$), and *Crossaster papposus* ($n = 4$), and the burrowing eyeless *Astropecten irregularis* ($n = 2$). We had previously observed that some starfish would retract the distal-most sensory tube feet when subjected to a passing shadow putatively as a predator avoidance response, and we tested the sensory basis behind this behavior. In room light, the arm tip including the sensory tube feet of a single arm was initially illuminated with a handheld torch for 1 min after which light was turned off. For *L. laevigata*, *M. glacialis*, and *C. papposus*, we tested both intact and eye-ablated specimens (number of test animals is the same as listed above in both cases). When illuminated, both the intact and eye-ablated animals of all species had the distal-most tube feet extended; typically moving them slowly from side to side (Fig. 4.5). The tube feet of *L. laevigata* were notably shorter/less extended than those of the other species (compare Fig. 4.5a, e, and i), which correlates with its coral habitat containing several fish species known to attack starfish including *Chaetodon* sp. (Cowan et al.

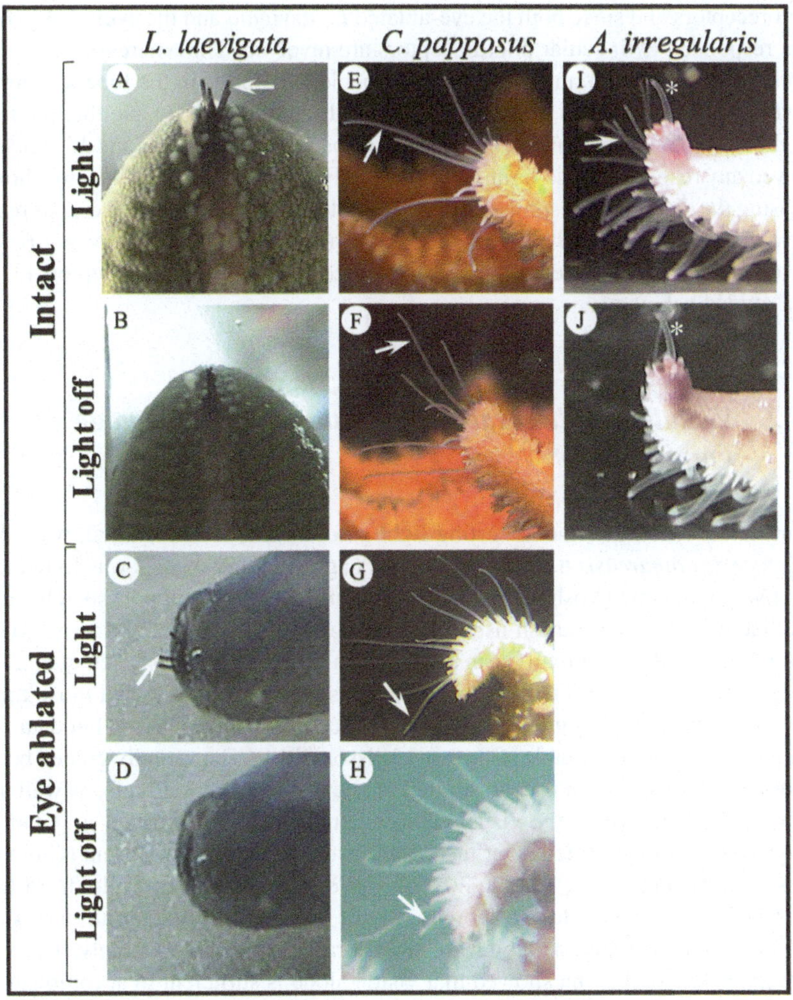

Fig. 4.5 Light-off response in the distal-most tube feet. (**a–d**) Both intact and eye-ablated *Linckia laevigata* respond to a shadow mimic (light-off) by retracting the distal-most black tube feet (arrows) showing that this behavior is at least partly guided by extraocular photoreceptors. (**e–h**) Neither intact nor eye-ablated *Crossaster papposus* responded to light-off and kept the distal-most tube feet extended at all times. (**i–j**) The eyeless *Astropecten irregularis* also respond to light-off by retracting the distal-most tube feet (arrows) adding further evidence to the presence of extraocular photoreceptors. Note that the unpaired terminal tube foot is not retracted (asterisk)

2017). The intact and eye-ablated *L. laevigata* and the eyeless *A. irregularis* all reacted in a similar way to the light-off stimulus mimicking a shadow passing the arm. With a latency of 0.7 +/−0.2 s for *L. laevigata* and 1.9 +/−0.6 s for *A. irregularis*, they rapidly withdraw all of the distal-most tube feet (Fig. 4.5). Interestingly, *A. irregularis* did not retract the terminal tube foot (Fig. 4.5j), which is of unknown function. This response to a pure light stimulation proved that it is governed by photoreceptors and since both the eye-ablated *L. laevigata* and the eyeless *A. irregularis* responded extraocular photoreceptors are involved. The whereabouts of these photoreceptors are still unknown, but a probable location is in the tube feet similar to what has been found in sea urchins (Ullrich-Lüter et al. 2011). Neither the intact nor the eye-ablated specimens of the last two species, *C. papposus* and *M. glacialis*, showed any response to the light-off stimulus (Fig. 4.5e–h, *M. glacialis* not shown), stressing the diversified response pattern. Despite the species differences, our results showed that similar to other echinoderms, some species of starfish do not rely on photoreception in their eyes alone but also utilize extraocular photoreceptors in their light guided behaviors.

4.4.2 Visually Guided Habitat Detection: Proof of Image Forming Eyes

Some of the first experiments testing light guided behaviors in starfish were done with *Asterias amurensis* a little over 50 years ago, and this species was found to be negative phototactic (Yoshida and Ohtsuki 1968). New data support such behavior and that at least some starfish use vision for negative phototaxis and move toward dark objects in their habitat (Garm and Nilsson 2014; Korsvig-Nielsen et al. 2019; Petie et al. 2016b). A limited number of species have been tested, and most knowledge about the visually guided behavior again comes from *A. planci*, but data from *L. laevigata* are also available (Garm and Nilsson 2014). The visually guided behavior of these two species has been studied in situ at tropical coral reefs, as well as in behavioral arenas where the visual environment could be controlled in detail and where no directional olfactory or mechanosensory stimuli were present (Garm and Nilsson 2014; Petie et al. 2016a, 2016b; Sigl et al. 2016). The results from these combined experiments clearly show that both species are attracted to dark structures, which in their natural habitat are large coral boulders. Interestingly, the results from the behavioral arena showed that vision alone is sufficient to accomplish this behavior. The results from the natural habitat are somewhat divergent. When tested close to the reef in weak non-laminar currents, eye-ablated *A. planci* and *L. laevigata* with assumed intact chemo- and mechanoreception walked randomly, whereas sham operated animals walked toward the coral reef, indicating that vision is required to accomplish the behavior (Fig. 4.6). Another study did indicate, though, that under stronger semi-laminar current conditions, blinded animals can navigate toward the reef possibly using rheotaxis in combination with olfaction (Sigl et al. 2016). Our previously unpublished data from the North Atlantic species *M. glacialis* support the inclusion of rheotactic information, since they were only attracted to a 37° large dark visual stimulus in the presence of a semi-laminar current (see Sect. 4.5 on "Multimodal control of Behavior" for details).

The results from the behavioral arena are well in line with the low spatial resolution in *A. planci* determined from eye morphology, since they were not attracted to

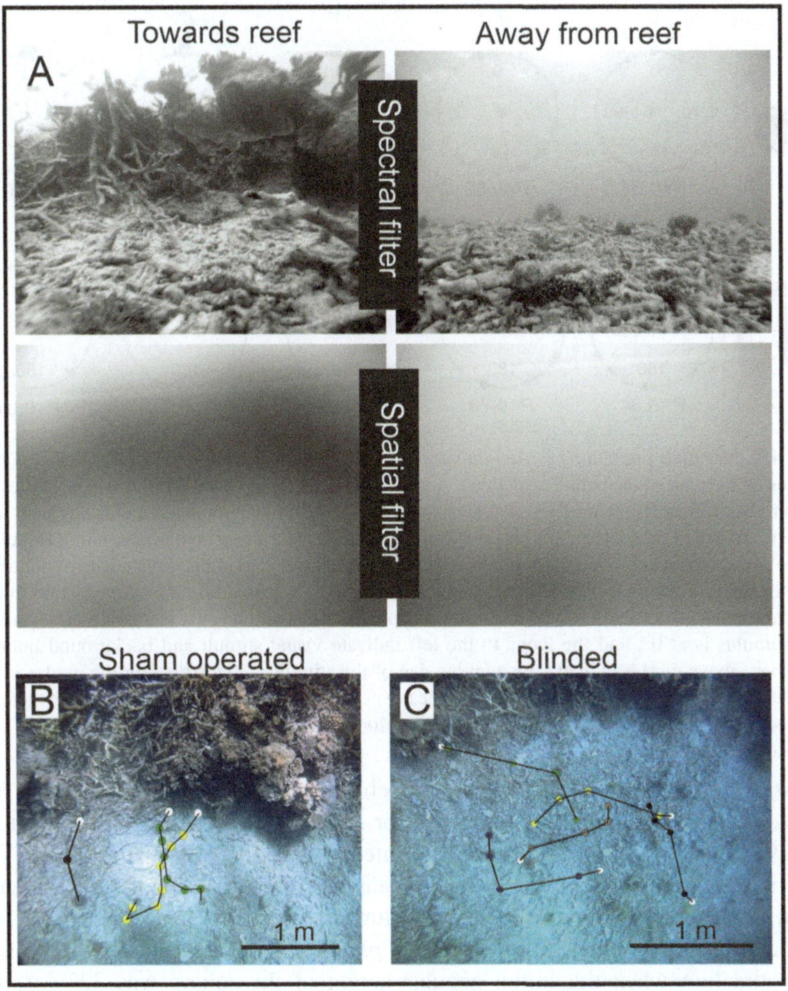

Fig. 4.6 Visually guided navigation in *Acanthaster planci*. (**a**) The visual scene toward and away from the coral habitat, when first spectrally filtered (blue opsin) and then spatially filtered (8 degrees resolution) to match the properties of vision in *A. planci*. Note that the coral reef is still visible as a dark area rising from the ocean floor. (**b**, **c**) Results from in situ behavioral experiments with *A. planci* showing that sham operated animals (vision intact, but a couple of locomotory tube feet removed) are able to detect the habitat whereas blinded animals (terminal tube foot with eye removed but other senses intact) walk randomly. There is 1 min between the dots; the white dots indicate the end positions. (Modified after Garm, 2017)

dark visual stimuli until they took up a visual angle of at least 14° (Fig. 4.7a) (Petie et al. 2016a). The results also supported the lower resolution in juveniles as they were not attracted until the stimuli had a visual angle of 27° or more (Korsvig-Nielsen et al. 2019). Being attracted to dark objects/areas can be accomplished without proper image formation, though, using simple directional intensity

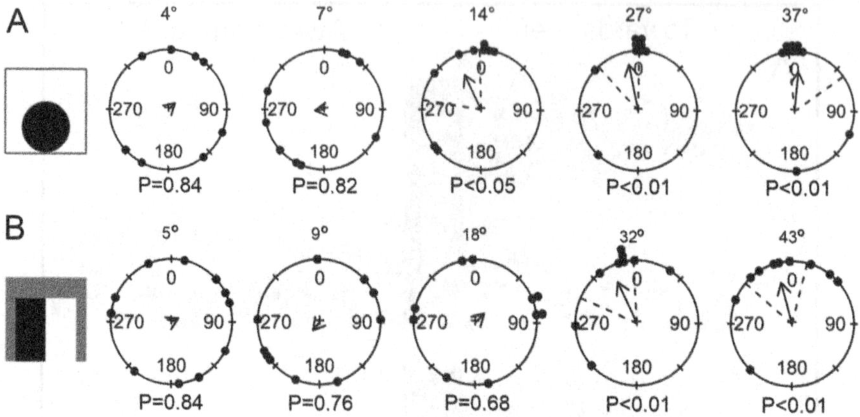

Fig. 4.7 Negative phototaxis is *Acanthaster planci*. (**a**) When presented with a series of black circles on the wall of the behavioral arena, *A. planci* ignores these visual stimuli when smaller than 14° in visual angle. If larger than 14°, *A. planci* displays negative phototaxis and is attracted to the circles. (**b**) When presented with visual stimuli detectable only using spatial resolution (black and white square on average intensity gray background), *A. planci* again displays negative phototaxis but not until the vertical part of the black stimulus is 32° in visual angle or larger. Note that the animals contact the stimuli on the left side (at approx. 340°), which is the black part. The center of the stimulus is at 0°, and the icons to the left indicate visual stimuli and background intensity. Numbers above circles indicate the angular size of the stimulus. Black dots on the circles = individual mean headings, central arrows = the mean vector, broken lines = 95% confidence interval of the mean vector, and P = result of Raleigh test. (Modified after Petie et al., 2016a)

measurements. Which mechanism lies behind the negative phototaxis has been tested through behavioral experiments for *A. planci* (Petie et al. 2016b). In the behavioral arena, the starfish were presented with a black and white square on a mid-intensity gray background, which means that just measuring the light intensity from a given area will not reveal the stimulus since it has the same average intensity as the gray background. Still, the adult *A. planci* showed clear directional walking and attraction when the initial angular height of the square was 32° or above (Fig. 4.7b). Importantly, they did not contact the center of the square but the black half of it (Petie et al. 2016a). These were the first direct behavioral proof of image forming eyes in starfish earlier suggested from the eye morphology. In contrast, juvenile *A. planci* were not able to discriminate the black and white stimulus from the gray background, even though they showed a strong tendency to be either attracted or repelled (Korsvig-Nielsen et al. 2019). This emphasizes that starfish vision changes (improves) with age and/or size of the animal, a phenomenon seen several places in the animal kingdom (Land and Nilsson 2012).

4.4.3 Eye Movements and Active Vision

An important part of the visual ecology of many animals is the ability to actively control the visual input through a variety of eye movements (Land 2019). Four functional types of eye movements exist: setting the gaze direction relative to own body position; compensation for self or object motion, thereby stabilizing the gaze; scanning movements; and fixational eye movements counteracting adaptation. Highly interesting, three of the four types of eye movements seem to be present in *A. planci* (Beer et al. 2016).

The individual eyes of *A. planci* have a broad horizontal visual field of about 100°, but it only spans approx. 20° vertically. Still, the terminal tube foot holding the eye is placed on a bendable nob on the arm tip allowing changes of the vertical part of the gaze direction. It was found that on the arms leading at the given time, the gaze direction was centered a few degrees above the horizon, which matches the desire to detect large coral boulders, their habitat, on the seafloor. The arms trailing at the given time, on the other hand, were held flat along the substrate but with the nob at the tip of the arm bent upward and the eye gazing approx. 45° above horizon possibly looking out for predators (Beer et al. 2016). Further, when traversing an obstacle movement of the nob would counteract 60–70% of the arm movements, thereby stabilizing the gaze direction (Beer et al. 2016). Lastly, *A. planci* raises and lowers the leading arms in a rhythmic way, which results in the eyes changing the vertical part of their gaze direction about 6° in a little more than 2 sec. This matches the temporal and spatial resolution and is, thus, putatively used as fixational eye movements refreshing the image on the retina and enhanced horizontal contrast lines without compromising spatial resolution (Beer et al. 2016; Petie et al. 2016b).

Even though active vision through eye movements in starfish has only been examined for *A. planci* so far, we predict that it is a common feature of starfish vision, since most visual tasks will benefit from this. The presence of these advanced aspects of vision stresses that starfish are highly dependent on visual information but also that complex neuronal circuitry is present providing the feedback needed to control the movements.

4.4.4 Other Starfish Behaviors Putatively Involving Vision

There are major gaps in our knowledge of starfish behavioral repertoires, and when it comes to which senses control the known behaviors, our knowledge is also limited (Garm 2017). As illustrated above, surprisingly few controlled experiments have tested their visually guided behaviors when considering that their eyes were discovered more than 200 years ago. To fully appreciate the starfish visual ecology, new hypothesis driven behavioral experiments are warranted. One of the highly interesting aspects of starfish ecology, which has only been studied in little details, is the bioluminescence found in some deep-sea species (Birk et al. 2018; Henning 1974). Especially members of the order Brisingida should be studied further, since they combine the ability to emit light with prominent eyes and relatively high spatial resolution compared to other starfish (Birk et al. 2018). A plausible explanation for this combination could be that they use the bioluminescence for visual communication in the darkness of the deep sea where finding a mate is often problematic. If true, this would take starfish vision and visually guided behaviors to a new level of complexity. The shallow water Antarctic species *Odontaster validus* does include photoreception in its reproductive behavior, but likely only extraocular photoreception measuring day length to coordinate gamete release (Pearse and Bosch 2002; Pearse et al. 1986). Several nonluminescent deep-sea starfish also have prominent eyes (Birk et al. 2018), and since many of them are predators, they might use vision to detect bioluminescent prey such as deep-sea corals. Vision could also be involved in prey detection in shallow water species, but considering the known spatial and temporal resolution, this would have to be rather large and stationary prey items. All these testable hypotheses are awaiting experimental proof, but several other hypotheses and ideas will undoubtedly emerge once more species and behaviors have been examined.

4.5 Multimodal Control of Behavior

It has been shown within the latest years that starfish can use a combination of rheotaxis, olfaction, and true vision to navigate the ocean floor (Petie et al. 2016b; Sigl and Laforsch 2016; Sigl et al. 2016). In a previously unpublished study, we investigated this further especially to gain information on the potential sensory hierarchy of such multimodal behavior. In a behavioral arena, *M. glacialis* was presented with either a rheotactic stimulus (semi-laminar flow, $n = 12$), visual stimulus (black area on white background, $n = 12$), olfactory stimulus (prey scent) in a semi-laminar flow ($n = 13$), combined rheotactic and visual stimuli (diverging 110°, $n = 14$), or combined olfactory, rheotactic, and visual stimuli (diverging 110°, $n = 10$). In the arena, *M. glacialis* showed a clear need for multimodal sensory input and walked randomly if only a single stimulus was present (Fig. 4.8). Interestingly, they also displayed a sensory hierarchy and were attracted to the visual stimulus when an odorless semi-laminar flow was present, and this is irrespective of the direction to the black area relative to the current. When a prey scent was added to the flow, this combined stimulus became attractive and overruled the visual input (Fig. 4.8). This supports that *M. glacialis* is mainly olfactory guided but also that other senses, like vision, take over if no reliable olfactory cue is present. This novel data are in contrast to *A. planci* where one study found that vision is not only necessary but also sufficient to navigate to their habitat in low current situation (Petie et al. 2016a). When a stronger directional current is present but with no visual cues, *A. planci* seems to use rheotaxis to find the coral boulders putatively including olfactory cues (Sigl et al. 2016). There is, thus, a high chance that starfish with ecologies differing from that of *M. glacialis* and *A. planci* will combine their senses in other ways with different hierarchal orders possibly also including other senses than vision, olfaction, and rheotaxis.

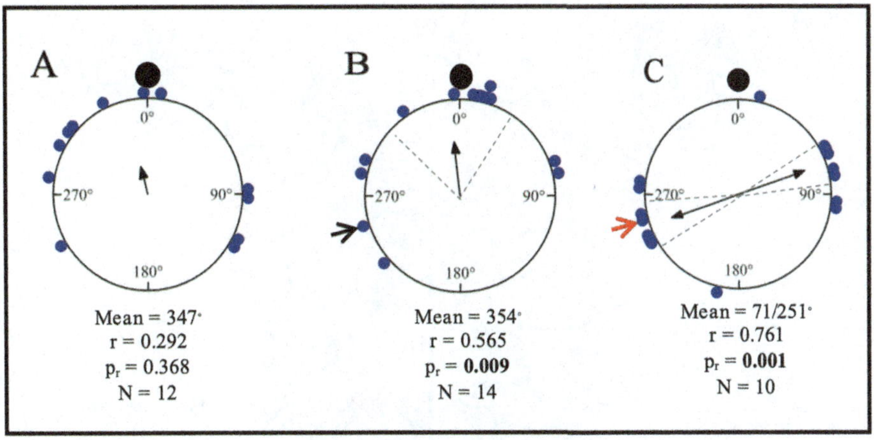

Fig. 4.8 Sensory hierarchy in *Marthasterias glacialis*. (**a**) When only presented with a visual stimulus (black circle at 0°), *M. glacialis* walks randomly in the behavioral arena. (**b**) When presented with both a visual stimulus and a semi-laminar flow (large black arrow indicates flow direction), *M. glacialis* showed negative phototaxis and walked toward the black visual cue. (**c**) When a chemical stimulus (prey odor) was added to the semi-laminar flow (red arrow), *M. glacialis* abandoned negative phototaxis and displayed an axial response with either positive or negative chemotaxis. Central black arrow = mean vector of all individual headings (blue dots), r = length of mean vector, p_r = result of Raleigh test, broken lines = 95% confidence interval, N = number of test animals

4.6 Processing of the Visual Information

Due to their secondary radial symmetry, adult starfish have no cephalization with a single body region, where nervous tissue and most senses are concentrated. Nonetheless, detailed morphological and physiological studies have proven that these animals still possess a central nervous system (CNS) or "brain" – it is just organized differently as what could be called a dispersed CNS. Besides the recognition of a dispersed CNS, though, we are still in the very early days of experimentally testing its functionality.

When a starfish explores the seafloor, they have no preset leading arm and all arms take turns leading the animal (Pearse et al. 1987). This is likely one of the reasons for the senses in starfish, including the eyes, being dispersed with a repetition on each arm. As every arm around the body is equally receptive to environmental stimuli, the sensory processing and integration seemingly need to follow the same pattern. Starfish do have a well-defined CNS including a nerve ring encircling the mouth, but the major part of the CNS is the radial nerve cords (RNC), dispersed with a repetition in every arm (Fig. 4.9b), and this is putatively where the visual information is processed. The nerve ring appears to function at least in part as a means of communication between the RNCs, which is backed by our unpublished results from *A. planci*. When bisecting the nerve ring on opposite sides of the animals, they became quiescent at first. After about 10 min, the animals started

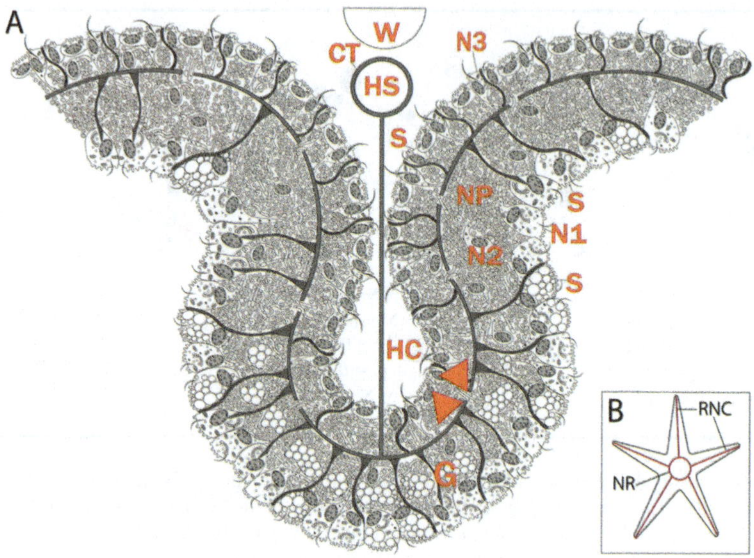

Fig. 4.9 The starfish central nervous system (CNS). (**a**) A schematic drawing of a cross section through the radial nerve cord (RNC) in starfish – not to scale. Arrowheads show neural connection between the ectoneural and hyponeural area. Aggregations of giant neurons (**g**) are drawn as seen in *Acanthaster planci*. Three other types of neurons are indicated: N1 is a neuron located in the neuroepithelium in the ectoneural area, N2 is a neuron with the cell body located in the neuropil, and N3 is a putative sensory neuron. (**b**) Schematic drawing of the starfish central nervous system comprising a RNC in each arm connected to a circumoral nerve ring (NR). CT = connective tissue, HC = hyponeural canal, HS = haemal sinus, NP = neuropil, S = supporting cells, W = water vascular system

moving, but it was obvious that the two halves of the animals could not communicate and coordinate locomotion. On several occasions, the result was the starfish pulling itself apart with the two halves afterward walking "normally" around the tank (Petie and Garm, unpublished results).

4.6.1 Structure of the Starfish CNS

The echinoderm CNS consists of a hyponeural area and an ectoneural area, and in starfish, the entire CNS (nerve ring and RNC) has this division, which is not the case for all echinoderm classes (Engle 2013; Mashanov et al. 2006; Märkel and Röser 1991; Viehweg et al. 1998; Zueva et al. 2018). The ectoneural area makes up the oral side part of the RNC and has by far the most neurons. The hyponeural area overlays the aboral surface of the ectoneural area (Viehweg et al. 1998) (Fig. 4.9). It was initially thought that the two areas were completely separated by a sheath of connective tissue including a basal lamina, but recent data from several echinoderm classes (Holothuroidea, Ophiuroidea, and Asteroidea) have found neural bridges in the connective tissue connecting the two areas (Mashanov et al. 2006; Zueva et al. 2018) (Fig. 4.9a). The starfish CNS lies orally to the water vascular and the haemal systems. In the arms, the connective tissue surrounding the haemal sinus continues orally and separates the hyponeural area in two also forming a two parted hyponeural canal (Fig. 4.9a). Little is known about the function of this canal, but it might be involved in maintenance of the RNC removing waste products and supplying nutrients. In other echinoderm classes, an additional ectoneural canal is present, and oddly enough in echinoids it is open to the external environment, giving access for seawater to enter (Märkel and Röser 1991).

The CNS is connected by peripheral nerves to different areas of the body. Peripheral nerves from the RNC in starfish innervates the spines, the spine-free zone of the interambulacrum, and all the tube feet including the distal-most sensory tube foot with the eye (Formery et al. 2021). Little is known from starfish, but in brittle stars, the peripheral nerves can originate from either the ectoneural area, the hyponeural area, or a mix of both (Zueva et al. 2018). Across echinoderms, the size of the hyponeural system is correlated with the amount of muscle tissue (Mashanov et al. 2016), and it is believed to be involved in motor control only (Cobb 1995; Cobb and Stubbs 1981). This is questioned by the presence of putative sensory cilia extending from the neuroepithelium and into the hyponeural canal (Fig. 4.9a), which has been observed in several echinoderm species (Cobb 1987; Mashanov et al. 2006; Viehweg et al. 1998). Based on their ultrastructure and position inside the canals, these cilia have been proposed to be proprioceptors or chemoreceptors. Still, the ectoneural system comprises by the far the most neurons of which many are sensory neurons (Cobb 1987) and the ectoneural area is the only part of the RNC that continues all the way to the distal-most tube foot at the tip of the arm where it directly contacts the compound eye (Fig. 4.10) (Moss et al. 1998). In total, morphological evidence strongly suggests that visual processing happens in the ectoneural part of each RNS.

4.6.2 The Ectoneural Part of the RNC

The space between the neuroepithelia is filled with neurites (the existing morphological data does not allow us to differentiate between axons and dendrites) and cell bodies of neurons, and this neuropil is partially divided into separate compartments by processes from the supporting cells (see below). In both the hypo- and the ectoneural area, most of the neurites run parallel with the RNC, but some run transversally (Fig. 4.11c). Synapses are omnipresent in the ectoneural part of the RNC as indicated by the putative synaptic vesicles, which again supports that this is an area of information processing. There are three types of synaptic vesicles all 100–150 nm in diameter – electron lucent vesicles, electron dense vesicles, and dense-cored vesicles – and in some neurites, more than one type of vesicle is present (Fig. 4.11d).

Most neurites found in the ectoneural part of the starfish RNC are quite small typically between 100 and 600 nm in diameter depending on the species (Fig. 4.11). This small diameter is a general echinoderm trait, and it has made it hard to

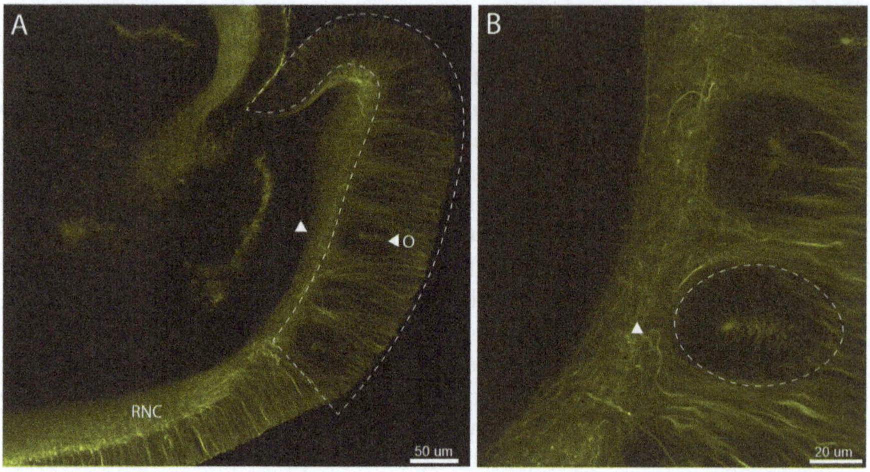

Fig. 4.10 Nervous system in the eye of *Asterias rubens* visualized with anti-tubulin immunostaining. (**a**) The eye (encircled by broken white line) sits as a direct extension of the radial nerve cord (RNC). Nerves from the ommatidia (**o**) collect beneath the eye (arrowhead) and contact the RNC. (**b**) Close-up of a couple of ommatidia and the nerve beneath the eye. Broken line indicates an ommatidium; the dark part is where the screening pigment is situated. Arrowhead points at nerve connecting to the eye

accomplish any form of electrophysiological work on the CNS. To our knowledge, the only published recordings from echinoderm CNS so far come from the larger (giant) neurons of brittle stars (Cobb and Moore 1989). These neurons are 10–20 μm in diameter and are of two types based on the electrophysiological data. The largest neurons alerted the whole animal, and their activity resulted in a "freezing" state of the body with the arms stiffening. The other group of giant neurons was active during behaviors such as escape and feeding. Importantly, the work showed that depending on the type of stimuli, the sensory integration happened either throughout the entire CNS or locally in a particular RNC (Cobb and Moore 1989). Whether this is a general echinoderm feature or specific to brittle stars is not known though.

An open question about the functional organization of the starfish RNC is if there are regional specializations where a certain type of information is processed. To look for such regions, we compared the morphology of the basal (close to the nerve ring), middle, and distal part of the RNC of three species: *A. planci* ($n = 3$), *A. rubens* ($n = 3$), and *A. irregularis* ($n = 3$) (Fig 4.12). Not surprisingly, we found a correlation between the absolute number of neurites and the size of the animal, with the largest species, *A. planci*, displaying the most and the small *A. irregularis* the least neurites. There was an interesting difference between species, though, in that the two eye carrying species, *A. planci* and *A. rubens*, displayed very similar patterns. They had a close to linear decline in the number of neurites from the basal to the distal part of the RNC (Fig. 4.12d, g). In the eyeless *A. irregularis*, on the other hand, the ectoneural area in the distal part of the RNC seemed swollen and had the highest number of neurites (Fig. 4.12a–c). We do not know the exact number of

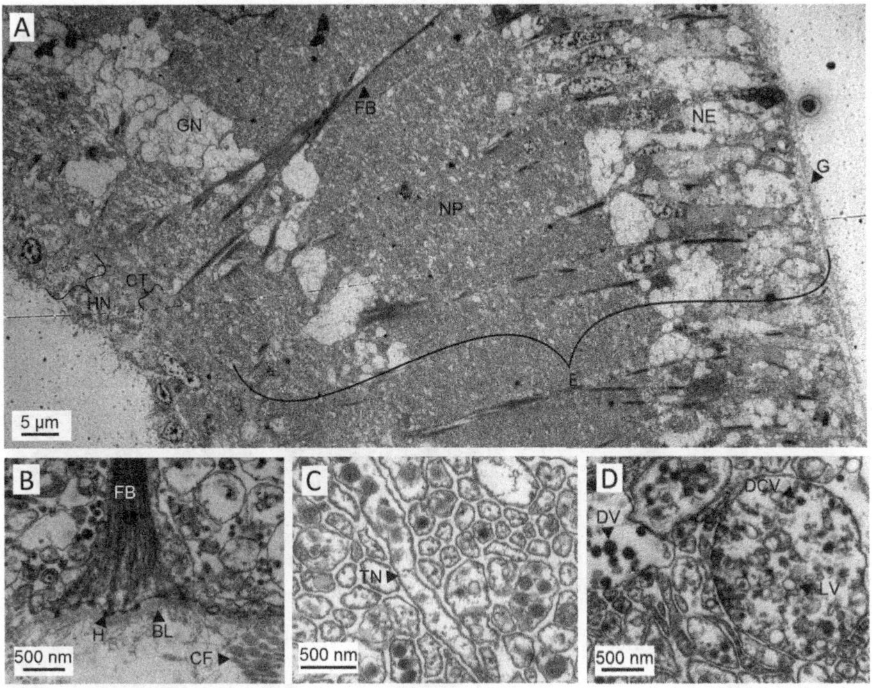

Fig. 4.11 TEM micrographs showing the organization of the distal part of the radial nerve cord (RNC) in *Acanthaster planci*. (**a**) Most of the approx. 80-µm-thick RNC is filled with ectoneural (**e**) neuropil (NP) occasionally transversed by fiber bundles (FB) from the supporting cells. Note that toward the hyponeural part, there is an aggregation of giant neurons (GN). (**b**) The FB extend from the neuropil to the basal lamina (BL) where they attach with hemidesmosomes (H). (**c**) Most neurons run parallel with the longitudinal axis of the RNC (cross-sectioned here), but some transverse the neuropil area (TN). (**d**) The neuropil displays a high number of synapses as indicated by putative synaptic vesicles. Different types of vesicles are seen: electron lucent vesicles (LV), dense vesicles (DV), and dense-cored vesicle (DCV). G = glycocalyx, NE = neuroepithelium, HN = hyponeural area, CT = connective tissue, CF = collagen fiber

neurons as the neurites can be branches of dendrites and/or axons, but it is likely significantly less than the millions of neurites we count. Still, there is putatively a correlation between the relative number of neurites and the amount of processing indicating a processing hotspot distally in the RNC of *A. irregularis*. This came as a surprise since *A. rubens* and *A. planci* have the same putative chemosensory and tactile tube feet distally on the arm as *A. irregularis*, and adding visual information thus appears to reduce the need for processing power. A part of the explanation could be that most of the visual information is processed already in the distal-most tube foot carrying the eye. This tube foot does hold a specialized nervous system laying in direct extension of the RNC as seen by our antibody stains (Fig. 4.10). Further, when compared to the eyeless distal tube feet in sea cucumber and sea urchin, the eye carrying starfish tube foot has a much denser nerve plexus with circumferential and longitudinal condensation (Formery et al. 2021).

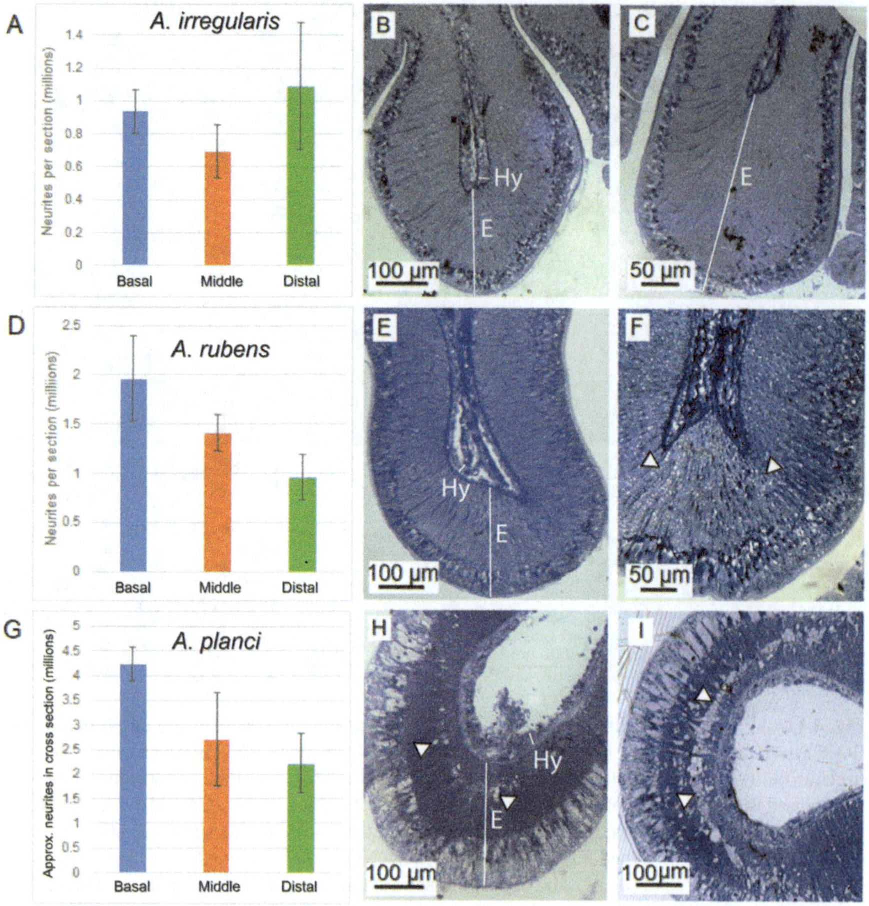

Fig. 4.12 Distribution of neurites along the radial nerve cord (RNC). (**a–c**) In *Astropecten irregularis*, the highest number of neurites in the RNC is found in the distal part (**c**), and the middle part (**b**) has the lowest number. (**d–f**) *Asterias rubens* in general has a higher number of neurites in the RNC than *A. irregularis*, and the highest number is here found in the basal part (**e**). The distal part has the lowest number (**f**). (**g–i**) Following its larger size, *Acanthaster planci* has the most neurites in the RNC of the three species (more than 4 million basally), and as in *A. rubens*, the highest number is found basally (**i**) and the lowest distally close to the eye (**j**). Arrowhead shows aggregation of giant neurons, which appear as lighter areas of the neuropil. E = ectoneural area, Hy = hyponeural area. Error bars in (**a, d**) and (**g**) indicate SD

As in the RNC of brittle stars, we also found giant neurons in the ectoneural part of the starfish RNC, though somewhat smaller (approx. 3–6 μm in diameter). In both *A. planci* and *A. rubens*, the number of giant neurons increases toward the distal part of the RNC (Fig. 4.12h, i). Furthermore, in the middle and distal parts of the RNC, the giant neurons were concentrated in subsystems in the central area, which is the area contacting the distal-most tube foot and the eye (Fig. 4.12f). As this pattern was not found in *A. irregularis*, it could indicate that the giant neurons play a part in the visual processing and integration. Still, we are only just beginning to understand the functional organization of the starfish CNS, and particularly electrophysiological data is warranted before any conclusions can be made about where and how the visual processing happens.

4.6.3 Supporting Cells

Both the ectoneural and hyponeural area have a neuroepithelium that surrounds the neuropil with intermingled supporting cells. The supporting cells have a fiber bundle in the cell body extending into a basal process and for the bipolar cells an additional apical process. The processes run through the neuropil area, where the apical process runs into the neuroepithelium and the basal process toward the basal lamina where they attach with hemidesmosomes (Figs. 4.9 and 4.11). We found no differences between *A. planci*, *A. rubens*, and *A. irregularis*, and the organization is similar in other echinoderms (Mashanov et al. 2010).

The function of the supporting cells is not fully understood, but they seem important for several processes in the RNS and thus potentially also for the visual processing. Bargmann and Behrens first suggested that the supporting cells in *A. rubens* are glial-like cells, but this was subsequently disputed (Bargmann and Behrens 1963; Cobb 1995; Cobb and Stubbs 1981). However, it has been shown in both starfish and sea cucumbers that they produce a material similar to Reissner's substance found in secretory glial cells in chordates. They also share morphological features with the radial glia of chordates found in the embryonic CNS, which for some chordates persist into adulthood (Mashanov et al. 2009; Viehweg et al. 1998). Additionally, the supporting cells of holothurians play important roles in both RNC regeneration after injury and in nervous tissue growth in general. During regeneration, the supporting cells near the lesion lose the processes and dedifferentiate. The processes are phagocytosed by the surrounding cells, and the remaining cell body becomes highly proliferative and an important source of new supporting cells and neurons (Mashanov et al. 2008, 2015).

Even if the supporting cells in some aspect resemble glial cell, they do not encircle the neurons as seen in chordates, and it is thus hard to imagine that they have the same function supporting the neurotransmission (Hartline and Colman 2007). Cnidarians also appear to lack glial sheaths in their CNS, and this lack allows neurons running in parallel in the hydrozoan nerve ring to enhance the signal by exciting each other, a phenomenon known as piggybacking (Mackie 2004). Whether this also happens in the starfish RNC is currently unknown.

4.6.4 *Specializations in* A. planci*: Neural Bulbs on the RNC*

It was recently discovered that *A. planci* has bulbous structures protruding from the ectoneural surface of the RNC (Smith 2018) (Fig. 4.13). These structures have so far not been observed in other species of echinoderms. The neural bulbs are found every approx. 50 μm along the entire length of the RNC, and they are up to 170 μm long. The function of these neural bulbs is still not completely understood, but they seem to be integrated parts of the RNCs and thus potentially involved in sensory processing. The center of the neural bulb is mostly filled with neurites connected to the RNC, but they are slightly larger than the average RNC neurites, 0.5–1.1 μm in diameter (Fig. 4.13d). The neuroepithelium of the ectoneural area also covers the neural bulb, and a high number of putative mechanosensory cilia with a collar of microvilli are found in this part of the neuroepithelium (Fig. 4.13b). Besides these putative sensory cells, the neuroepithelium of the bulbs contains many secretory cells (Smith 2018) (Fig 4.13c).

Interestingly, juveniles of *A. planci* do not have the neural bulbs, and their emergence seems correlated with a shift in diet. Juveniles feed on algae, but when they

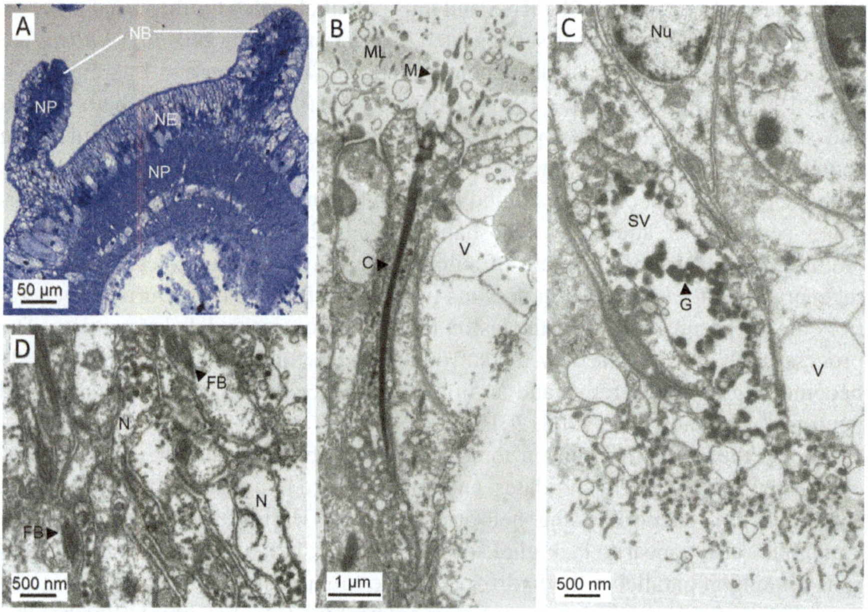

Fig. 4.13 Structure of the neural bulbs on the RNC of *Acanthaster planci*. (**a**) Cross section of the RNC showing two neural bulbs (NB). (**b**) Putative sensory cell in the NB epithelium displaying a ciliary rootlet (C) and a collar of microvilli (M) around the cilium. (**c**) Secretory cell with vesicles (SV) filled with granular material (G) secreted into the mucus layer on the outside of the NB epithelium. (**d**) The center of the neural bulb is filled with neurites (N) and fiber bundles (FB) crossing the bulb in different directions. ML = mucus layer, NP = neuropil, NE = neuroepithelium, Nu = nucleus, V = vacuole

reach a size of 5–8 cm in diameter, they start feeding on hard corals and this is when the bulbs start to appear. It has thus been suggested that the sensory system of the bulbs along with the high secretory activity is used in counteractions against the cnidocytes and other defense mechanisms of the corals (Smith 2018).

4.7 Concluding Remarks

Most starfish have a prominent compound eye at the tip of each arm, and recent evidence suggests that they are not as strictly olfactory guided as previously suggested (Dale 1997; Garm and Nilsson 2014). Even in the deep sea, the eyes persist in many species indicating that detection of bioluminescence plays an important role also for some starfish. Their visual capacity seems to be closely linked to their ecology and behavioral needs, with low pass spatial and temporal filtering removing most except for large and stationary objects from sight. The behavioral evidence points out negative phototaxis and habitat detection as important visually guided behaviors but also that vision is part of sensory hierarchy that can be dominated by olfaction. A great enigma when it comes to starfish vision is where and how the visual information is processed. The radial symmetry of starfish results in a CNS where a central brain with distinct sensory specific centers is missing. Instead the evidence suggests that each eye carrying arm has its own processing center, the radial nerve cord (RNC). Here we have presented some of the first data on the structure of starfish RNC and shown how it differs between eye carrying and eyeless species, but the interpretation still suffers from a complete lack of physiological data.

Interestingly, the echinoderm CNS resembles the cnidarian CNS especially what is found in Cubomedusae. The cnidarian CNS is also radially symmetric, and a major part is a nerve ring encircling the mouth (Garm et al. 2006; Mackie 2004). Furthermore, Cubomedusae have four sensory structures called rhopalia, which each holds an additional part of the CNS, the rhopalial nervous system (RNS), connected to the ring nerve. The rhopalia also carries a set of six eyes each, and the visual information processing putatively happens in the RNS (Bielecki et al. 2013; Garm and Mori 2009; Nilsson et al. 2005). This arrangement has a strong resemblance to starfish with each rhopalium putatively paralleling an arm. The combined physiological, behavioral, and modeling data from cubomedusae shows that behavioral control is not accomplished through a collaboration between the four rhopalia but rather as a competition. At any given time, the rhopalium receiving the strongest stimulation overrules the others through a resetting mechanism and becomes the sole control unit (Satterlie and Nolen 2001; Satterlie and Spencer 1979; Stöckl et al. 2011). Whether the same is the case for starfish is still to be tested, but it is currently the most likely hypothesis, and, thus, starfish are in many ways a collection of semi-independent arms!

Acknowledgments The authors appreciate the input from the editors and the anonymous reviewer. We would also like to acknowledge our coworkers from the results presented here: Dr. Ronald Petie, Prof. Dan-E Nilsson, Prof. Mike Hall, Dr. Martin F. Blicher, Mrs. Marie-Helene Birk, Mrs. Sabrina Beer, and Mrs. Camilla Wentzel.

References

Bargmann W, Behrens B (1963) Über den Feinbau des Nervensystems des Seesterns (Asterias rubens L.). Zeitschrift für Zellforschung und Mikroskopische Anatomie 59:746–770

Beer S, Wentzel C, Petie R, Garm A (2016) Active control of the visual field in the starfish *Acanthaster planci*. Vision Res 127:28–34

Bielecki J, Nachman G, Garm A (2013) Swim pacemaker response to bath applied neurotransmitters in the cubozoan *Tripedalia cystophora*. J Comp Physiol A 199:785–795

Birk M, Blicher ME, Garm A (2018) Deep-sea starfish from the Arctic have well developed eyes in the dark. Proc R Soc London B Biol Sci 285:20172743. https://doi.org/10.1098/rspb.2017.2743

Boivin Y, Larrivée D, Himmelman JH (1986) Reproductive cycle of the subarctic brooding asteroid *Leptasterias polaris*. Mar Biol 92:329–337

Castilla J, Crisp D (1973) Responses of *Asterias rubens* to water currents and their modification by certain environmental factors. Netherlands J Sea Res 7:171–190

Chia F, Koss R (1994) Asteroidea. In: Harrison FW, Chia F-S (eds) *Echinodermata*, vol 14, pp 169–246

Clark EG, Kanauchi D, Kano T, Aonuma H, Briggs DEG, Ishiguro A (2019) The function of the ophiuroid nerve ring: how a decentralized nervous system controls coordinated locomotion. J Exp Biol 222:jeb192104. https://doi.org/10.1242/jeb.192104

Clements M, Wolfe K, Schwartz K, Bryne M (2019) Forever fissiparous: asexual propagation and stable demography in a tropical and geographically isolated asterinid sea star. Mar Biol 166. https://doi.org/10.1007/s00227-019-3518-0

Cobb JLS (1987) Neurobiology of the echinodermata. In: Nervous systems in invertebrates, Springer, Boston, pp 483–525

Cobb JLS (1995) The nervous systems of Echinodermata: recent results and new approaches. In: The nervous systems of invertebrates: an evolutionary and comparative approach, Birkhäuser, Basel, pp 407–424

Cobb JLS, Moore A (1989) Studies on the integration of sensory information by the nervous system of the brittlestar *Ophiura ophiura*. Mar Freshw Behav Physiol 14:211–222

Cobb JLS, Stubbs TR (1981) The giant neurone system in ophiuroids I. The general morphology of the radial nerve cords and circumoral ring. Cell Tiss Res 219:197–207

Cowan Z-L, Pratchett M, Messmer V, Ling S (2017) Known predators of crown-of-thorns starfish (Acanthaster spp.) and their role in mitigating, if not preventing, population outbreaks. Diversity 9. https://doi.org/10.3390/d9010007

Dale J (1997) Chemosensory search behavior in the starfish *Asterias forbesi*. Biol Bull 193:210–212

D'Aniello S, Delroisse J, Valero-Garcia A, Lowe EK, Byrne M, Cannon JT, Halanych KM, Elphick MR, Mallefet J, Kaul-Strehlow S et al (2015) Opsin evolution in the Ambulacraria. Mar Genom 24:177–183

De'ath G, Moran PJ (1998) Factors affecting the behaviour of crown-of-thorns starfish (*Acanthaster planci* L.) on the Great Barrier Reef 2: feeding preferences. J Exp Mar Biol Ecol 220:107–126

Döring CC, Kumer S, Tumu SC, Kourtesis I, Hausen H (2020) The visual pigment xenopsin is widespread in protostome eyes and impacts the view on eye evolution. eLife 9:e55193. https://doi.org/10.7554/eLife.55193

Engle S (2013) Ultrastructure and development of the body cavities in Antedon bifida (Pennant, 1777) (Comatulida, Crinoida), vol. PhD

Fabricius KE, Okaji K, De'ath G (2000) Three lines of evidence to link outbreaks of the crown-of-thorns seastar *Acanthaster planci* to the release of larval food limitation. Coral reefs 29:593–605
Fenchel T (1965) Feeding biology of the sea star *Luidia sarsi* Düben and Koren. Ophelia 2:223–236
Formery L, Orange F, Formery A, Yaguchi S, Lowe CJ, Schubert M, Croce JC (2021) Neural anatomy of echinoid early juveniles and comparison of nervous system organization in echinoderms. J Comp Neurol 529:1135–1156
Fritsches KA, Brill RW, Warrant EJ (2005) Warm eyes provide superior vision in swordfishes. Curr Biol 15:55–58
Garm A (2017) Sensory biology of starfish - with emphasis on recent discoveries in their visual ecology. Integrat Comp Biol. https://doi.org/10.1093/icb/icx086
Garm A, Mori S (2009) Multiple photoreceptor systems control the swim pacemaker activity in box jellyfish. J Exp Biol 212:3951–3960
Garm A, Nilsson DE (2014) Visual navigation in starfish: first evidence for use of vision and eyes in starfish. Proc R Soc London B Biol Sci 281:20133011. https://doi.org/10.1098/rspb.2013.3011
Garm A, Ekström P, Boudes M, Nilsson DE (2006) Rhopalia are integrated parts of the central nervous system in box jellyfish. Cell Tiss Res 325:333–343
Govardovskii VI, Fyhrquist N, Reuter T, Kuzmin DG, Donner K (2000) In search of the visual pigment template. Vis Neurosci 17:509–528
Hall M, Kocot KM, Baughman KW, Fernandez-Valverde SL, Gauthier MEA, Hatleberg WL, Krishman A, McDougall C, Motti CA, Shoguchi E et al (2017) The crown-of-thorns starfish genome as a guide for biocontrol of this coral reef pest. Nature 544:231–234
Hamel J, Mercier A (1994) Influence of environmental factors on prespawnign behaviour, spawning and developmental biology of the brooding starfish *Leptasterias Polaris*. In: David B, Guille A, Feral J, Roux M (eds) *Echinoderms through times*. Balkema, Rotterdam, p 310
Hartline DK, Colman DR (2007) Rapid conduction and the evolution of giant axons and myelinated fibers. Curr Biol 17:R29–R35
Henning PJ (1974) New observations on the bioluminescence of echinoderms. L Zool London 172:401–418
Jourdain S (1865) On the eyes of *Asteracanthion rubens*, Müll. & Trosch. (*Uraster rubens*, Forbes). Annals Magazine Nat His 15:238–240
Kalmus H (1929) Versuche fiber die Bewegungen der Seesterne, besonders von *Asterina gibbosa*. Zool Vergleich Phsiol 9:703–733
Kirwan JD, Bok MJ, Smolka J, Foster JJ, Hermandez JC, Nilsson DE (2018) The sea urchin Diadema africanum uses low resolution vision to find shelter and deter enemies. J Exp Biol 221. https://doi.org/10.1242/jeb.176271
Korsvig-Nielsen C, Hall M, Motti CA, Garm A (2019) Eyes and negative phototaxis in juvenile crown-of-thorns starfish, *Acanthaster* species complex. Biol Open 8. https://doi.org/10.1242/bio.041814
Land MF (2019) Eye movements in man and other animals. Vision Res 192:1–7
Land MF, Nilsson DE (2012) Animal eyes. Oxford University Press, Oxford
Lascalli T (2000) Larval budding, metamorphosis, and the evolution of life-history patterns in echinoderms. Invert Biol 119:234–241
Lesser MP, Carleton KL, Böttger SA, Barry TM, Walker CW (2011) Sea urchin tube feet are photosensory organs that express a rhabdomeric-like opsin and PAX6. Proc R Soc London B Biol Sci 278:3371–3379
Lowe EK, Garm A, Ullrich-Lüter EM, Cuomo C, Arnone MI (2018) The crowns have eyes: Multiple opsins found in the eyes of the Crown-of-Thorns Starfish *Acanthaster planci* including the first r-opsin utilized by a deuterostome eye. BMC Evol Biol Accepted for publication
Lucas JS (2013) Crown-of-thorns starfish. Curr Biol 23:945–946
Lythgoe JN (1979) The ecology of vision. Clarendon Press, Oxford
Mackie GO (2004) Central neural circuitry in the jellyfish *Aglantha*: a model "simple nervous system". Neuro-Signals 13:5–19

Märkel K, Röser U (1991) Ultrastructure and organization of the epineural canal and the nerve cord in sea urchins (Echinodermata, Echinoida). Zoomorphol 110:267–279x

Mashanov VS, Zueva OR, Heinzeller T, Dolmatov IY (2006) Ultrastructure of the circumoral nerve ring and the radial nerve cords in holothurians (Echinodermata). Zoomorphol 125:27–38

Mashanov VS, Zueva OR, Heinzeller T (2008) Regeneration of the radial nerve cord in a holothurian: a promising new model system for studying post-traumatic recovery in the adult nervous system. Tiss Cell 40:351–375

Mashanov VS, Zueva OR, Heinzeller T, Aschauser B, Neumann WW, Grondoda JM, Cifuentes M, Garcia-Arrarás JE (2009) The central nervous system of sea cucumbers (Echinodermata: Holothuroidea) shows positive immunostaining for a chordate glial secretion. Front Zool 6. https://doi.org/10.1186/1742-9994-6-11

Mashanov VS, Zueva OR, Garcia-Arrarás JE (2010) Organization of glial cells in the adult sea cucumber central nervous system. Glia 58:1581–1593

Mashanov VS, Zueva OR, Garcia-Arrarás JE (2015) Myc regulates programmed cell death and radial glia dedifferentiation after neural injury in an echinoderm. BMC Dev Biol 15:1–9

Mashanov VS, Zueva OR, Rubilar T, Epherra L, Garcia-Arrarás JE (2016) Echinodermata. In: Schmidt-Rhaesa A, Harzsch S, Purschke G (eds) *Structure and Evolution of Invertebrate Nervous Systems*. Oxford University Press, Oxford

McFarland WN, Munz FW (1975) Part II The photic environment of clear tropical seas during the day. Vision Res 15:1063–1070

Moore S, Thorndyke MC (1993) Immunocytochemical mapping of the novel echinoderm neuropeptide SALMFamide 1 (S1) in the starfish *Asterias rubens*. Cell Tiss Res 274:605–618

Moran PJ (1986) The *Acanthaster* phenomenon. Oceanograph Mar Biol Annu Rev 24:379–480

Moss C, Jackie Hunter A, Thorndyke MC (1998) Patterns of bromodeoxyuridine incorporation and neuropeptide immunoreactivity during arm regeneration in the starfish *Asterias rubens*. Philos Trans R Soc London 353:421–436

Motti CA, Bose U, Roberts RE, McDougall C, Smith MK, Hall M, Cummins SF (2018) Chemical ecology of chemosensation in Asteroidea: insights towards management strategies of pest species. J Chem Ecol 44:147–177

Mueller B, Bos AR, Graf G, Gumanao GS (2011) Size-specific locomotion rate and movement pattern of four common Indo-Pacific sea stars (Echinodermata; Asteroidea). Aquat Biol 12:157–164

Nilsson DE, Gislén L, Coates MM, Skogh C, Garm A (2005) Advanced optics in a jellyfish eye. Nature 435:201–205

Pearse JS, Bosch I (2002) Photoperiodic regulation of gametogenesis in the Antarctic sea star *Odontaster validus* Koehler: evidence for a circannual rhythm modulated by light. Invert Repro Dev 41:73–81

Pearse JS, Eernisse DT, Pearse VB, Bauchamp KA (1986) Photoperiodic regulation of gametogenesis in sea stars, with evidence for an annual calendar independent of fixed daylength. American Zool 26:417–431

Pearse V, Pearse J, Buchsbaum M, Buchsbaum R (1987) Living invertebrates. Blackwell Science Inc.

Penn PE, Alexander CG (1980) Fine structure of the optic cusion in the asteroid *Nepanthia belcheri*. Mar Biol 58:251–256

Petie R, Garm A, Hall M (2016a) Crown-of-thorns starfish have true image forming vision. Front Zool 13:41. https://doi.org/10.1186/s12983-016-0174-9

Petie R, Hall M, Hyldahl M, Garm A (2016b) Visual orientation by the crown-of-thorns starfish (*Acanthaster planci*). Coral reefs 35:1139–1150. https://doi.org/10.1007/s00338-016-1478-0

Satterlie RA, Nolen TG (2001) Why do cubomedusae have only four swim pacemakers? J Exp Biol 204:1413–1419

Satterlie RA, Spencer AN (1979) Swimming control in a cubomedusan jellyfish. Nature 231:141–142

Scheibling RE (1981) Optimal foraging movements of *Oreaster reticulatus* (L.) (Echinodermata: Asteroidea). J Exp Mar Biol Ecol 51:173–185

Sigl R, Laforsch C (2016) The influence of water currents on movement patterns on sand in the crown-of-thorns seastar (*Acanthaster cf. solaris*). Diversity 8:25. https://doi.org/10.3390/d8040025

Sigl R, Streibl S, Laforsch C (2016) The role of vision for navigation in the crown-of-thorns seastar *Acanthaster planci*. Sci Rep 6:30834. https://doi.org/10.1038/srep30834

Sloan NA (1980) The arm curling and terminal tube-foot response of the ateroid *Crossaster papposus*. J Nat Hist 14:469–482

Smith MK (2018) Molecular investigations of the neural system in the Crown-of-Thorns Seastar, Acanthaster planci species-complex, vol. PhD. University of Queensland

Stöckl AL, Petie R, Nilsson DE (2011) Setting the pace: new insights into central pattern generator interactions in box jellyfish swimming. PLoS ONE 6:1–11

Sumner-Rooney L, Kirwan JD, Lowe EK, Ullrich-Lüter EM (2020) Extraocular vision in a brittle star Is mediated by chromatophore movement in response to ambient light. Curr Biol 30:319–327

Ullrich-Lüter EM, Dupont S, Arboleda E, Hausen H, Arnone MI (2011) Unique system of photoreceptors in sea urchin tube feet. In: Proceedings of the National Academy of Sciences USA, vol 108, pp 8367–8372

Ullrich-Lüter EM, D'Aniello S, Arnone MI (2013) C-opsin Expressing Photoreceptors in Echinoderms. Integrat Comp Biol 53:27–38

Valentincic T (1975) Amino-acid chemoreception and other releasing factors in the feeding response of the sea star *Marthasterias glacialis* (D.). In: Barnes H (ed) *Proceedings from the ninth European marine biological symposium*. University of Aberdeen press, Aberdeen, pp 693–705

Viehweg J, Khoury M, Heinzeller T, Mashanova D, Mashanov VS (1998) Secretory radial glia in the ectoneural system of the sea star Asterias rubens (Echinodermata). Acta Zool 92:119–131

Vöcking O, Kourtesis I, Tumu SC, Hausen H (2017) Co-expression of xenopsin and rhabdomeric opsin in photoreceptors bearing microvilli and cilia. eLife 6:e23435

Yoshida M, Ohtsuki H (1968) The phototactic behavior of the starfish *Asterias amurensis* Lütken. Biol Bull, pp 516–532

Zhang R, Wang C, Zhou Y, Zhang H (2019) Morphology and molecular phylogeny of two new species in genus Freyastera (Asteroidea: Brisingida: Freyellidae), with a revised key to close species and ecological remarks. Deep-Sea Res 1:154. https://doi.org/10.1016/j.dsr.2019.103163

Zueva OR, Khoury M, Heinzeller T, Mashanova D, Mashanov VS (2018) The complex simplicity of the brittle star nervous system. Front Zool 15:1–26

Chapter 5
Distributed Visual Systems in Pteriomorphian Bivalves

Daniel I. Speiser, Daniel R. Chappell, Jorge A. Audino, Alexandra C. N. Kingston, and Jeanne M. Serb

Abstract Pteriomorphia includes bivalves such as scallops, file clams, oysters, mussels, and ark clams. Like other bivalves, pteriomorphians do not have heads and lack complex anterior sensory organs. Instead, they have sensory organs, such as eyes and tentacles, distributed along their mantles at the edges of their valves. At least five separate lineages of pteriomorphians have evolved distributed visual systems that include dozens to hundreds of mantle eyes. Pteriomorphia is a valuable group in which to study distributed visual systems because species within the group show considerable variation in their eye morphology, ecology, locomotory abilities, and neuroanatomy. In the following chapter, we will introduce pteriomorphian bivalves, describe the structure and function of their mantle eyes, present what is known about their visual ecology, and detail their neuroanatomy. We will conclude by asking questions about how and why distributed visual systems have evolved in pteriomorphian bivalves.

Keywords Neuroethology · Neuroanatomy · Eye evolution · Visual ecology · Photoreception · Light-influenced behavior · Scallop · Mollusca

5.1 Introduction

Bivalvia is an ecologically and morphologically diverse class of mollusks (Stanley 1975; Bieler et al. 2013). Bivalves are distinguished by their hinged two-part shells (or valves) made of calcium carbonate ($CaCO_3$). Most bivalves can shut their shells so that their bodies are fully shielded, an action that serves as the primary means of

D. I. Speiser (✉) · D. R. Chappell
Department of Biological Sciences, University of South Carolina, Columbia, SC, USA

J. A. Audino · J. M. Serb
Department of Ecology, and Organismal Biology, Iowa State University, Ames, IA, USA

A. C. N. Kingston
Department of Biological Science, University of Tulsa, Tulsa, OK, USA

E. Buschbeck, M. Bok (eds.), *Distributed Vision*, Springer Series in Vision Research, https://doi.org/10.1007/978-3-031-23216-9_5

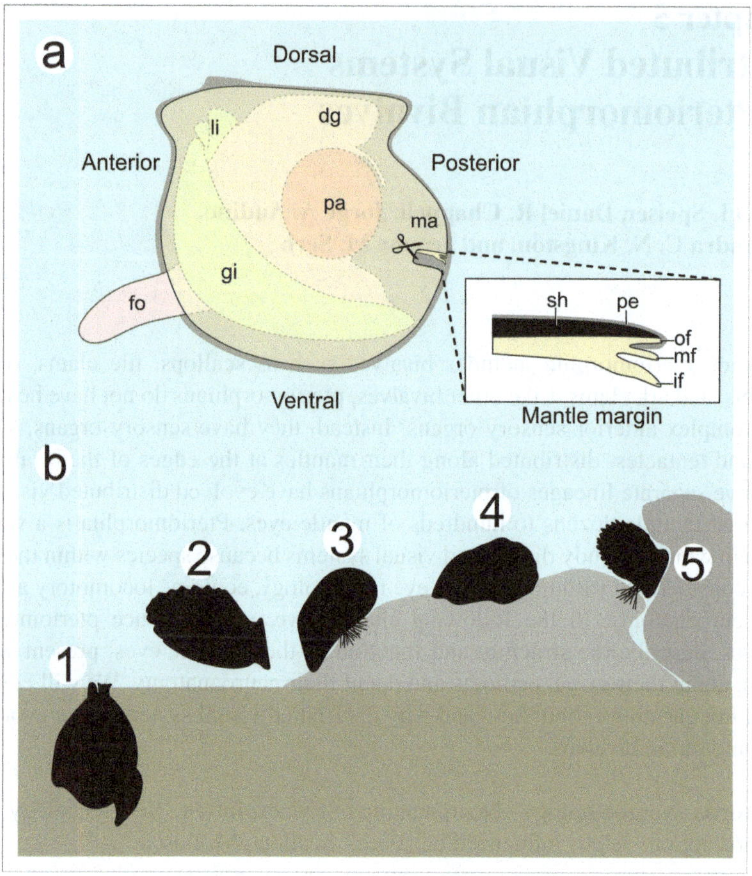

Fig. 5.1 Generalized morphology (**a**) and ecology (**b**) of pteriomorphian bivalves. (**a**) Schematic representation of the bivalve body after removal of the left valve. Lateral view of some organs (transparency; dotted line) covered by the left mantle lobe. The inset shows the mantle margin, including the outer, middle, and inner mantle folds. (**b**) General modes of life according to occupation of the substrate and mobility. Infaunal lifestyle is indicated in (**1**) and different epifaunal habits in (**2–5**), including (**2**) swimming (high mobility), (**3**) byssal attachment (low mobility), (**4**) cementing (sessile), and (**5**) byssal attachment in crevices (low mobility). Abbreviations: *dg* digestive gland, *fo* foot, *gi* gill, *if* inner fold, *li* lips, *ma* mantle, *mf* middle fold, *of* outer fold, *pa* posterior adductor muscle, *pe* periostracum, *sh* shell

defense for many species. Shell formation is guided by the mantle, a body wall that encloses the visceral organs (Fig. 5.1a). Bivalves do not have heads, and they lack the complex anterior sensory organs found in many other mollusks. Instead, bivalves often have numerous sensory organs, such as tentacles and eyes, lining the mantle margins at the edges of their valves (Fig. 5.1a). A muscular foot is used by many species for movements that can include crawling and burrowing. Most bivalves filter-feed using their gills, and many of the remaining species are deposit feeders. Bivalves live in aquatic habitats ranging from freshwater to saltwater, they are found at latitudes from pole to pole, and they can inhabit depths from the intertidal to the abyssal. As adults, most bivalves are either infaunal (i.e., living amidst the substrate) or epifaunal (i.e., living above the substrate), and, with the exception of a few groups that can swim, they are either slow-moving or sessile (Fig. 5.1b).

Extant bivalves, numbering nearly 10,000 described species, are divided into 6 major clades: Anomalodesmata, Archiheterodonta, Imparidentia, Palaeoheterodonta, Protobranchia, and Pteriomorphia (Bieler et al. 2014; González et al. 2015; Combosch et al. 2017). Pteriomorphia, the focus of this chapter, includes approximately 2,000 recognized species divided between 5 monophyletic orders (Fig. 5.2): Pectinida, scallops and relatives; Limida, file clams; Ostreida, oysters and relatives; Mytilida, mussels; and Arcida, ark clams and relatives (Combosch and Giribet 2016; Lemer et al. 2016; Audino et al. 2020). These orders diverged long ago: the fossil records of Arcida and Mytilida appear to extend, respectively, to the lower Ordovician (~480 mya) (Cope 1997) and the Devonian (~420 mya) (Distel 2000). Deeply diverged evolutionary histories indicate pteriomorphian bivalves, despite superficial similarities, can differ from each other in substantial ways.

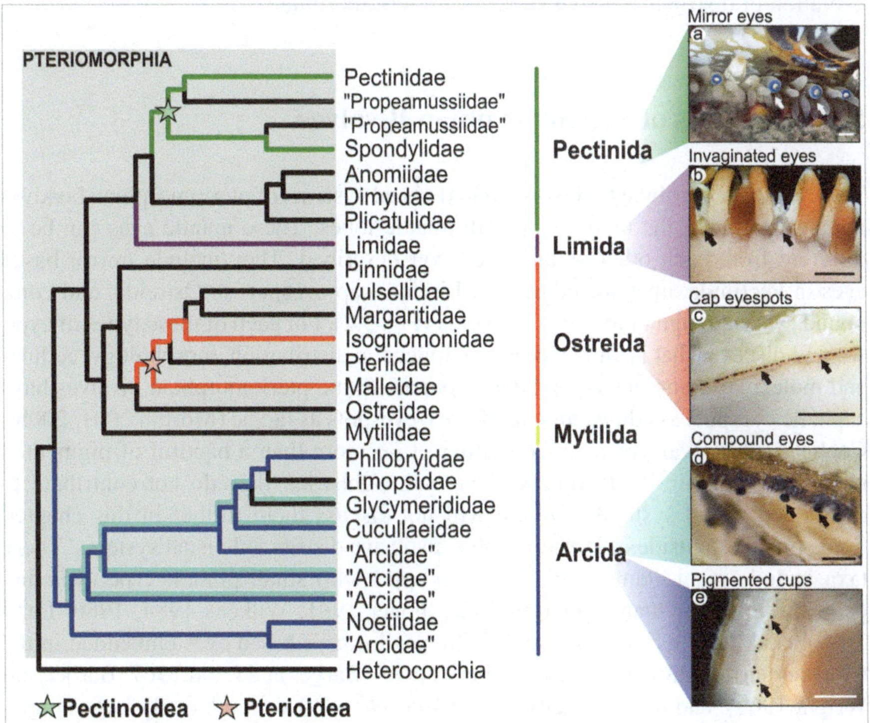

Fig. 5.2 Phylogenetic relationships within Pteriomorphia (gray square) and ancestral state estimations of different types of mantle eyes contributing to distributed visual systems within the group (redrawn after Audino et al. 2020). These mantle eyes include (**a**) mirror-based eyes (green), (**b**) invaginated eyes (purple), (**c**) cap eyespots (red), (**d**) compound eyes (light blue), and (**e**) pigmented cups (blue). The photographs display the mantle eyes of (**a**) *Spondylus tenuis* (Spondylidae), (**b**) *Lima lima* (Limidae), (**c**) *Isognomon radiatus* (Isognomonidae), (**d**) *Arca noae* (Arcidae), and (**e**) *Barbatia domingensis* (Arcidae). Scale bars = 1 mm. All photographs by J.A. Audino

Pteriomorphia is a valuable group in which to study distributed visual systems that can include anywhere from dozens to hundreds of eyes. A compelling reason to study these distributed visual systems is the morphological diversity of the mantle eyes of pteriomorphians, which range in complexity from simple pigmented cups to image-forming eyes with mirror-based optics (Audino et al. 2020). A second reason is the behavioral and ecological diversity of pteriomorphians. Mobile species, for example, may use vision for behavioral tasks, such as habitat selection, that are not options for sessile species. Neuroanatomical differences between species are a third reason: pteriomorphians with mantle eyes may not use similar approaches, or even the same neural structures, to process visual information. Finally, from a phylogenetic perspective, pteriomorphians have gained mantle eyes at least five times and may have lost them even more times than that (Audino et al. 2020). Frequent, independent gains and losses of eyes make pteriomorphian bivalves an intriguing group in which to ask why and how distributed visual systems evolve. In the following chapter, we will introduce pteriomorphians, describe their distributed visual systems, present what is known about their visually influenced behaviors, and compare their neuroanatomical structures. We will conclude by asking questions about the evolution of distributed visual systems in Pteriomorphia.

5.2 The Eyes of Pteriomorphian Bivalves

The eyes that contribute to the distributed visual systems of pteriomorphian bivalves are arrayed along the margins of both mantle lobes. These mantle eyes can be as different from each other as any eyes ever described. They include mirror-based eyes in Pectinida, invaginated eyes in Limida, cap eyespots in Ostreida, and compound eyes and pigmented cups in Arcida (Fig. 5.2). For each of these types of eyes, we will discuss what is known of its phylogenetic distribution, morphology, cellular and molecular components, and development. Some pteriomorphian bivalves have cephalic eyespots as adults, and many have eyespots as larvae (Morton 2001, 2008). Cephalic and larval eyespots, consisting of no more than a handful of pigmented photoreceptors, are interesting in their own rights, but they do not contribute to distributed visual systems, and we will not discuss them further in this chapter. Some bivalves outside of Pteriomorphia also have distributed visual systems. These taxa include giant clams (e.g., *Tridacna* spp.) with small pinhole-type eyes dispersed along their mantle margins (Fankboner 1981; Wilkens 1984, 1986; Land 2003); cockles (e.g., *Cardium* spp.) with small mirror-based eyes embedded in the tentacles surrounding the bases of their siphons (Barber and Land 1967; Barber and Wright 1969); and lantern shells (e.g., *Laternula* spp.) with relatively large single-chambered eyes, perhaps with camera-like optics, on the tips of tentacles surrounding the openings of their siphons (Adal and Morton 1973). These eyes, while morphologically disparate, are limited to two superorders (Imparidentia and Anomalodesmata).

5.2.1 *Pectinida: Mirror-Based Eyes*

5.2.1.1 Phylogenetic Distribution and General Description

Pectinida includes jingle shells (Anomiidae), micro-scallops (Cyclochlamydidae), dimyarian oysters (Dimyidae), scallops (Pectinidae), kittenpaws (Plicatulidae), glass scallops (Propeamussiidae), and thorny oysters (Spondylidae). Scallops, the best known pectinids, are able to swim using a form of jet propulsion (Cheng et al. 1996; Alejandrino et al. 2011; Tremblay et al. 2015). Other pectinids, including anomiids and spondylids, cement to the substrate and are sessile as adults. Distributed visual systems based on dozens of mirror-based eyes (up to 1 mm in diameter) are found in scallops, spondylids, and some propeamussids (Dakin 1928; Speiser and Johnsen 2008a; Smedley et al. 2019). Ancestral state reconstruction indicates mirror-based eyes were present in the last common ancestor of Pectinoidea (Fig. 5.2) and then lost in at least two lineages of deep-dwelling *Propeamussium* (Audino et al. 2020). The eyes of scallops (Fig. 5.2a) are positioned on the middle mantle fold, and they ring the edges of the right and left valves from one side of the hinge to the other. Scallops add eyes as they grow, which helps explain why individuals can have eyes of varying sizes and why conspecifics can have different numbers of eyes (Whoriskey et al. 2014; Audino et al. 2022). The eyes of scallops are found at the ends of short stalks, and they are interspersed with sensory tentacles. The eyestalks are mobile, due to longitudinal muscle fibers, but they are less flexible than the sensory tentacles, which have both longitudinal and transverse fibers (Audino et al. 2015a).

5.2.1.2 Morphology and Optics of Mirror-Based Eyes

It has long been recognized that the eyes of scallops (Fig. 5.3a) each contain a cornea, a biconvex lens, two separate retinas (referred to by their relative positions as "distal" and "proximal"), and a concave mirror (Krohn 1840; Patten 1886; Hesse 1901; Dakin 1910a). The outer surface of the scallop eye is a single continuous layer of epithelial cells. The cornea is formed by transparent epithelial cells, and the rest of the eye is surrounded by pigmented epithelial cells. Differences between these pigmented epithelial cells can cause the eyes of scallops to vary in color. For example, the photonic nanostructures that make the eyes of *Argopecten irradians* bright blue are absent in the black-eyed *Placopecten magellanicus* (Harris et al. 2019). Directly beneath the cornea, an optic vesicle made of connective tissue encloses the lens, retinas, and mirror (Dakin 1910a). The lenses of scallops are unusual in several ways: they are nonspherical in shape, they are less dense than most biological lenses, and they appear to be composed of metabolically active cells (Land 1965; Barber et al. 1967).

The distal and proximal retinas in the eyes of scallops are composed of ciliary and rhabdomeric photoreceptors, respectively. The photoreceptive cilia of the distal

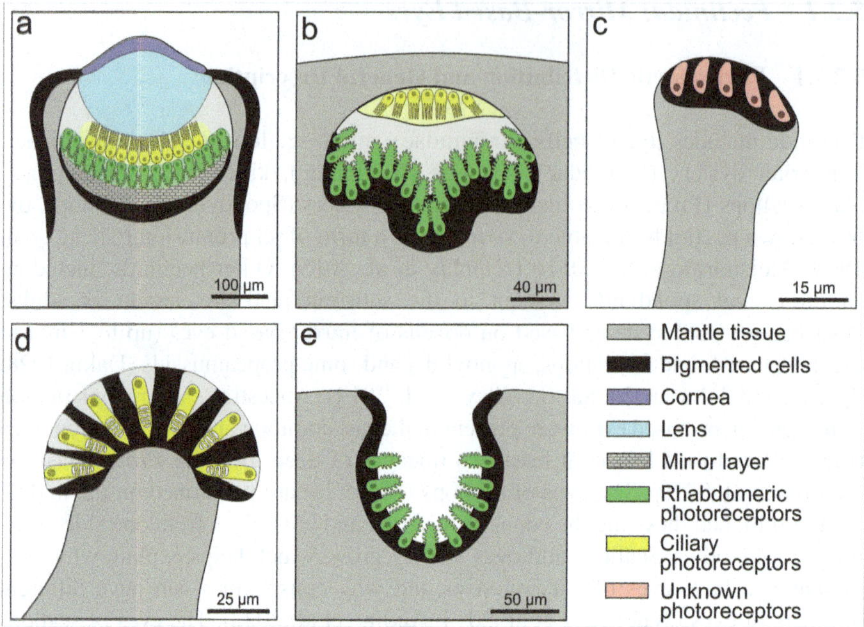

Fig. 5.3 Schematic representations of the mantle eyes contributing to the distributed visual systems of pterimorphian bivalves. (**a**) Mirror-based eyes of scallops and relatives (Pectinoidea), drawn after *Placopecten magellanicus* (Speiser and Johnsen 2008a). (**b**) Invaginated eyes of file clams (Limida), drawn after *Ctenoides scaber* (Bell and Mpitosos 1968; Mpitosos 1973; McReynolds 1976; Nasi 1991). (**c**) Cap eyespots of tree oysters and hammer oysters (Pterioidea), drawn after *Isognomon bicolor* (Audino et al. 2020). (**d**) Compound eyes of ark clams and relatives (Arcida), drawn after *Barbatia domingensis* (Nilsson 1994). (**e**) Pigmented cups of ark clams and relatives (Arcida), drawn after *Anadara notabilis* (Nilsson 1994)

photoreceptors extend toward the lens, and the photoreceptive microvillar structures of the proximal photoreceptors extend toward the mirror at the back of the eye (Fig. 5.3a). The microvilli of adjacent proximal photoreceptors interdigitate to form rhabdoms (Barber et al. 1967). Visual processing does not appear to occur within the eyes: synaptic connections have not been observed between photoreceptors from the same retina, between photoreceptors from different retinas, or between photoreceptors and the glial cells that lie between the two retinas (Miller 1958; Barber et al. 1967). Axons from the distal and proximal photoreceptors leave the optic vesicle separately and then join beneath the eye to form a single optic nerve that travels down the eyestalk (Dakin 1910a; Hartline 1938; Barber et al. 1967; Malkowsky and Jochum 2015).

The concave mirror at the back of the scallop eye follows the curve of the proximal hemisphere of the optic vesicle (Fig. 5.3a). The mirror is a multilayer reflector (or Bragg reflector) that is highly reflective because its many internal surfaces produce multiple specular reflections that are in phase with each other (Land 1966a).

These internal surfaces come from dozens of thin (~75 nm) layers of guanine crystals alternating with equally thin layers of cytoplasm (Land 1966a). The guanine crystals are square in shape, and they are arranged like tiles in a mosaic (Palmer et al. 2017). A clear viscous substance, termed the "rod matrix" (Dakin 1910a), separates the mirror from the proximal photoreceptors. A thin layer of red-pigmented cells lies between the mirror and the optic vesicle (Barber et al. 1967).

The eyes of scallops were the first eyes discovered to form images using a concave mirror (Land 1965), and scallops remain among the few animals known to have single-chambered eyes with mirror-based optics (Land 2000; Wagner et al. 2009). Light passes through the cornea, lens, and retinas and is then reflected by the mirror as a high-quality, inverted image (Land 1965; Palmer et al. 2017). It is thought that this image falls in the vicinity of the distal retina (Land 1965). The lens has a low refractive index and so contributes little to the focusing power of the eye, but its unusual nonspherical shape may help it correct for spherical aberration caused by the mirror (Land 1965). Scallop eyes have complicated optics that have yet to be fully understood (Speiser et al. 2016; Palmer et al. 2017). For example, the function of the proximal retina remains unclear. It has been argued the proximal retina lies too close to the mirror to receive focused light (Land 1965), yet each proximal retina can contain as many as 10,000 photoreceptors (Dakin 1910a), and the proximal photoreceptors outnumber the distal photoreceptors by about 10:1. It would be unprecedented for such retinas to not be associated with the acquisition of spatial information. Several possible solutions to this mystery have been proposed. One option is that scallops use the muscles in their eyestalks to change the shapes of their eyes so that one retina or the other receives focused light (Speiser et al. 2016). Another option is that the distal and proximal retinas receive focused light from the center and periphery, respectively, of an eye's field of view (Palmer et al. 2017).

5.2.1.3 Cellular and Molecular Components of Mirror-Based Eyes

Like most biological lenses, the lenses of scallops largely consist of water-soluble proteins known as crystallins. These proteins are so named because they pack together in a crystal-like manner at high concentrations, thereby allowing lenses to have high refractive indices while remaining transparent (Wistow and Piatigorsky 1988). The lenses of scallops contain a single type of Ω-crystallin, a repurposed aldehyde dehydrogenase that lacks enzymatic function (Piatigorsky et al. 2000; Carosa et al. 2002; Horwitz et al. 2006; Piatigorsky 2008). The scallop Ω-crystallin forms homotetramers in vivo and its expression is regulated, at least in part, by transcription factors (e.g., Pax 6) associated with eye development in other taxa (Piatigorsky et al. 2000; Carosa et al. 2002; Horwitz et al. 2006; Piatigorsky 2008).

The ciliary photoreceptors of the distal retina and the rhabdomeric photoreceptors of the proximal retina have opposite physiological responses to light. The distal photoreceptors hyperpolarize when light levels increase, and they depolarize when light levels decrease; in other words, they are inhibited by light and excited by the removal of light (Hartline 1938; McReynolds and Gorman 1970a, 1970b). Their hyperpolarizing response is caused by the opening of potassium channels and an increased outward flow of K^+ (Gorman and McReynolds 1978; Cornwall and Gorman 1983). Their depolarizing response is caused by the closing of potassium channels and a decreased outward flow of K^+, as well as the opening of voltage-gated calcium channels and an increased inward flow of Ca^{2+} (Cornwall and Gorman 1979). In contrast, the proximal photoreceptors depolarize when light levels increase due to the opening of sodium channels and an increased inward flow of Na^+ (McReynolds and Gorman 1970a, 1970b; Gomez and Nasi 1996).

The distal and proximal photoreceptors detect light using different molecular components. Unlike other well-characterized ciliary photoreceptors, the distal photoreceptors respond to light using G_o-opsins that interact with G_o-type G-proteins (Kojima et al. 1997; Kingston et al. 2017). The activation of this G_o-mediated phototransduction cascade leads to increased activity by guanylate cyclase and an increase in intracellular cGMP concentration (Gomez and Nasi 1995, 2000). Other ciliary photoreceptors, such as the rods and cones of vertebrates, respond to light using c-opsins and G_i- or G_t-type G-proteins (Fain et al. 2010). The proximal photoreceptors are more conventional: like the rhabdomeric photoreceptors found in the cephalic eyes of many invertebrates, they respond to light through a phototransduction cascade involving G_q-opsin (= r-opsin) and a G_q-type G-protein (Kojima et al. 1997; Kingston et al. 2017).

5.2.1.4 Development of Mirror-Based Eyes

The eyes of scallops develop unlike any other eyes yet described. In *Nodipecten nodosus*, mantle eyes begin developing as pigmented papillae after metamorphosis (Audino et al. 2015b). The optic vesicle forms first and is followed by the mirror, which begins as loosely packed plates of guanine crystals that compact over the course of development. The remaining structures of the eye begin as an undifferentiated mass of cells within the optic vesicle. From this mass of cells, the proximal retina develops first, the distal retina second, and the lens third. The lens appears to develop from retinal precursor cells, a surprise because the lenses of other single-chambered eyes, like those of vertebrates and cephalopods, develop from populations of ectodermal cells separate from those that give rise to the retinas. The optic nerves form early in development, with serotonergic neurons appearing in the eyestalks of juveniles while the retinas are still differentiating. Scallops add eyes as they grow, and the process of eye development in juveniles and adults appears to be similar (Butcher 1930; Audino et al. 2015b). Scallops also regenerate their mantle eyes. When eyes were removed from the mantle margins of *Argopecten* (= *Pecten*) *gibbus*, new eyes developed within 40 days (Butcher 1930). Eyes in *A. gibbus* also continued developing and extending optic nerves when they were transplanted from the mantle to other locations, including the surface of the gonad (Butcher 1930).

5.2.2 *Limida: Invaginated Eyes*

5.2.2.1 Phylogenetic Distribution and General Description

Limida includes file clams (Limidae), a group best known for species with brightly colored mantle tissues and many long tentacles extending from their mantle margins (Mikkelsen and Bieler 2003; Dougherty et al. 2019). File clams tend to be epifaunal and to attach voluntarily and reversibly to their substrate using byssal threads. Many species can swim by rowing with their tentacles and clapping with their valves (Donovan et al. 2004). Some species produce nests of byssal threads that can accumulate enough sediment over time to form biogenic reefs (Hall-Spencer and Moore 2000). Some limids, including species of *Ctenoides* and *Lima*, have invaginated eyes between the long tentacles on their middle mantle fold (Fig. 5.2b). Species of *Acesta*, *Limaria*, and *Limatula* lack mantle eyes, but ancestral state reconstruction (Audino et al. 2020) indicates mantle eyes were present in the last common ancestor of Limida (Fig. 5.2).

5.2.2.2 Morphology of Invaginated Eyes

The invaginated eyes of *Ctenoides mitis* (= *C. floridanus*) are 140–180 μm in diameter (transverse section), and ~18 of them are present along the margins of each mantle lobe (Morton 2000). In *C. scaber* (= *Lima scabra*), invaginated eyes are described as 600 μm long, 200 μm wide, and 150 μm deep (Bell and Mpitosos 1968). The eyes of *C. mitis* and *C. scaber* are embedded within connective tissue beneath the translucent epithelium of the mantle (Mpitosos 1973; Morton 2000). In *C. scaber*, these eyes include two separate retinas referred to by their relative positions as "distal" and "proximal" (Fig. 5.3b). The distal retina is composed of round, transparent photoreceptors which extend bundles of cilia and neuronal processes (Bell and Mpitosos 1968; Mpitosos 1973; McReynolds 1976; Nasi 1991). The proximal retina lies against the back of the eye and is composed of a single layer of rhabdomeric photoreceptors interspersed with red-pigmented cells (Mpitosos 1973; Nasi 1991). Like the eyes of *C. scaber*, the eyes of *Lima lima* appear to contain ciliary photoreceptors, rhabdomeric photoreceptors, and pigmented cells (Salvini-Plawen 2008). Characterizing the invaginated eyes of limids has been challenging because of conflicting interpretations of their morphology and limited taxonomic sampling. We strongly encourage a comprehensive, in-depth survey of the enigmatic eyes of file clams.

5.2.2.3 Cellular and Molecular Components of Invaginated Eyes

Like the distal and proximal photoreceptors in the eyes of scallops, the photoreceptors of the distal and proximal retinas in the eyes of file clams demonstrate opposite responses to light. The ciliary photoreceptors from the invaginated eyes of *C. scaber* hyperpolarize in response to increased amounts of light and depolarize in response to the dimming of light (Mpitosos 1973; Nasi 1991; Gomez and Nasi 1994). Like the rhabdomeric photoreceptors found in the cephalic eyes of many invertebrates, those from the mantle eyes of *C. scaber* depolarize in response to increased light levels (Mpitosos 1973). The two sets of photoreceptors in the eyes of file clams have similar spectral sensitivities (Mpitosos 1973), but their opposite physiological responses to light suggest they employ different phototransduction cascades.

5.2.3 Ostreida: Cap Eyespots

Ostreida includes foam oysters (Gryphaeidae), tree oysters (Isognomonidae), hammer oysters (Malleidae), pearl oysters (Margaritidae), true oysters (Ostreidae), pen shells (Pinnidae), feather oysters (Pteriidae), and relatives (Vulsellidae). Except for pen shells and some species of hammer oysters, which are semi-infaunal, most oysters are epifaunal (Fig. 5.1b). True oysters cement to their substrate, and their epifaunal relatives use byssal threads to attach to a variety of substrates (Tsubaki et al. 2011). Many epifaunal species of tree oyster (Morton 2001, 2008 Tëmkin 2006) and hammer oyster (Audino et al. 2020) have distributed visual systems consisting of hundreds of cap eyespots. These cap eyespots, located on the outer mantle folds, are small (around 20 μm in diameter), densely packed, and present along both valves from one side of the hinge to the other (Fig. 5.2c). Cap eyespots (Fig. 5.3c) are formed by clusters of pigmented cells and photoreceptors, and they appear to lack lenses or other light-focusing structures (Morton 2001, 2008; Audino et al. 2020). Ancestral state reconstruction (Audino et al. 2020) indicates cap eyespots were present in the last common ancestor of Pterioidea and then lost in Margaritidae, Pteriidae, and Vulsellidae (Fig. 5.2).

5.2.4 Arcida: Compound Eyes and Pigmented Cups

5.2.4.1 Phylogenetic Distribution and General Description

Arcida includes the ark clams (Arcidae) and relatives (including Cucullaeidae, Glycymerididae, Limopsidae, Noetiidae, and Philobryidae). This ecologically diverse group includes epifaunal species that attach to rocks using byssal threads and infaunal species that burrow in soft sediment (Audino and Marian 2018). Two distinct types of mantle eyes contribute to distributed visual systems in Arcida: compound eyes (Fig. 5.2d) and pigmented cups (Fig. 5.2e). Both types of eyes are located on the outer mantle fold beneath recently secreted periostracum (the transparent, organic, outermost layer of the shell). Ancestral state reconstruction (Fig. 5.2) indicates the last common ancestor of Arcida was epifaunal, shallow-dwelling, and had both types of eyes (Audino et al. 2019). Subsequent losses of one or both types of eyes have occurred in several lineages that have become infaunal (burrowers) or transitioned to deeper habitats (Audino et al. 2019). This has led to complex distributions of eye types in Arcida. For example, species of *Arca* have both types of eyes, species in Glycymerididae only have compound eyes, species of *Anadara* only have pigmented cups, and species in Limopsidae are eyeless (Waller 1980; Nilsson 1994; Audino et al. 2019). Given such an intriguing distribution of eye types, Arcida is a promising lineage in which to ask how and why animals gain and lose eyes (Sumner-Rooney et al. 2016).

5.2.4.2 Morphology and Components of Compound Eyes

The compound eyes of arcids (Fig. 5.3d) resemble slightly flattened globes composed of dozens of ommatidia (Waller 1980). In the compound eyes of *Barbatia domingensis* (= *B. cancellaria*) and *Arca zebra*, each ommatidium is a cone of pigmented cells with a single ciliary photoreceptor at its base (Nilsson 1994). Within each of these photoreceptors, sensory cilia project into a central vacuole and form a stack of flattened sacs (Nilsson 1994). Consequently, the ciliary photoreceptors of arcids have a very different morphology than the ciliary photoreceptors found in the eyes of pectinids and limids. The ommatidia of the compound eyes of arcids do not contain lenses or other light-focusing structures (Nilsson 1994). Compound eyes are found all along the mantle margins, but they tend to be more numerous toward the posterior (Audino and Marian 2018). The compound eyes of arcids are highly variable in number and size. In *B. domingensis*, for example, individuals can have up to 300 compound eyes with diameters from 180 to 300 μm and ommatidia numbering from 100 to 160. In comparison with those of *B. domingensis*, the compound eyes of *A. zebra* are fewer in number (90–100 per individual) and smaller in size (~40 ommatidia) (Waller 1980; Nilsson 1994; Audino and Marian 2018). With 79 ommatidia, the compound eyes of *Glycymeris bimaculata* appear to have sizes that fall between those of *B. domingensis* and *A. zebra* (Morton and Puljas 2016). The largest compound eyes described in Arcida are those of *A. noae*, which can reach 1 mm in diameter (Morton and Peharda 2008).

5.2.4.3 Morphology and Components of Pigmented Cups

The pigmented cups of arcids (Fig. 5.3e) are small pits lined with rhabdomeric photoreceptors and cells packed with screening pigment (Morton 1987). The interior of each cup is filled by densely packed microvilli extending from the photoreceptors (Nilsson 1994). The sizes and numbers of pigmented cups vary between individuals and species. Within individuals of *Barbatia candida*, an epifaunal species, there is a continuous gradation from smaller cups to larger, with the largest cups reaching widths of 120 μm (Audino and Marian 2018). These pigmented cups tend to be larger, more densely packed, and more heavily pigmented toward the posteriors of animals (Audino and Marian 2018). In *B. domingensis*, also an epifaunal species, individuals can have as many as 2000 pigmented cups, with the largest reaching 80 μm in diameter (Nilsson 1994). Compared to epifaunal species of *Barbatia*, burrowing species of *Anadara* have pigmented cups that are fewer in number (only ~40 in *A. notabilis*), similar in size, and concentrated toward the anterodorsal region of the mantle (Nilsson 1994; Audino et al. 2019).

5.3 Visual Ecology of Pteriomorphian Bivalves

Pteriomorphian bivalves tend to respond to the sudden dimming of light by engaging in defense responses such as withdrawing their mantle tissue and closing their valves. In eyeless bivalves, defensive "shadow responses" can be initiated by extraocular photoreceptors located in the mantle (Kennedy 1960; Wiederhold et al. 1973). Distributed visual systems may enhance the defensive abilities of bivalves in several ways. For example, they may make it possible for animals to detect moving objects that do not alter light levels (Nilsson 1994). They may also make the detection of predators more reliable by allowing bivalves to distinguish between the appearances of objects and uniform changes in light levels. Can the distributed visual systems of pteriomorphians help inform other types of behavior? Habitat selection is an option for mobile species, such as scallops and file clams, which live in visually distinct habitats, such as rock crevices or seagrass beds, from which they may become dislodged and to which they may seek to return (Fig. 5.1b). For each group of pteriomorphian bivalves with distributed visual systems, we will discuss what is known about the functional properties of its eyes and its visually influenced behaviors. Where possible, we will also comment on how differences in visual performance between species may relate to ecological factors such as locomotory abilities and habitat depth.

5.3.1 Visual Ecology of Scallops (Pectinida)

5.3.1.1 Visual Performance of the Mirror-Based Eyes of Scallops

The eyes of scallops provide fine-grained spatial resolution compared to the eyes of most non-cephalopod mollusks. Physiological recordings indicate the eyes of scallops respond to moving stripes with angular widths as narrow as 2° (Land 1966b). Ray-tracing analyses (Land 1965; Speiser et al. 2016; Palmer et al. 2017) and behavioral trials (Buddenbrock and Moller-Racke 1953; Speiser and Johnsen 2008b; Chappell et al. 2021) provide similar estimates of spatial resolution. Relative to those of any animal (Petie et al. 2016), the eyes of scallops sample slowly in time. Electroretinography indicates the eyes of *Ylistrum* (=*Amusium*) *japonicum* have a maximum temporal sampling rate of ~5 Hz (Kanmizutaru et al. 2005). Thus, the eyes of scallops may be able to resolve relatively small objects, but only if these objects are moving slowly. The ecological importance of vision to scallops is emphasized by their eyes demonstrating pupillary responses that appear to balance trade-offs between spatial resolution and sensitivity (i.e., light-gathering power) under varying light conditions (Miller et al. 2019).

It remains unclear if scallops have color vision, which is defined as the ability of a viewer to distinguish between different wavelengths of light with equivalent intensities. Color vision generally requires an animal to have eyes with two or more types of photoreceptors with different spectral sensitivities and a neural mechanism for comparing the responses of these photoreceptors to light. In behavioral experiments, *P. maximus* demonstrated spectral response peaks at 480 and 540 nm (Cronly-Dillon 1966). Microspectrophotometry (MSP) indicates the proximal and distal photoreceptors of *A. irradians* absorb maximally at 506 and 535 nm, respectively, while those of *P. magellanicus* absorb maximally at 488 and 513 nm (Speiser et al. 2011). However, differences in spectral sensitivity between the proximal and distal photoreceptors will only confer color vision if scallops compare activity between their two retinas and there is no evidence that they do so (Speiser et al. 2011). Unlike MSP readings, electrophysiological recordings indicate proximal and distal photoreceptors from the eyes of *A. irradians* both have peak sensitivities at ~500 nm (McReynolds and Gorman 1970a). Inconsistencies between the results of these studies may be due to interspecific variation, the use of different wavelength intervals in experiments, or other factors. If scallops have color vision, transcriptome sequencing suggests a straightforward mechanism: the eyes of *P. magellanicus* express at least two G_q-opsins (Pairett and Serb 2013), and the eyes of *A. irradians* express at least four (Porath-Krause et al. 2016), indicating the proximal retinas of scallops may contain multiple spectral classes of photoreceptors.

Differences between the distal and proximal retinas suggest scallops use them for separate visual tasks. The distal retina appears well-suited for detecting the edges of moving objects, an ability relevant to predator detection (Land 1966b). Light primes the distal photoreceptors to respond to edges by removing inhibition from the voltage-gated calcium channels whose openings enhance the excitatory responses of these photoreceptors to the dimming of light (Cornwall and Gorman 1979, 1983). Unlike the phasic photoreceptors of the distal retina, the tonic photoreceptors of the proximal retina demonstrate persistent activity in the light and thus may be better suited for monitoring changes in ambient light intensity and evaluating static environmental features, such as those relevant to habitat selection (Land 1966b).

Functional properties of the eyes of scallops appear to correlate with ecological factors. For example, the distal and proximal photoreceptors of *A. irradians*, a species that tends to live in shallower, greener water, are maximally sensitive to longer (greener) wavelengths of light than the corresponding photoreceptors from the eyes of *P. magellanicus*, a species that tends to live in deeper, bluer water (Speiser et al. 2011). The functional properties of scallop eyes may also be associated with the locomotory abilities of species. For example, the eyes of mobile scallops are more likely to express two or four copies of G_q-opsin (= r-opsin) than the eyes of sessile species, which tend to express a single G_q-opsin (Serb et al. 2013). Consistent with this pattern, the eyes of swimming pectinids (e.g., *A. irradians* and *P. magellanicus*) tend to be larger and have longer focal lengths than the eyes of sessile species (e.g., *Crassadoma gigantea* and *Spondylus americanus*), implying they provide finer-grained spatial resolution (Speiser and Johnsen 2008a).

5.3.1.2 Visually Influenced Behaviors of Scallops

Scallops may use their distributed visual systems to inform at least three different types of behavior: defensive responses, exploration of objects with their sensory tentacles, and habitat selection (Buddenbrock and Moller-Racke 1953). First, like other bivalves, scallops respond to the dimming of light by retracting their mantle tissue and closing their valves (Gutsell 1930; Buddenbrock and Moller-Racke 1953). Spatial resolution may contribute to these defensive responses: moving objects elicit stronger reactions from scallops than uniform changes in light intensity (Gutsell 1930). Second, scallops such as *Euvola* (=*Pecten*) *ziczac* track the movements of approaching predators, such sea stars, with their sensory tentacles (Wilkens 1981; Speiser and Wilkens 2016). It appears they do so, at least in part, using spatial information acquired by their distributed visual system. In controlled laboratory trials, *A. irradians* directed their sensory tentacles toward static visual stimuli and tracked isoluminant rotating visual stimuli with rotating waves of tentacle extension (Chappell et al. 2021). By using spatial vision to direct their chemosensory tentacles toward moving objects, scallops may distinguish threats from non-threats more efficiently. Third, scallops may use their distributed visual system to locate preferred habitats. In support of this possibility, *Mimachlamys varia* (=*Pecten varius*) have been observed visually locating rock crevices (Buddenbrock and Moller-Racke 1953), and *A. irradians* have been shown to visually detect the beds of eelgrass (*Zostera*) in which they tend to seek shelter (Hamilton and Koch 1996).

5.3.2 Visual Ecology of File Clams (Limida)

File clams, like many bivalves, respond to the dimming of light by withdrawing their mantle tissue and closing their valves (Morton 2000). The invaginated eyes of file clams respond physiologically to decreases in light intensity and so may contribute to these defensive behaviors (Mpitosos 1973). It is unclear, however, if distributed visual systems based on invaginated eyes provide file clams with spatial vision. This is worth exploring because file clams display novel behaviors to which spatial vision may contribute. First, file clams cannot fully withdraw their large tentacles into their valves. Instead of hiding in their shells, species such as *Limaria* (= *Lima*) *hians* defend themselves by autotomizing and releasing noxious mucus from their tentacles (Gilmour 1967). Second, some file clams, such as *C. ales*, produce flashing displays in which they furl and unfurl strips of reflective mantle tissue

(Dougherty et al. 2014). In an intriguing study, *C. ales* were found to increase their flash rates in response to changing light levels, an indication their flashing display may have an aposematic function (Dougherty et al. 2017). Third, most file clams can swim, so they may use spatial vision to locate and retreat into crevices (Morton 1979). We hope future studies on file clams will explore the evolutionary relationships between their distributed visual systems and their distinctive behaviors such as tentacle autotomy, flashing displays, and swimming.

5.3.3 Visual Ecology of Oysters (Ostreida)

The functions of the distributed visual systems of oysters have yet to be investigated. The structures of cap eyespots suggest each may provide low-resolution spatial vision (Fig. 5.3c). Cap eyespots are numerous and densely packed on the mantle margins of oysters, so they may provide a considerable amount of spatial information if they function collectively (Fig. 5.2c). If distributed visual systems based on cap eyespots provide spatial information, they may help epibyssate species orient their bodies relative to the substrate (Audino et al. 2020). Many oysters can detach from their substrate, which allows minor adjustments in position and location, with subsequent byssal reattachment (Stanley 1972). Spatial information provided by a distributed network of cap eyespots may contribute to this behavior, particularly in crevice-dwellers (Fig. 5.1b), such as species of tree oyster (*Isognomon*) and hammer oyster (*Malleus*).

5.3.4 Visual Ecology of Ark Clams (Arcida)

The distributed visual systems of ark clams provide coarse spatial information. Based on morphological estimates (Nilsson 1994), the ommatidia of the compound eyes have acceptance angles of ~30° (*B. domingensis*), and the pigmented cups have acceptance angles ranging from 20° (*A. notabilis*) to 40° (*B. domingensis*). In behavioral tests of spatial vision, *B. domingensis* reliably demonstrated defensive responses to a moving stripe with an angular width of 6° (Nilsson 1994). Distributed visual systems that provide coarse-grained spatial information may help ark clams survive by making it possible for them to detect movements of predators that are not accompanied by changes in light levels (Nilsson 1994). Ark clams with distributed visual systems tend to be epibenthic and to be able to crawl using their foot, so these bivalves may also use their distributed visual systems for habitat selection or to position their bodies on the substrate (Audino et al. 2019). Visual performance in ark clams may be associated with ecological factors such as habitat depth. In Glycymerididae, for example, shallow-dwelling species have compound eyes that are larger than those of deeper-dwelling species (Morton and Puljas 2016).

5.4 Neuroanatomy and Visual Processing in Pteriomorphia

The nervous systems of pteriomorphians, like those of other bivalves, tend to include three pairs of ganglia: the cerebral (or cerebro-pleural) ganglia, the pedal ganglia, and the visceral (or parietal-visceral) ganglia. These distinct neural structures have cortices of neuronal somata that surround cores of neuropil (Richter et al. 2010). The paired ganglia of bivalves may be fused, linked transversally by nerves known as commissures (e.g., cerebral commissure, visceral commissure, or pedal commissure), or joined longitudinally by nerves known as connectives (e.g., cerebral-pedal connective, cerebral-visceral connective, etc.). Numerous nerves (e.g., pallial nerves) emanate from ganglia in a radial fashion. These nerves connect ganglia to the peripheral nervous system and the organs of the body. In addition to ganglia, many pteriomorphians have prominent neural structures termed circumpallial nerves. These innervate the organs along the mantle margins, including any sensory organs that might be present. Despite their name, circumpallial nerves are medullary cords rather than nerves. Like ganglia, the circumpallial nerves consist of a core of neuropil surrounded by a cortex of neuronal somata (Audino et al. 2015a). To begin exploring how pteriomorphian bivalves process visual information they acquire with their mantle eyes, we will compare their ganglia, commissures, connectives, and circumpallial nerves.

5.4.1 Neuroanatomy of Scallops (Pectinida)

5.4.1.1 Neuroanatomical Structures of Pectinids

The nervous systems of pectinids include cerebral ganglia, pedal ganglia, a fused and elaborated visceral ganglion, and a circumpallial nerve (Fig. 5.4a). The cerebral and pedal ganglia are paired structures whose halves are joined by commissures (Drew 1906; Dakin 1910b). In scallops and spondylids, the fused visceral ganglion has multiple lobes (Drew 1906; Dakin 1910b, 1928). These lobes include a single ventrocentral lobe (VCL), a pair of dorsocentral lobes (DCL), and a pair of crescent-shaped lateral lobes (LL). The visceral ganglion is also associated with two small accessory ganglia (AG), one located near the dorsal ends of each lateral lobe. Each lobe and accessory ganglion has a cortex of small, densely packed neuronal somata and a core of neuropil, making each of these substructures as complex as most whole ganglia in bivalves (Dakin 1910b; Croll et al. 1995). The ganglia of pectinids are linked longitudinally by connectives, but the connectives of scallops and spondylids do not match. In scallops, the pedal ganglia are connected to the cerebral ganglia, but not to the visceral ganglion; in spondylids, the pedal ganglia are connected to the visceral ganglion, but not to the cerebral ganglia (Dakin 1928). Lastly, pectinids have a prominent circumpallial nerve that innervates the sensory structures along their mantle margins, including the eyes and tentacles (Drew 1906; Audino et al. 2015a).

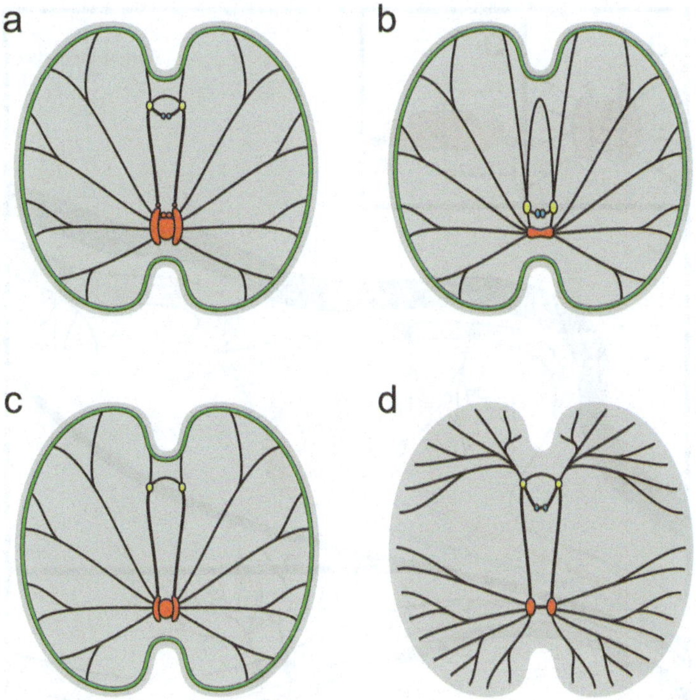

Fig. 5.4 Schematic representations of the nervous systems of pterimorphian bivalves with distributed visual systems. (**a**) Scallops and relatives (Pectinoidea), drawn after *Placopecten magellanicus* (Drew 1906). (**b**) File clams (Limida), drawn after *Lima* spp. (Pelseneer 1911). (**c**) Oysters and relatives (Ostreida), drawn after *Crassostrea virginica* (Galtsoff 1964). (**d**) Ark clams and relatives (Arcida), drawn after *Arca* spp. and *Barbatia* spp. (Heath 1941; Audino et al. 2019). The diagrams are oriented with anterior up, posterior down, and the ventral side of the animal facing the viewer. Mantles are shown divided into left and right lobes. Structures: yellow = cephalic ganglia; blue = pedal ganglia; red = visceral ganglia; green = circumpallial nerve; black = commissures, connectives, and pallial nerves; gray = mantle lobes

5.4.1.2 Visual Processing in Scallops

Optic nerves from nearly all of the mantle eyes of scallops travel to the lateral lobes of the visceral ganglion (Fig. 5.5). Autoradiographic experiments indicate each of these optic nerves exits an eyestalk, joins the circumpallial nerve, crosses the mantle via one of several pallial nerves, and then enters a lateral lobe (Spagnolia and Wilkens 1983). Every photoreceptor in an eye appears to be represented by its own axon in the optic, circumpallial, and pallial nerves (Spagnolia and Wilkens 1983). The circumpallial nerves not only serve as conduits for the axons of sensory receptors and motor neurons, but they also have cortices of interneurons. These features raise the possibility that scallops process sensory-motor information peripherally in

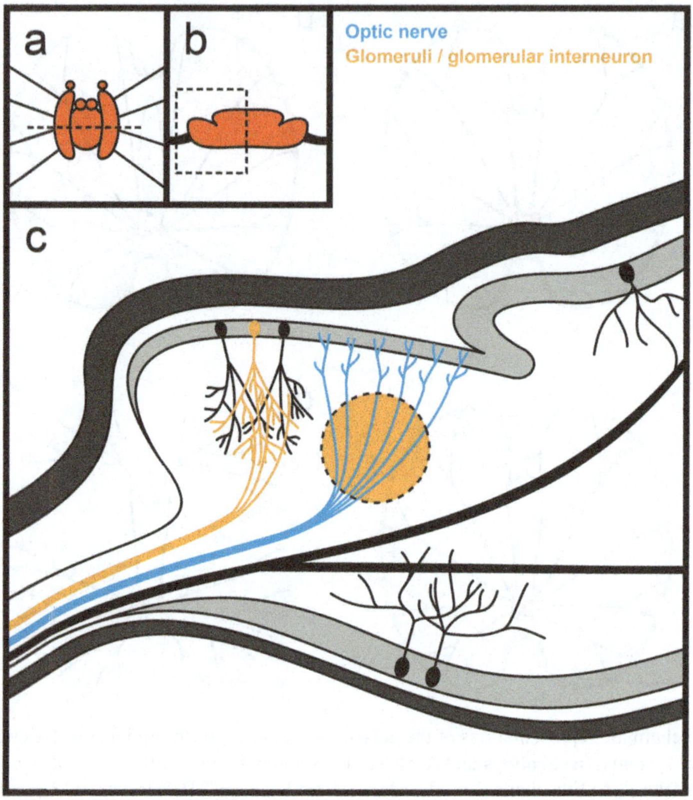

Fig. 5.5 Schematic representation of a provisional visual-motor circuit in the lateral lobe of the visceral ganglion of the bay scallop *Argopecten irradians* (adapted from Spagnolia and Wilkens 1983). The insets (**a**) and (**b**) show the sectioning plane and restricted field of view, respectively, of the visceral ganglion (red) represented in (**c**). (**c**) Optic nerves (one nerve shown in blue) enter a lateral lobe, pass through a layer of glomerular neuropil (one glomerulus shown as an orange circle), and then make synaptic contact with the cortex of the lateral lobe (light gray). The cortex is composed of interneurons (one interneuron shown in orange) that project multiple types of neurites. Those with bushy arborizations contribute to the glomeruli inside the lateral lobes, and others function as efferents that travel to the mantle lobes via the pallial nerves. Also shown here are the capsule of connective tissue surrounding the visceral ganglion (dark grey) and other interneurons and neural tracts (black)

their circumpallial nerves. Evidence suggests, however, that the optic nerves remain anatomically and physiologically segregated from other types of fibers in the circumpallial nerves (Spagnolia and Wilkens 1983). Optic nerves from the dozen or so mantle eyes closest to the anterior and posterior sides of the hinge have optic nerves that project to the cerebral ganglion instead of the visceral ganglion (Wilkens 1981). The functional significance of separate groups of mantle eyes sending optic nerves to different ganglia has yet to be explored.

Behavioral trials indicate *A. irradians* can locate and track visual stimuli with their sensory tentacles, a demonstration of spatial vision that indicates scallops retain within their sensory-motor circuits the fine-grained visual information acquired by their eyes (Chappell et al. 2021). Consistent with this finding, scallops appear to process visual information from their mantle eyes using glomerular neuropil with a somatotopic organization (Wilkens and Ache 1977; Spagnolia and Wilkens 1983). After entering a lateral lobe, optic nerves (Fig. 5.5, blue) pass through a layer of glomerular neuropil before innervating the cortex of the lobe (Spagnolia and Wilkens 1983). At the cortex, optic nerves make synaptic connections with interneurons (Fig. 5.5, orange). These cortical interneurons project multiple types of neurites: those with bushy arborizations contribute to the glomeruli inside the lateral lobes, whereas others function as efferents that travel to the mantle lobes via the pallial nerves (Spagnolia and Wilkens 1983).

Glomeruli within the lateral lobes have positions that appear to correspond point-for-point with the positions of the eyes with which they share synaptic connections (Spagnolia and Wilkens 1983). Electrophysiological recordings support the somatotopic arrangement of the lateral lobes: levels of electrical activity measured from the anterior to posterior regions of lateral lobes corresponded to the amount of light presented to eyes positioned anterior to posterior along the mantle margins (Wilkens and Ache 1977). From what we have learned so far, we can conclude scallops process visual information in unique ways. They are the only bilaterians in which somatotopic visual processing occurs outside of a brain, and they appear to use glomerular structures to integrate visual information. The latter comes as a surprise because animals typically use glomerular neuropil to integrate olfactory information rather than visual information (Ache and Young 2005).

5.4.2 Neuroanatomy of File Clams (Limida)

The nervous systems of file clams include paired cerebral ganglia, paired pedal ganglia, a fused visceral ganglion, and a circumpallial nerve (Fig. 5.4b). The cerebral, pedal, and visceral ganglia of pteriomorphian bivalves tend to be relatively distant from each other, but the cerebral and pedal ganglia of some file clams are close enough to the visceral ganglion to form a dense neural complex (Pelseneer 1911). Intriguingly, different species of file clam display varying degrees of neural condensation. In *Acesta* (= *Lima*) *excavata*, the cerebral and pedal ganglia are distinct from the visceral ganglion, but in *L. lima* (= *L. squamosa*), the cerebral, pedal, and visceral ganglia are highly condensed (Pelseneer 1911). The visceral ganglia of file clams appear to have multiple lobes, but it is not known if these lobes represent distinct inner regions of neuropil (as they do in scallops) (Pelseneer 1911). Due to the condensation of the cerebral and pedal ganglia toward the visceral ganglion in some species, the cerebral commissures of file clams can be much longer than those of other bivalves (Pelseneer 1911). The ganglia of file clams are linked by connectives including cerebral-visceral connectives and cerebral-pedal connectives, but visceral-pedal connectives are absent.

5.4.3 Neuroanatomy of Oysters (Osterida)

The nervous systems of oysters include paired cerebral ganglia, a fused visceral ganglion, and a circumpallial nerve, but pedal ganglia are absent (Fig. 5.4c). The lack of pedal ganglia may be attributed to post-metamorphic oysters losing their foot, the organ innervated by pedal ganglia in other bivalves. The fused visceral ganglion is subdivided into a central lobe and two lateral lobes (Galtsoff 1964). The circumpallial nerves of oysters, like those of other pteriomorphian bivalves, are medullary cords with cortices of neuronal somata and cores of neuropil (Duvernoy 1853).

5.4.4 Neuroanatomy of Ark Clams (Arcida)

Arcids have nervous systems with cerebral ganglia, pedal ganglia, and visceral ganglia, but they lack a circumpallial nerve (Fig. 5.4d). Unlike the visceral ganglia of other pteriomorphian bivalves, those of arcids are not fused; instead, they are paired structures in which the halves are joined by a short visceral commissure (Heath 1941). The ganglia are linked longitudinally by connectives including cerebral-visceral connectives and cerebral-pedal connectives, but visceral-pedal connectives are absent (Heath 1941). Unlike other pteriomorphians, arcids lack circumpallial nerves. Instead of being innervated by a circumpallial nerve, the mantle margin of an arcid is innervated by a plexus formed by branching pallial nerves (Heath 1941; Audino et al. 2019). Half of these pallial nerves emanate from the paired cerebral ganglia, and the other half emanate from the paired visceral ganglia (Heath 1941). Consequently, pallial nerves from the cerebral ganglia make a larger contribution to the innervation of the mantle in arcids (Fig. 5.4d) than in other pteriomorphians (Fig 5.4a–c). Optic nerves from the mantle eyes of arcids appear to join with the pallial nerves, but neural structures associated with visual processing have yet to be identified (Morton 1987).

5.5 Evolution of Distributed Visual Systems in Pteriomorphia

5.5.1 Why Do Some Pteriomorphian Bivalves Have Eyes When Many Do Not?

Ancestral state reconstruction (Audino et al. 2020) indicates pteriomorphian bivalves have gained mantle eyes at least five times and lost them many more times than that (Fig. 5.2). A first step toward learning why pteriomorphians gain and lose mantle eyes is identifying the light-influenced behaviors to which these eyes contribute. Many animals, including many mollusks, detect light using extraocular photoreceptors (Ramirez et al. 2011; Kingston and Cronin 2016). Thus, the presence of eyes in pteriomorphian bivalves is probably not explained by their contributions to tasks associated with nondirectional photoreception such as circadian entrainment or shadow detection (Nilsson 2013). Pteriomorphians with distributed visual systems, including scallops and ark clams, demonstrate behaviors consistent with spatial resolution, i.e., the ability to detect objects with relatively small angular sizes (Nilsson 1994; Speiser and Johnsen 2008b). Spatial resolution helps animals respond to potential threats by making it possible for them to detect the movements of objects and to distinguish shadows cast by objects from uniform changes in light conditions. We have also learned that scallops demonstrate spatial vision, i.e., they are able to locate objects within their visual environment (Chappell et al. 2021). Spatial vision may help pteriomorphians direct their sensory tentacles toward objects (Chappell et al. 2021), and it may help mobile species locate preferred habitats, such as rock crevices or seagrass beds (Hamilton and Koch 1996).

A second step toward understanding the evolution of distributed visual systems in pteriomorphian bivalves will involve identifying ecological factors that correlate with the presence of mantle eyes. Epifaunal species, for example, are more likely to have mantle eyes than infaunal species (Audino et al. 2020). Further, eyes may be lost when lineages make transitions from shallow to deep habitats. For example, at least two lineages of deep-dwelling glass scallops (*Propeamussium* spp.) have lost the mirror-based eyes present in the last common ancestor of pectinids (Audino et al. 2020). Changes in locomotory abilities also may influence the evolution of distributed visual systems in pteriomorphians. The origin of swimming, for example, may have opened possibilities for new visually influenced behaviors in pectinids, such as habitat selection, that could have introduced new selective pressures on the visual abilities of these bivalves.

5.5.2 Why Do Pteriomorphian Bivalves Have so Many Mantle Eyes?

Pteriomorphian bivalves with distributed visual systems can have dozens to hundreds of eyes. Visual systems can incur high metabolic costs (Niven and Laughlin 2008; Moran et al. 2015), so it is curious to see bivalves invest in numerous eyes with overlapping fields of view. Large numbers of eyes may benefit pteriomorphian bivalves by providing near-complete coverage of their visual surroundings at all times. Further, eyes with overlapping fields of view may make distributed visual systems more reliable by increasing their signal-to-noise ratios (Nilsson 1994). Distributed visual systems may also function as coincidence detectors if animals only initiate defense responses when several eyes with overlapping fields of view detect motion simultaneously. Consistent with this prediction, calculations suggest the ark clam *B. domingensis* samples every point in its visual field with 755 ommatidia (Nilsson 1994) and the scallop *P. maximus* samples every point in its visual field with at least 1 photoreceptor from 17 separate eyes (Land 1968). Enhancing the reliability of distributed visual systems through oversampling may be an efficient strategy if these visual systems involve limited amounts of neural processing. Sensory-motor processing in scallops, for example, appears to involve only one or two layers of interneurons (Fig. 5.5), which suggests bivalves may pay a low marginal metabolic cost to add new eyes to their distributed visual systems.

5.5.3 Why Do the Eyes of Pteriomorphians Tend to Include Two Different Types of Photoreceptors?

The mantle eyes of pteriomorphian bivalves tend to include two very different types of photoreceptors: rhabdomeric photoreceptors that are excited by light and ciliary photoreceptors that are excited by the dimming of light. In pectinids and limids, these photoreceptors are divided between separate retinas within the same eyes (Hartline 1938; Mpitosos 1973). In ark clams and relatives, these photoreceptors are divided between different types of eyes: the compound eyes have ciliary photoreceptors, and the pigmented cups have rhabdomeric photoreceptors (Nilsson 1994). The rhabdomeric photoreceptors in the eyes of pteriomorphians are morphologically and physiologically similar to the rhabdomeric photoreceptors found in the cephalic eyes of other invertebrates, and they appear to function using homologous molecular components (Barber et al. 1967; Kojima et al. 1997; Kingston et al. 2017). In contrast, the ciliary photoreceptors in the eyes of pteriomorphians are dissimilar to other well-characterized ciliary photoreceptors, such as the rods and cones in the eyes of vertebrates (Wilkens 2008). Learning more about these unusual ciliary photoreceptors will help us more accurately reconstruct the evolutionary history of light-sensitive cell types in Metazoa (Plachetzki et al. 2007; Arendt 2008; Nilsson and Arendt 2008). To learn why two very different types of photoreceptors tend to be associated with the distributed visual systems of pteriomorphians, we must ask whether they represent two sources of input for a single visual pathway or if they represent parallel visual pathways associated with different sets of visually influenced behaviors.

5.5.4 How Do Pteriomorphian Bivalves Process Visual Information?

When we compare neuroanatomical structures across Pteriomorphia, we find a great deal of diversity. Ganglia can have different relative positions, they can be paired or fused, they can be elaborated with lobes and other accessory structures, or they can be absent. Given these differences, pteriomorphians with distributed visual systems may not use the same approaches, or even the same neural structures, to process visual information. For example, pteriomorphians may divide visual processing between their cephalic and visceral ganglia in different ways. Scallops appear to process visual information from most of their mantle eyes in the lateral lobes of their visceral ganglia, whereas ark clams appear to split visual processing more evenly between their cerebral and visceral ganglia. Pteriomorphians may also process visual information from their mantle eyes in a centralized manner, a decentralized manner, or through a combination of these two approaches. With this in mind, we see what may be another meaningful neuroanatomical difference between ark clams and other pteriomorphians: ark clams lack the circumpallial nerve that is present in pectinids, file clams, and oysters (Fig. 5.4). The circumpallial nerve is a medullary cord, so ark clams may have less capacity for peripheral processing in their mantle lobes than other pteriomorphians. The nervous systems of arcids resemble those of non-pteriomorphian bivalves (i.e., they contain three pairs of unfused ganglia and no circumpallial nerve), suggesting the diversification of pteriomorphians (excluding arcids) may be associated with innovations that include the enlargement of the posterior muscular system (e.g., the posterior adductor) over the anterior system (e.g., the anterior adductor), the consolidation of centralized processing toward a fused visceral ganglion, and the enhancement of peripheral processing in the mantle lobes (through the origin and evolution of the circumpallial nerve).

5.5.5 *Future Directions*

It will be fascinating to explore the coevolution of distributed visual systems, ecological traits, and neural structures in pteriomorphian bivalves. For example, do changes in ecology, such as transitions to epifaunal or infaunal lifestyles, respond to changes in sensory systems, or do they drive these changes? Likewise, how are the origins of novel locomotory abilities, such as swimming, reflected by changes in distributed visual systems? Do changes in neural architecture make distributed visual systems possible? Or do origins of novel sensory structures along the mantle margins drive changes to nervous systems? These questions may be pursued using integrative approaches (e.g., behavior, physiology, anatomy) within a phylogenetic framework so that ancestral states may be reconstructed and the timing of evolutionary events estimated. For the many reasons discussed in this chapter, such as their morphological diversity, global distribution, and rich fossil record, pteriomorphian bivalves are a fascinating group in which to study how distributed visual systems function and how they may coevolve with light-influenced behaviors, locomotory abilities, neural architectures, and other traits.

References

Ache BW, Young JM (2005) Olfaction: Diverse Species, Conserved Principles. Neuron 48(3):417–430

Adal MN, Morton B (1973) The fine structure of the pallial eyes of *Laternula truncata* (Bivalvia: Anomalodesmata: Pandoracea). J Zool 170(4):533–556

Alejandrino A, Puslednik L, Serb JM (2011) Convergent and parallel evolution in life habit of the scallops (Bivalvia: Pectinidae). BMC Evol Biol 11(1):1–9

Arendt D (2008) The evolution of cell types in animals: emerging principles from molecular studies. Nature Rev Genet 9(11):868–882

Audino J, Marian JEAR (2018) Comparative and functional anatomy of the mantle margin in ark clams and their relatives (Bivalvia: Arcoidea) supports association between morphology and life habits. J Zool 305(3):149–162

Audino JA, Marian JEAR, Wanninger A, Lopes SGBC (2015a) Anatomy of the pallial tentacular organs of the scallop *Nodipecten nodosus* (Linnaeus, 1758) (Bivalvia: Pectinidae). Zool Anz 258:39–46

Audino JA, Marian JEAR, Wanninger A, Lopes SGBC (2015b) Development of the pallial eye in *Nodipecten nodosus* (Mollusca: Bivalvia): insights into early visual performance in scallops. Zoomorphology 134(3):403–415

Audino JA, Serb JM, Marian JEAR (2019) Ark clams and relatives (Bivalvia: Arcida) show convergent morphological evolution associated with lifestyle transitions in the marine benthos. Biol J Linn Soc 126(4):866–884

Audino JA, Serb JM, Marian JEAR (2020) Hard to get, easy to lose: Evolution of mantle photoreceptor organs in bivalves (Bivalvia, Pteriomorphia). Evolution 74(9):2105–2120

Audino JA, Adams DC, Serb JM (2022) Variation in eye abundance among scallops reveals ontogenetic and evolutionary convergence associated with life habits. Evolution 76(7):1607–1618

Barber VC, Evans EM, Land MF (1967) The fine structure of the eye of the mollusc *Pecten maximus*. Z Zellforsch 76(3):295–312

Barber VC, Land MF (1967) Eye of the cockle, *Cardium edule*: Anatomical and physiological investigations. Experientia 23(8):677–678

Barber VC, Wright DE (1969) The fine structure of the eye and optic tentacle of the mollusc *Cardium edule*. J Ultrastruct Res 26(5-6):515–528

Bell A, Mpitosos GJ (1968) Morphology of the eye of the flame fringe clam, *Lima scabra*. Biol Bull 135(2):414–415

Bieler R, Mikkelsen PM, Collins TM, Glover EA, González VL, Graf DL, Harper EM, Healy J, Kawauchi GY, Sharma PP (2014) Investigating the Bivalve Tree of Life–an exemplar-based approach combining molecular and novel morphological characters. Invertebr Syst 28(1):32–115

Bieler R, Mikkelsen PM, Giribet G (2013) Bivalvia—A Discussion of Known Unknowns. Am Malacol Bull 31(1):123–133

Von BW, Moller-Racke I (1953) Uber den lichtsinn von *Pecten*. Pubbl Staz Zool Napoli 24:217–245

Butcher EO (1930) The formation, regeneration, and transplantation of eyes in *Pecten* (*Gibbus borealis*). Biol Bull 59(2):154–164

Carosa E, Kozmik Z, Rall JE, Piatigorsky J (2002) Structure and expression of the scallop Ω-crystallin gene: evidence for convergent evolution of promoter sequences. J Biol Chem 277(1):656–664

Chappell DR, Horan TM, Speiser DI (2021) Panoramic spatial vision in the bay scallop *Argopecten irradians*. Proc R Soc Lond B Biol Sci 288(1962):20211730

Cheng J, Davison I, DeMont M (1996) Dynamics and energetics of scallop locomotion. J Exp Biol 199(9):1931–1946

Combosch DJ, Collins TM, Glover EA, Graf DL, Harper EM, Healy JM, Kawauchi GY, Lemer S, McIntyre E, Strong EE, Taylor JD (2017) A family-level tree of life for bivalves based on a Sanger-sequencing approach. Mol Phylogenetics Evol 107:191–208

Combosch DJ, Giribet G (2016) Clarifying phylogenetic relationships and the evolutionary history of the bivalve order Arcida (Mollusca: Bivalvia: Pteriomorphia). Mol Phylogenetics Evol 94:298–312

Cope JC (1997) The early phylogeny of the class Bivalvia. Palaeontology 40:713–746

Cornwall MC, Gorman ALF (1979) Contribution of calcium and potassium permeability changes to the off response of scallop hyperpolarizing photoreceptors. J Physiol 291(1):207–232

Cornwall MC, Gorman ALF (1983) The cation selectivity and voltage dependence of the light-activated potassium conductance in scallop distal photoreceptor. J Physiol 340(1):287–305

Croll RP, Too CK, Pani AK, Nason J (1995) Distribution of serotonin in the sea scallop *Placopecten magellanicus*. Invertebr Reprod Dev 28(2):125–135

Cronly-Dillon JR (1966) Spectral sensitivity of the scallop *Pecten maximus*. Science 151(3708):345–346

Dakin WJ (1910a) The eye of *Pecten*. Q J Microsc Sci 55:49–112

Dakin WJ (1910b) The visceral ganglion of *Pecten*, with some notes on the physiology of the nervous system, and an inquiry into the innervation of the osphradium in the Lamellibranchiata. Mitt Zool Stn Neapel 20:1–40

Dakin WJ (1928) The eyes of *Pecten*, *Spondylus*, *Amussium* and allied Lamellibranchs, with a short discussion on their evolution. Proc R Soc Lond B Biol Sci 103(725):355–365

Distel DL (2000) Phylogenetic relationships among Mytilidae (Bivalvia): 18S rRNA data suggest convergence in mytilid body plans. Mol Phylogenetics Evol 15:25–33

Donovan DA, Elias JP, Baldwin J (2004) Swimming behavior and morphometry of the file shell *Limaria fragilis*. Mar Freshwater Behav Physiol 37(1):7–16

Dougherty LF, Dubielzig RR, Schobert CS, Teixeira LB, Li J (2017) Do you see what I see? Optical morphology and visual capability of 'disco'clams (*Ctenoides ales*). Biol Open 6(5):648–653

Dougherty LF, Johnsen S, Caldwell RL, Marshall NJ (2014) A dynamic broadband reflector built from microscopic silica spheres in the 'disco' clam *Ctenoides ales*. J R Soc Interface 11(98):20140407

Dougherty LF, Niebergall AK, Broeckling CD, Schauer KL, Li J (2019) Brightly coloured tissues in limid bivalves chemically deter predators. R Soc Open Sci 6(10):191298

Drew GA (1906) The habits, anatomy, and embryology of the giant scallop (*Pecten tenuicostatus*, Mighels). Univ Maine Stud 6:1–71

Duvernoy GL (1853) Mémoires sur le système nerveux des mollusques acéphales. Mémoires de l'Académie des sciences de l'Institut de France 24:1–210

Fain GL, Hardie R, Laughlin SB (2010) Phototransduction and the evolution of photoreceptors. Curr Biol 20(3):R114–R124

Fankboner P (1981) Siphonal eyes of giant clams (Bivalvia: Tridacnidae) and their relationship to adjacent zooxanthellae. Veliger 23(3):245–249

Galtsoff PS (1964) The American oyster *Crassostrea virginica* Gmelin. US Fish Wildl Serv Fish Bull 64:1–480

Gilmour THJ (1967) The defensive adaptations of *Lima hians* (Mollusca, Bivalvia). J Mar Biol Assoc UK 47(1):209–221

Gomez MP, Nasi E (1994) The light-sensitive conductance of hyperpolarizing invertebrate photoreceptors: a patch-clamp study. J Gen Physiol 103(6):939–956

Gomez MP, Nasi E (1995) Activation of light-dependent K+ channels in ciliary invertebrate photoreceptors involves cGMP but not the IP3/Ca2+ cascade. Neuron 15(3):607–618

Gomez MP, Nasi E (1996) Ion permeation through light-activated channels in rhabdomeric photoreceptors. Role of divalent cations. J Gen Physiol 107(6):715–730

Gomez MP, Nasi E (2000) Light transduction in invertebrate hyperpolarizing photoreceptors: possible involvement of a Go-regulated guanylate cyclase. J Neurosci 20(14):5254–5263

González VL, Andrade SC, Bieler R, Collins TM, Dunn CW, Mikkelsen PM, Taylor JD, Giribet G (2015) A phylogenetic backbone for Bivalvia: an RNA-seq approach. Proc R Soc Lond B Biol Sci 282(1801):20142332

Gorman ALF, McReynolds JS (1978) Ionic effects on the membrane potential of hyperpolarizing photoreceptor in scallop retina. J Physiol 275(1):345–355

Gutsell JS (1930) Natural history of the bay scallop (*Pecten irradians*). Bull US Bur Fish 46:569–632

Hall-Spencer J, Moore P (2000) *Limaria hians* (Mollusca: Limacea): a neglected reef-forming keystone species. Aquat Conserv 10(4):267–277

Hamilton PV, Koch KM (1996) Orientation toward natural and artificial grassbeds by swimming bay scallops, *Argopecten irradians* (Lamarck, 1819). J Exp Mar Biol Ecol 199(1):79–88

Harris OK, Kingston AC, Wolfe CS, Ghoshroy S, Johnsen S, Speiser DI (2019) Core–shell nanospheres behind the blue eyes of the bay scallop Argopecten irradians. J R Soc Interface 16(159):20190383

Hartline HK (1938) The discharge of impulses in the optic nerve of *Pecten* in response to illumination of the eye. J Cell Comp Physiol 11(3):465–478

Heath H (1941) The anatomy of the pelecypod family Arcidae. Am Phil Soc Trans 31(5):287–319

Hesse R (1901) Untersuchungen über die Organe der Lichtempfindung bei niederen Thieren. VI. Die Augen einiger Mollusken. Z wiss Zool 68:379–477

Horwitz J, Ding L, Vasiliou V, Cantore M, Piatigorsky J (2006) Scallop lens Ω-crystallin (ALDH1A9): A novel tetrameric aldehyde dehydrogenase. Biochem Biophys Res Commun 348(4):1302–1309

Kanmizutaru T, Anraku K, Toyoda S (2005) Light perception capability of pallial eyes in Japanese moon scallop *Amusium japonicum* as determined by electroretinogram. Nippon Suisan Gakkaishi 71(6):928–934

Kennedy D (1960) Neural photoreception in a lamellibranch mollusc. J Gen Physiol 44(2):277–299

Kingston ACN, Chappell DR, Miller HV, Lee SJ, Speiser DI (2017) Expression of G proteins in the eyes and parietovisceral ganglion of the bay scallop *Argopecten irradians*. Biol Bull 233(1):83–95

Kingston ACN, Cronin TW (2016) Diverse distributions of extraocular opsins in crustaceans, cephalopods, and fish. Integr Comp Biol 56(5):820–833

Kojima D, Terakita A, Ishikawa T, Tsukahara Y, Maeda A, Shichida Y (1997) A novel Go-mediated phototransduction cascade in scallop visual cells. J Biol Chem 272(37):22979–22982

Krohn A (1840) Über augenähnliche Organe bei *Pecten* und *Spondylus*. Arch Anat Physiol Wiss Med 7:371–386

Land MF (1965) Image formation by a concave reflector in the eye of the scallop, *Pecten maximus*. J Physiol 179(1):138–153

Land MF (1966a) A multilayer interference reflector in the eye of the scallop, *Pecten maximus*. J Exp Biol 45(3):433–447

Land MF (1966b) Activity in the optic nerve of *Pecten Maximus* in response to changes in light intensity, and to pattern and movement in the optical environment. J Exp Biol 45(1):83–99

Land MF (1968) Functional aspects of the optical and retinal organization of the mollusc eye. Symp Zool Soc Lond 23:75–96

Land MF (2000) Eyes with mirror optics. J Opt A: Pure Appl Opt 2(6):R44–R50

Land MF (2003) The spatial resolution of the pinhole eyes of giant clams (*Tridacna maxima*). Proc R Soc Lond B Biol Sci 270(1511):185–188

Lemer S, González VL, Bieler R, Giribet G (2016) Cementing mussels to oysters in the pteriomorphian tree: a phylogenomic approach. Proc R Soc Lond B Biol Sci 283(1833):20160857

Malkowsky Y, Jochum A (2015) Three-dimensional reconstructions of pallial eyes in Pectinidae (Mollusca: Bivalvia). Acta Zool 96(2):167–173

McReynolds J (1976) Hyperpolarizing photoreceptors in invertebrates. In: Neural principles in vision, Springer, pp 394–409

McReynolds JS, Gorman ALF (1970a) Membrane conductances and spectral sensitivities of *Pecten* photoreceptors. J Gen Physiol 56(3):392–406

McReynolds JS, Gorman ALF (1970b) Photoreceptor potentials of opposite polarity in the eye of the scallop, *Pecten irradians*. J Gen Physiol 56(3):376–391

Mikkelsen PM, Bieler R (2003) Systematic revision of the western Atlantic file clams, *Lima* and *Ctenoides* (Bivalvia: Limoida: Limidae). Invertebr Syst 17(5):667–710

Miller HV, Kingston ACN, Gagnon YL, Speiser DI (2019) The mirror-based eyes of scallops demonstrate a light-evoked pupillary response. Curr Biol 29(9):R313–R314

Miller WH (1958) Derivatives of Cilia in the Distal Sense Cells of the Retina of *Pecten*. J Cell Biol 4(2):227–228

Moran D, Softley R, Warrant EJ (2015) The energetic cost of vision and the evolution of eyeless Mexican cavefish. Sci Adv 1(8):e1500363

Morton B (1979) A comparison of lip structure and function correlated with other aspects of the functional morphology of *Lima lima*, *Limaria* (*Platilimaria*) *fragilis*, and *Limaria* (*Platilimaria*) *hongkongensis* sp. nov. (Bivalvia: Limacea). Can J Zool 57(4):728–742

Morton B (1987) The pallial photophores of *Barbatia virescens* (Bivalvia: Arcacea). J Molluscan Stud 53(2):241–243

Morton B (2000) The pallial eyes of *Ctenoides floridanus* (Bivalvia: Limoidea). J Molluscan Stud 66(4):449–455

Morton B (2001) The evolution of eyes in the Bivalvia. Oceanogr Mar Biol 39:165–205

Morton B (2008) The evolution of eyes in the Bivalvia: New insights. Am Malacol Bull 26:35–45

Morton B, Peharda M (2008) The biology and functional morphology of *Arca noae* (Bivalvia: Arcidae) from the Adriatic Sea, Croatia, with a discussion on the evolution of the bivalve mantle margin. Acta Zool 89(1):19–28

Morton B, Puljas S (2016) The ectopic compound ommatidium-like pallial eyes of three species of Mediterranean (Adriatic Sea) *Glycymeris* (Bivalvia: Arcoida). Decreasing visual acuity with increasing depth? Acta Zool 97(4):464–474

Mpitosos G (1973) Physiology of vision in the mollusk *Lima scabra*. J Neurophysiol 36(2):371–383

Nasi E (1991) Electrophysiological properties of isolated photoreceptors from the eye of *Lima scabra*. J Gen Physiol 97(1):17–34

Nilsson D-E (1994) Eyes as optical alarm systems in fan worms and ark clams. Philos Trans R Soc Biol Lond 346(1316):195–212

Nilsson D-E (2013) Eye evolution and its functional basis. Vis Neurosci 30(1-2):5–20

Nilsson D-E, Arendt D (2008) Eye evolution: the blurry beginning. Curr Biol 18(23):R1096–R1098

Niven JE, Laughlin SB (2008) Energy limitation as a selective pressure on the evolution of sensory systems. J Exp Biol 211(11):1792–1804

Pairett AN, Serb JM (2013) De novo assembly and characterization of two transcriptomes reveal multiple light-mediated functions in the scallop eye (Bivalvia: Pectinidae). PLOS One 8(7):e69852

Palmer BA, Taylor GJ, Brumfeld V, Gur D, Shemesh M, Elad N, Osherov A, Oron D, Weiner S, Addadi L (2017) The image-forming mirror in the eye of the scallop. Science 358(6367):1172–1175

Patten W (1886) Eyes of molluscs and arthropods. Mitt Zool Stn Neapel 6:542–756

Pelseneer P (1911) Les lamellibranchs de l'expedition du Siboga, partie anatomique. Siboga Expedition Monograph 53:1–126

Petie R, Hall MR, Hyldahl M, Garm A (2016) Visual orientation by the crown-of-thorns starfish (*Acanthaster planci*). Coral Reefs 35(4):1139–1150

Piatigorsky J (2008) Evolution of mollusc lens crystallins: Glutathione S-transferase/S-crystallins and aldehyde dehydrogenase/Ω-crystallins. Am Malacol Bull 26:73–81

Piatigorsky J, Kozmik Z, Horwitz J, Ding L, Carosa E, Robison WG Jr, Steinbach PJ, Tamm ER (2000) Ω-Crystallin of the scallop lens: a dimeric aldehyde dehydrogenase class 1/2 enzyme-crystallin. J Biol Chem 275(52):41064–41073

Plachetzki DC, Degnan BM, Oakley TH (2007) The origins of novel protein interactions during animal opsin evolution. PLOS One 2(10):e1054

Porath-Krause AJ, Pairett AN, Faggionato D, Birla BS, Sankar K, Serb JM (2016) Structural differences and differential expression among rhabdomeric opsins reveal functional change after gene duplication in the bay scallop, *Argopecten irradians* (Pectinidae). BMC Evol Biol 16(1):1–15

Ramirez MD, Speiser DI, Pankey MS, Oakley TH (2011) Understanding the dermal light sense in the context of integrative photoreceptor cell biology. Vis Neurosci 28(4):265–279
Richter S, Loesel R, Purschke G, Schmidt-Rhaesa A, Scholtz G, Stach T, Vogt L, Wanninger A, Brenneis G, Döring C (2010) Invertebrate neurophylogeny: suggested terms and definitions for a neuroanatomical glossary. Front Zool 7(1):1–49
Salvini-Plawen LV (2008) Photoreception and the polyphyletic evolution of photoreceptors (with special reference to Mollusca). Am Malacol Bull 26:83–100
Serb JM, Porath-Krause AJ, Pairett AN (2013) Uncovering a gene duplication of the photoreceptive protein, opsin, in scallops (Bivalvia: Pectinidae). Integr Comp Biol 53(1):68–77
Smedley GD, Audino JA, Grula C, Porath-Krause A, Pairett AN, Alejandrino A, Lacey L, Masters F, Duncan PF, Strong EE, Serb JM (2019) Molecular phylogeny of the Pectinoidea (Bivalvia) indicates Propeamussiidae to be a non-monophyletic family with one clade sister to the scallops (Pectinidae). Mol Phylogenetics Evol 137:293–299
Spagnolia T, Wilkens LA (1983) Neurobiology of the scallop. II. Structure of the parietovisceral ganglion lateral lobes in relation to afferent projections from the mantle eyes. Mar Behav Physiol 10(1):23–55
Speiser DI, Gagnon YL, Chhetri RK, Oldenburg AL, Johnsen S (2016) Examining the effects of chromatic aberration, object distance, and eye shape on image-formation in the mirror-based eyes of the bay scallop *Argopecten irradians*. Integr Comp Biol 56(5):796–808
Speiser DI, Johnsen S (2008a) Comparative morphology of the concave mirror eyes of scallops (Pectinoidea). Am Malacol Bull 26:27–33
Speiser DI, Johnsen S (2008b) Scallops visually respond to the size and speed of virtual particles. J Exp Biol 211(13):2066–2070
Speiser DI, Loew ER, Johnsen S (2011) Spectral sensitivity of the concave mirror eyes of scallops: potential influences of habitat, self-screening and longitudinal chromatic aberration. J Exp Biol 214(3):422–431
Speiser DI, Wilkens LA (2016) Neurobiology and behaviour of the scallop. In: Elsevier S (ed) Scallops: Biology, Ecology, Aquaculture, and Fisheries, SE Shumway and GJ Parsons. Diego, CA, pp 219–251
Stanley SM (1972) Functional morphology and evolution of byssally attached bivalve mollusks. J Paleontol 46:165–212
Stanley SM (1975) Adaptive Themes in the Evolution of the Bivalvia (Mollusca). Annu Rev Earth Planet Sci 3(1):361–385
Sumner-Rooney L, Sigwart JD, McAfee J, Smith L, Williams ST (2016) Repeated eye reduction events reveal multiple pathways to degeneration in a family of marine snails. Evolution 70(10):2268–2295
Tëmkin I (2006) Morphological perspective on the classification and evolution of Recent Pterioidea (Mollusca: Bivalvia). Zool J Linn Soc 148(3):253–312
Tremblay I, Samson-Dô M, Guderley HE (2015) When behavior and mechanics meet: scallop swimming capacities and their hinge ligament. J Shellfish Res 34(2):203–212
Tsubaki R, Kameda Y, Kato M (2011) Pattern and process of diversification in an ecologically diverse epifaunal bivalve group Pterioidea (Pteriomorphia, Bivalvia). Mol Phylogenetics Evol 58:97–104
Wagner H-J, Douglas RH, Frank TM, Roberts NW, Partridge JC (2009) A novel vertebrate eye using both refractive and reflective optics. Curr Biol 19(2):108–114
Waller TR (1980) Scanning electron microscopy of shell and mantle in the order Arcoida (Mollusca: Bivalvia). Smithson Contrib Zool 313:1–58
Whoriskey K, Whoriskey S, Whoriskey FG (2014) Systematic differences in eye numbers between the left and right valves of the sea scallop *Placopecten magellanicus*: an evolutionary response for visualizing the water column? J Shellfish Res 33(2):337–341
Wiederhold ML, MacNichol EF, Bell AL (1973) Photoreceptor spike responses in the hardshell clam, *Mercenaria mercenaria*. J Gen Physiol 61(1):24–55

Wilkens LA (1981) Neurobiology of the scallop. I. Starfish-mediated escape behaviours. Proc R Soc Lond B Biol Sci 211(1184):341–372

Wilkens LA, Ache BW (1977) Visual responses in the central nervous system of the scallop *Pecten ziczac*. Experientia 33(10):1338–1340

Wilkens LA (1984) Ultraviolet sensitivity in hyperpolarizing photoreceptors of the giant clam *Tridacna*. Nature 309(5967):446–448

Wilkens LA (1986) The visual system of the giant clam *Tridacna*: behavioral adaptations. Biol Bull 170(3):393–408

Wilkens LA (2008) Primary inhibition by light: a unique property of bivalve photoreceptors. Am Malacol Bull 26:101–109

Wistow GJ, Piatigorsky J (1988) Lens crystallins: the evolution and expression of proteins for a highly specialized tissue. Annu Rev Biochem 57(1):479–504

Chapter 6
Distributed Light-Sensing Systems in Chitons

Daniel R. Chappell, Daniel I. Speiser, Douglas J. Eernisse, and Alexandra C. N. Kingston

Abstract Chitons (Mollusca; Polyplacophora) have thousands of sensory organs, termed aesthetes, embedded in their eight dorsal shell plates. Aesthetes are thought to be sensitive to light, as well as other types of external stimuli. In some chitons, the aesthetes are associated with clusters of photoreceptors and pigmented cells (eyespots). In other chitons, the aesthetes are interspersed with eyes that have lenses made of shell material (shell eyes). Chitons, as a group, thus demonstrate at least three distinct types of distributed light-sensing systems: aesthetes alone, aesthetes plus eyespots, and aesthetes plus shell eyes. Differences between their distributed light-sensing systems may help explain why chitons show differences between their light-influenced behaviors. Species that only have aesthetes, for example, do not show evidence of spatial resolution, whereas species with eyespots use spatial information about light to influence their locomotory behaviors, and species with shell eyes use spatial information about light to inform their defensive responses. In the following chapter, we will introduce chitons; detail the structures and functions of their aesthetes, eyespots, and shell eyes; present what is known of their light-influenced behaviors; and describe their cord-based nervous systems. We will conclude by asking questions about how chitons acquire, process, and act upon information about light and how and why their distributed light-sensing systems have evolved.

Keywords Neuroethology · Neuroanatomy · Eye evolution · Visual ecology · Photoreception · Light-influenced behavior · Mollusca · Polyplacophora

---

D. R. Chappell · D. I. Speiser
Department of Biological Sciences, University of South Carolina, Columbia, SC, USA

D. J. Eernisse
Department of Biological Science, California State University, Fullerton, CA, USA

A. C. N. Kingston (✉)
Department of Biological Science, University of Tulsa, Tulsa, OK, USA

E. Buschbeck, M. Bok (eds.), *Distributed Vision*, Springer Series in Vision Research, https://doi.org/10.1007/978-3-031-23216-9_6

6.1 Introduction to Chitons

Chitons (Mollusca; Polyplacophora) are distinguished from other mollusks by their eight dorsal shell plates (Fig. 6.1) (Eernisse 2007). These armored plates are mineralized with calcium carbonate, as aragonite, and they overlap so that chitons are well-protected from predators and yet still flexible enough to wedge into crevices or roll into balls (Connors et al. 2012; Sigwart et al. 2019). In living chitons, the shell plates (also termed valves) include a porous upper layer, termed the tegmentum, and a nonporous lower layer, termed the articulamentum (Carter and Hall 1990). The shell plates are encircled by a ring of muscular mantle tissue termed the girdle. The girdles of chitons may be smooth, adorned with hairs or bristles, or embedded with

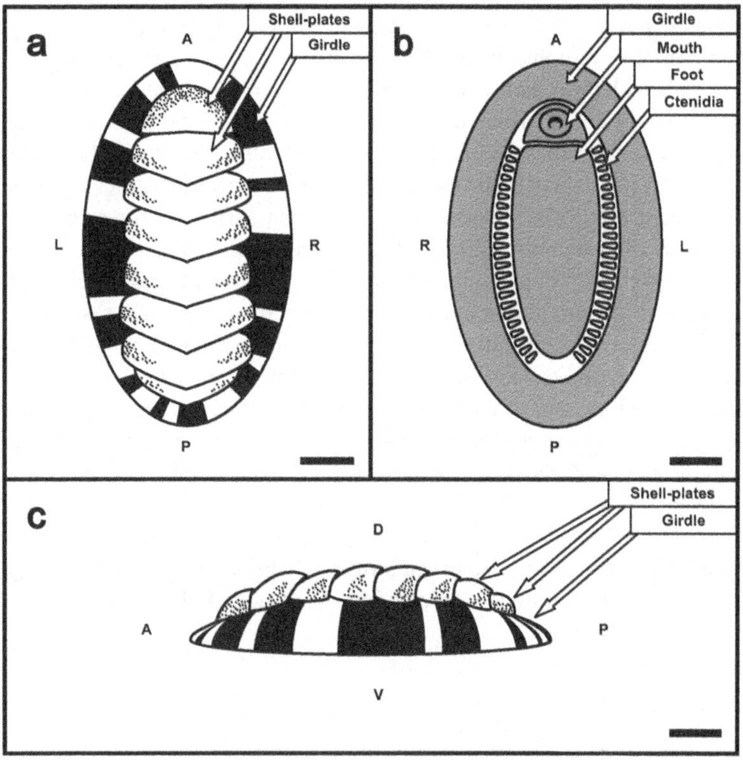

Fig. 6.1 An illustration of the external anatomy of the chiton *Acanthopleura granulata*. (**a**) Dorsal view showing the eight overlapping shell plates and the black and white striped girdle tissue that characterizes this species. The small black dots on the shell plates represent shell eyes. (**b**) Ventral view showing the centrally located foot, the pallial cavity surrounding the foot in which the ctenidia (gills) are located, the anteriorly located mouth, and the underside of the girdle. (**c**) Lateral view showing the eight overlapping shell plates and the surrounding girdle. Body axes are indicated as anterior-posterior (A-P), dorsal-ventral (D-V), and right-left (R-L). Scale bars (**a**–**c**) 500μm

calcified structures such as spines or overlapping scales (Treves et al. 2003; Connors et al. 2019).

Chitons, together with the worm-like aplacophorans (Solenogastres and Caudofoveata), form the clade Aculifera (Sigwart and Sutton 2007; Kocot et al. 2011; Smith et al. 2011; Vinther et al. 2012). Conchifera, the sister clade to Aculifera, includes all remaining mollusks (bivalves, cephalopods, gastropods, monoplacophorans, and scaphopods). Living chitons belong to a single subclass, Neoloricata, divided currently into the orders Lepidopleurida, Callochitonida, and Chitonida, the latter divided into the suborders Acanthochitonina and Chitonina (Fig. 6.2) (Sirenko 2006; Irisarri et al. 2020). The fossil record of Polyplacophora extends back to the Ordovician, but fossil records of extant genera extend back to the Cretaceous at the earliest (Sirenko 2006; Puchalski et al. 2008).

All chitons are benthic and marine, and most species live on rocks or other hard substrates. The majority of species live in intertidal or shallow subtidal habitats, but some are found in deep sea habitats including some of the deepest known trenches (Eernisse 2007). Chitons crawl using waves of muscle contractions that propagate longitudinally along their broad foot, and they respire using rows of gills (or ctenidia) that hang down from the roof of the pallial groove that separates their foot from their girdle (Fig. 6.1) (Yonge 1939; Eernisse and Reynolds 1994). When disturbed, chitons can clamp to their substrate by lowering their girdle, creating a seal, and then raising part of their foot to create suction (Crozier 1921). Like many mollusks, chitons feed using an organ termed a radula. The radulas of chitons have many separate rows of teeth (25–150 rows, usually with 17 pairs of teeth per row). Chitons mineralize the cores of their teeth with calcium phosphate, as apatite, and coat the cusps of their largest pairs of teeth (the major laterals) with iron oxides, including magnetite (Lowenstam 1962; Kirschvink and Lowenstam 1979; Brooker and Shaw 2012). Chitons use their iron-coated, self-sharpening teeth to scrape edible material from the hard surfaces on which they live and feed (Bullock 1988; Shaw et al. 2010).

Chitons have thousands of innervated, multicellular organs termed aesthetes (or esthetes) embedded within the porous tegmental layers of their shell plates (Fig. 6.3a) (Moseley 1885). In many species, the aesthetes fill a substantial volume of the tegmentum (Vendrasco et al. 2008). Aesthetes are found in the shell plates of all extant and fossil chitons (Puchalski et al. 2008), and they are absent from the valves of any extant or fossil conchiferan mollusks, such as bivalves or gastropods. Consequently, aesthetes are considered a synapomorphic trait of Polyplacophora.

Aesthetes have long been thought to have a variety of secretory and sensory functions, including sensitivity to light (Moseley 1885; Blumrich 1891; Nowikoff 1907; Arey and Crozier 1919). Some chitons, including those in two distantly related genera, *Chiton* (Nowikoff 1909; Haas and Kriesten 1978) and *Callochiton* (Baxter et al. 1990; Sturrock and Baxter 1995), have clusters of photoreceptors and pigmented cells ("eyespots") attached to their aesthetes (Fig. 6.3b). Other chitons have eyes with lenses made of shell material ("shell eyes") interspersed with their aesthetes (Fig. 6.3c) (Moseley 1885; Boyle 1969a; Speiser et al. 2011; Li et al. 2015). Species with shell eyes include those in the subfamilies Acanthopleurinae

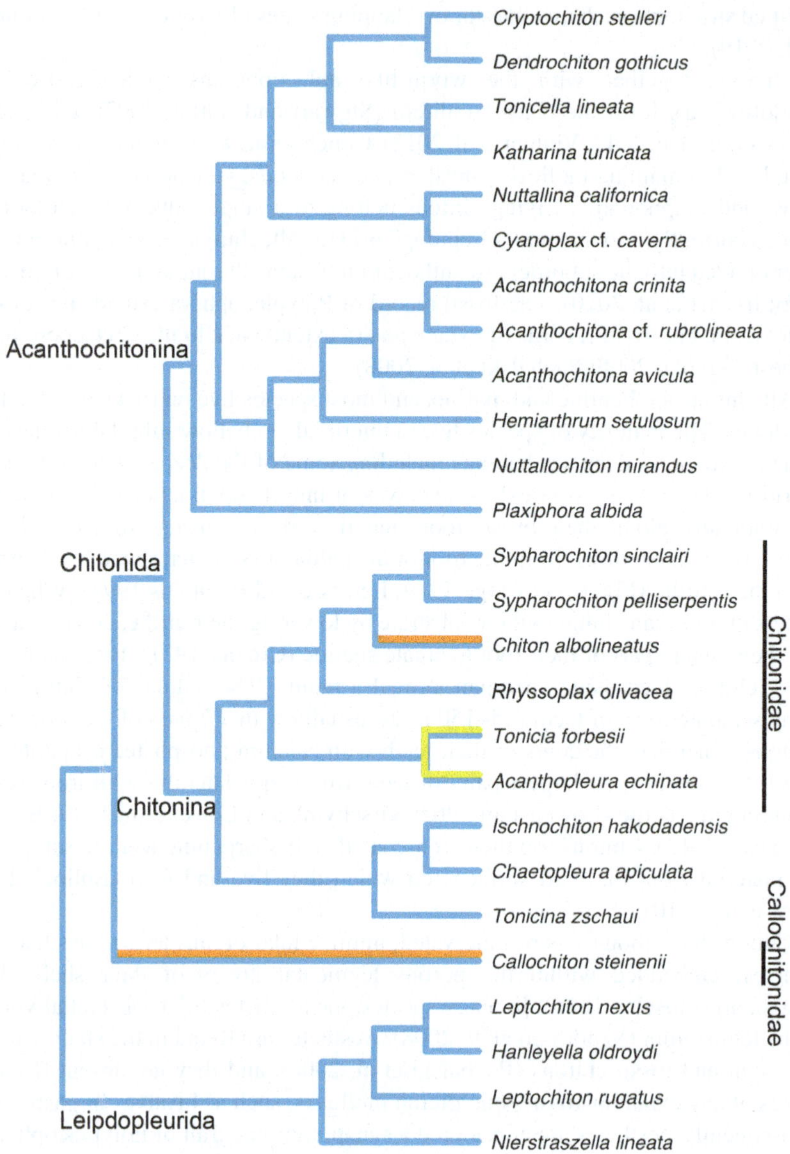

Fig. 6.2 A phylogeny of chitons (adapted from Irisarri et al. 2020) illustrating the distribution of aesthetes, eyespots, and eyes across the group. Blue terminal branches indicate species known to have aesthetes (which includes all fossil and extant chitons). Orange terminal branches indicate species from genera known to have eyespots. Yellow terminal branches indicate species from genera known to have shell eyes. From the phylogenetic relationships shown here, we infer aesthetes are an ancestral trait of living chitons, eyespots evolved independently from aesthetes at least twice (once in Chitonidae and once in Callochitonidae), and eyespots and shell eyes independently evolved from aesthetes in Chitonidae. Shell eyes are also present in *Schizochiton* (not represented in the phylogeny), but the phylogenetic position of this genus has been difficult to resolve due to its highly divergent DNA sequences

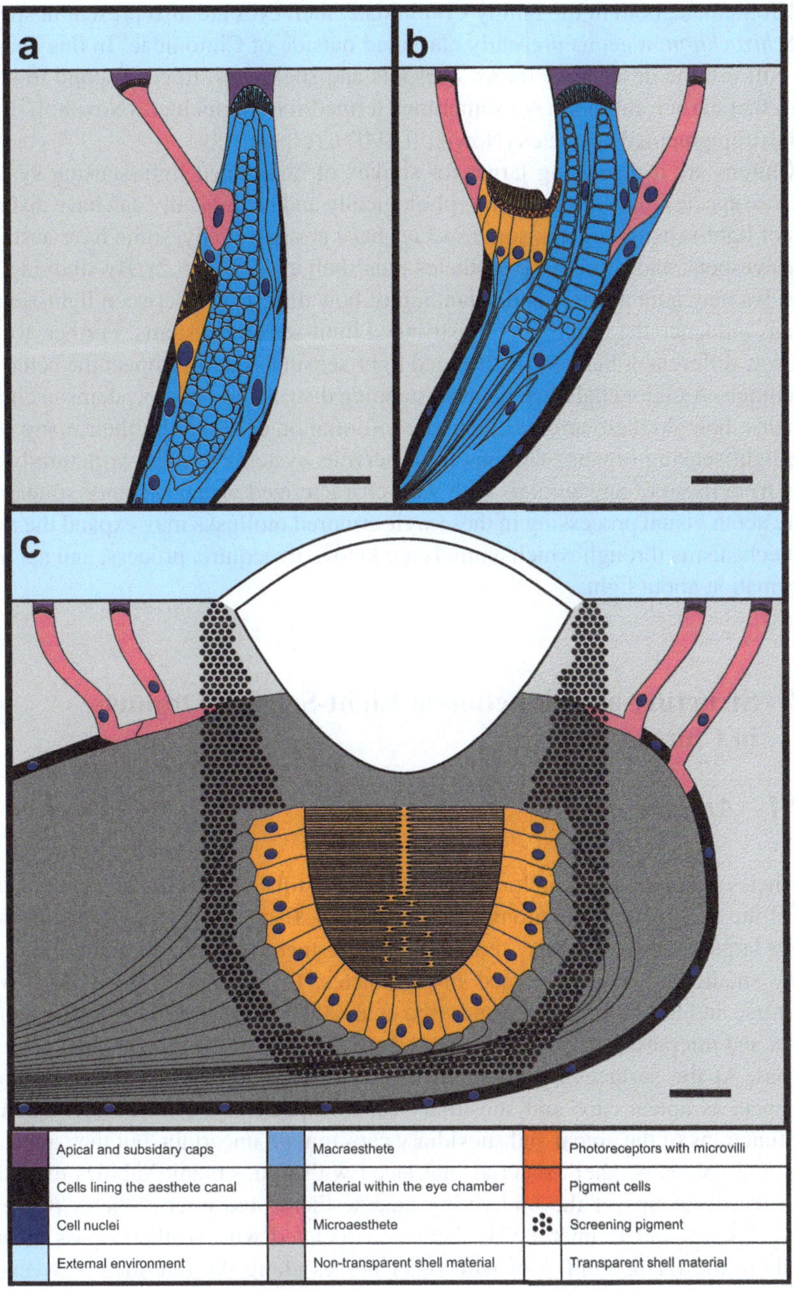

Fig. 6.3 Sensory structures embedded within the shell plates of chitons. (**a**) An aesthete similar to those found in all chitons, drawn after *Lepidochitona cinerea* (adapted from Boyle 1974). (**b**) An aesthete with an attached eyespot (= intrapigmented aesthete) drawn after *Chiton marmoratus* (adapted from Haas and Kriesten 1978). (**c**) A shell eye (= extrapigmented aesthete) drawn after *Acanthopleura granulata* (adapted from Speiser et al. 2011 and Li et al. 2015). The structures of these sensory organs are color-coded as indicated in the figure legend. Scale bars (**a–c**) 10 μm

and Toniciinae, both in the family Chitonidae. Shell eyes are also present in species of *Schizochiton*, a genus presently classified outside of Chitonidae. In this chapter, we will use the descriptive terms, eyespots and shell eyes, to correspond to structures that earlier authors have sometimes termed intrapigmented (Nowikoff 1909) and extrapigmented aesthetes (Nowikoff 1907), respectively.

Chitons are a promising target for studies of distributed light-sensing systems because species that are similar morphologically and ecologically can have different sets of light-sensing organs: some species have aesthetes only, some have aesthetes plus eyespots, and some have aesthetes plus shell eyes (Fig. 6.2). By studying chitons, we may gain a better understanding of how differences between light-sensing organs influence the functions of distributed light-sensing systems. Further, we can ask how differences between distributed light-sensing systems impact the behaviors of animals. A major challenge to understanding distributed visual systems in chitons remains: how do these animals integrate information collected by their many separate light-sensing organs? Chitons have nervous systems that are structurally distinct from those of any animals with well-characterized visual systems, so learning more about visual processing in these well-armored mollusks may expand the range of mechanisms through which animals are known to acquire, process, and act upon information about light.

6.2 Structure and Function of Light-Sensing Organs in Chitons

6.2.1 Aesthetes

Aesthetes are innervated, multicellular organs that fill channels in the tegmental layers of the shell plates of chitons (Moseley 1885). Each aesthete consists of a relatively large central structure, termed a macraesthete (or megalaesthete), and the many smaller structures, termed micraesthetes, that branch from it (Fig. 6.3a) (Eernisse and Reynolds 1994; Vendrasco et al. 2008; Schwabe 2010). The macraesthetes and micraesthetes are unpigmented and found across all eight shell plates of chitons. At the surfaces of the shell plates, the macraesthetes and micraesthetes terminate as apical caps and subsidiary caps, respectively. The precise structures and functions of the apical and subsidiary caps remain uncertain, but they appear to be porous plugs of shell material interlaced with thin, organic tubules extending from the distal tips of the underlying macraesthetes and micraesthetes (Omelich 1967; Fischer and Renner 1978). Nerves associated with aesthetes pass through canals in the tegmentum, join with other nerves along the interface between the tegmentum and articulamentum, and then exit shell plates through lateral or terminal slits in the articulamentum (Moseley 1885; von Knorre 1925; Boyle 1974; Fischer 1988). Across species, aesthetes vary with regard to size, packing density, and patterns of distribution within shell plates (e.g., Boyle 1976; Sturrock and Baxter 1993; Fernandez et al. 2007; Vendrasco et al. 2008).

Morphological evidence suggests aesthetes are multifunctional organs (Haas and Kriesten 1978; Fischer 1979, 1988). The macraesthetes of chitons tend to contain long, thin cells that project microvilli (and sometimes cilia) toward the surfaces of the shell plates in which they are embedded (e.g., Boyle 1974). There is general consensus that these cells have sensory functions, but the stimuli to which they are sensitive have yet to be established experimentally. Authors have speculated they may function as mechanoreceptors (e.g., Moseley 1885), chemoreceptors (e.g., Baxter et al. 1987, 1990), or photoreceptors (e.g., Blumrich 1891; Boyle 1972; Fischer and Renner 1978). It is possible that some or all of these predictions are correct: aesthetes may vary in function between species and may respond to cues associated with one or more sensory modalities. Along with sensory cells, macraesthetes tend to contain club-shaped, vesicle-packed cells proposed to have secretory functions (Baxter et al. 1987, 1990).

In many species, one to several cells with photoreceptor-like morphology are attached to the sides of macraesthetes (Fischer and Renner 1978). These photoreceptors project microvilli and in some cases cilia as well (Fischer and Renner 1978; Eernisse and Reynolds 1994). Unlike the sensory cells in the central bodies of macraesthetes, the photoreceptors on the sides of macraesthetes lie beneath solid shell material so that their microvillar and ciliary projections are not exposed to the external environment. Photoreceptors are associated with at least some of the macraesthetes from some of the shell plates in all examined species that lack eyespots or eyes. These species span the phylogenetic diversity of living chitons: *Lepidopleurus cajetanus* (Fischer 1988) and *Leptochiton asellus* (Sturrock and Baxter 1993) in Lepidopleurida; *Acanthochitona fascicularis* (Fischer 1979), *Lepidochitona cinerea* (Boyle 1974), and *Tonicella marmorea* in Acanthochitonina (Baxter et al. 1987; Sturrock and Baxter 1993); and *Rhyssoplax olivacea* (as *Chiton olivaceus*) in Chitonina (Fischer and Renner 1978).

6.2.2 Eyespots

In some chitons, including species in Callochitonidae (Baxter et al. 1990; Sturrock and Baxter 1995) and Chitonidae (Haas and Kriesten 1978), a subset of macraesthetes have eyespots attached to them (Figs. 6.3b and 6.4a). Given the distant phylogenetic relationship between these families (Irisarri et al. 2020), it is likely that eyespots have evolved independently at least twice in chitons. The eyespots of chitons lie beneath 15–20μm of shell material, and they include two types of cells: photoreceptors and cells packed with pigment granules (Haas and Kriesten 1978; Baxter et al. 1990; Sturrock and Baxter 1995). The cell bodies of the photoreceptors lie underneath or alongside the pigmented cells. Bundles of microvilli (and

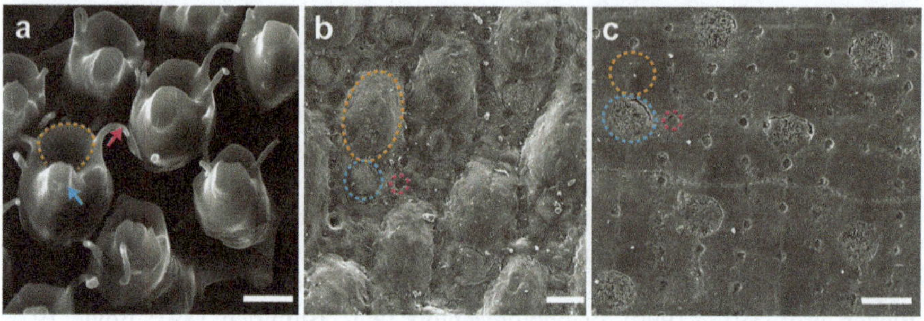

Fig. 6.4 Structures associated with the eyespots of *Chiton*. (**a**) Resin casts of aesthete canals in the shell plates of *Chiton viridis*. In (**a**), the blue arrow indicates the superficial end of a macraesthete canal, the pink arrow indicates the superficial end of a micraesthete canal, and the dashed orange circle indicates the position of an eyespot. (**b**) and (**c**) Scanning electron micrographs of the surfaces of shell plates from *Chiton tuberculatus* and *Chiton marmoratus*, respectively. In (**b**) and (**c**), the dashed blue circles indicate apical caps (associated with macraesthetes), the dashed pink circles indicate subsidiary caps (associated with micraesthetes), and the dashed orange circles indicate areas of transparent shell material overlying eyespots. Scale bars (**a**–**c**) 20μm

sometimes cilia) project from the photoreceptors and pass through gaps between the pigmented cells so that a layer of pigment separates the sensory regions of photoreceptors (their microvilli and/or cilia) from their cell bodies (Haas and Kriesten 1978; Baxter et al. 1990; Sturrock and Baxter 1995). The pigment cells associated with eyespots are influenced by light conditions: the eyespots of *Callochiton septemvalvis* (as *C. achatinus*), for example, become more heavily pigmented after animals are exposed continuously to bright light for several weeks (Sturrock and Baxter 1995).

The eyespots of chitons are tiny, measuring only ~20μm in *C. septemvalvis* (Baxter et al. 1990) and ~ 35μm in *Chiton tuberculatus* (Kingston et al. 2018). The number of cells per eyespot can vary between species: the eyespots of *C. septemvalvis* consist of 1 photoreceptor and several pigmented cells (Baxter et al. 1990; Sturrock and Baxter 1995), whereas those of *Chiton marmoratus* and *C. tuberculatus* contain around 20 cells in total, including multiple cells of each type (Haas and Kriesten 1978; Kingston et al. 2018). The eyespots of chitons tend to be numerous and densely packed. In *C. septemvalvis*, *C. marmoratus*, and *C. tuberculatus*, thousands of them are distributed across the shell plates (Sturrock and Baxter 1995; Kingston et al. 2018). In *C. tuberculatus* and *C. marmoratus*, they are as dense as 400 per mm^2 and separated on the curved shell plates by angles as narrow as 0.5° (Kingston et al. 2018).

The shell material overlying the eyespots may be modified to provide light to the photoreceptors below (Haas and Kriesten 1978; Baxter and Jones 1984; Kingston et al. 2018). In *C. marmoratus* and *C. tuberculatus*, the shell material above the eyespots is transparent and nonporous (Kingston et al. 2018). Intriguingly, the external curvature of this transparent shell material varies between species in ways that may influence the focusing of light (Kingston et al. 2018). In *C. tuberculatus*, the transparent regions of the shell have convex external surfaces and so may function as light-focusing structures (Fig. 6.4b); in *C. marmoratus*, however, the transparent regions have flat external surfaces and may only function as windows (Fig. 6.4c). Three-dimensional reconstruction followed by optical modeling will be necessary to tell if morphological differences between the lens-like structures in the shell plates of *C. tuberculatus* and *C. marmoratus* correspond to meaningful functional differences.

6.2.3 Shell Eyes

In some chitons, the aesthetes are interspersed with hundreds to thousands of shell eyes (Moseley 1885; Nowikoff 1907; Boyle 1969a; Speiser et al. 2011; Li et al. 2015). Species with shell eyes include those in the Chitonidae subfamilies Acanthopleurinae (Nowikoff 1907) and Toniciinae (Boyle 1969a), as well as those in the enigmatic genus *Schizochiton* (Moseley 1885). The shell eyes of chitons each includes a retina, a layer of screening pigment, and a lens (Fig. 6.3c). Under each lens, a retina resides in a chamber derived from an enlarged aesthete canal and filled partially with pillars of intrachamber calcified material (Li et al. 2015). These retinas, described in detail for *Acanthopleura granulata* (Speiser et al. 2011) and *Onithochiton neglectus* (Boyle 1969b), are composed of ~100 photoreceptors that project bundles of microvilli ~8μm long and sometimes cilia as well. The pigment associated with shell eyes is extracellular and incorporated into the shell material surrounding each retina. In *A. granulata*, the red-brown screening pigment associated with the shell eyes is a pheomelanin (Speiser et al. 2014). Unlike most biological lenses, which tend to be protein-based, these shell eye lenses are made of aragonite. Like other aragonitic structures, these lenses are birefringent (Speiser et al. 2011). In *A. granulata*, the axis at which lenses have a single refractive index (the *c* axis) does not align with the axis through which light travels to reach the retinas (the optical axis); as a result, the lenses focus double images on to the underlying retinas (Li et al. 2015).

Shell eyes vary in size and in their patterns of distribution across valves. For example, the shell eyes of *O. neglectus* and *Tonicia lebruni* have diameters of ~40–45μm (Boyle 1969b), those of *A. granulata* reach diameters of ~80μm (Speiser et al. 2011; Li et al. 2015), and those of *Schizochiton incisus* may be as large as ~145μm (Moseley 1885). In some genera, such as *Tonicia* and *Onithochiton*, the eyes of younger and smaller chitons are spaced regularly (Fig. 6.5a), and animals maintain a relatively ordered arrangement of eyes as they grow and add new eyes to their valves (Boyle 1969b; Sigwart and Sumner-Rooney 2021). In other species, such as *A. granulata*, the eyes are relatively ordered when animals are young (Fig. 6.5b) and then become less ordered as animals add new eyes to the edges of their growing shell plates. We have yet to learn how differences between the sizes and distributions of shell eyes impact the functions of the distributed visual systems to which they contribute.

Fig. 6.5 The distributions of shell eyes across the valves of (**a**) *Tonicia schrammi* and (**b**) a young specimen of *Acanthopleura granulata*. Both animals are displayed anterior up. The eyes on the anterior and posterior valves of both animals are ordered irregularly, while the eyes on the intermediate valves are ordered in a more regular manner. In (**a**) and (**b**), the white arrows point to shell eyes. Scale bars (**a**, **b**) 500μm

6.2.4 Other Light-Sensing Organs in Chitons

Aesthetes, eyespots, and shell eyes are not the only light-sensing structures reported from chitons. Larval chitons have a pair of ocelli with microvillous photoreceptors (Kowalevsky 1883; Heath 1904; Rosen et al. 1979; Fischer 1980; Henry et al. 2004). These larval ocelli have drawn interest from developmental biologists because they differ from those of other studied mollusks in both position and cell lineage. The larval ocelli of chitons are located posterior to the prototroch and develop from second-quartet micromeres, whereas the larval ocelli of conchiferan mollusks tend to be located anterior to the prototroch and develop from first-quartet micromeres (Henry et al. 2004). The larval ocelli persist through metamorphosis, eventually disappearing from view in juveniles as the overlying shell plates thicken (Eernisse 1988). Species in Lepidopleurida, such as *L. asellus*, may retain their larval ocelli into adulthood as small, pigmented sensory organs located near their mouths (Sigwart et al. 2014). Behavioral experiments indicate these structures, termed Schwabe organs, contribute to *L. asellus* detecting and avoiding artificial sources of upwelling light (Sumner-Rooney and Sigwart 2015). Additionally, diverse sensory structures, including photoreceptor-like cells, have been identified in the girdles of chitons (Fischer 1980; Leise 1986, 1988; Checa et al. 2017).

6.3 Light-Influenced Behaviors in Chitons

6.3.1 Light-Influenced Behaviors Observed Across Chitons

Most chitons demonstrate two types of behavior that require light sensitivity: light-guided locomotion and defensive clamping responses (Omelich 1967; Boyle 1972; Speiser et al. 2011). Many chitons crawl away from bright light, and aesthetes appear to contribute to this light-guided locomotory behavior (Omelich 1967). For example, *Ischnochiton maorianus*, a chiton that lacks eyespots or eyes, was found to crawl away from a light source more slowly when its shell plates were abraded, painted, or covered than when its shell plates were unaltered (Boyle 1972). Further, all chitons studied to date respond defensively to the sudden dimming of light by clamping to their substrate (Arey and Crozier 1919; Boyle 1972; Speiser et al. 2011; Kingston et al. 2018). Photoreceptors associated with aesthetes may contribute to these responses, as may the photoreceptor-like sensory cells that have been identified in the girdles of some species (Fischer 1980; Leise 1986, 1988; Checa et al. 2017).

6.3.2 Light-Influenced Behaviors in Chitons with Eyespots

Chitons with eyespots use spatial information about light to influence their locomotory behaviors. In behavioral trials, *C. tuberculatus* oriented their bodies to dark visual stimuli in their lateral field of view (Kingston et al. 2018). By doing so, *C. tuberculatus* demonstrated behavior consistent with spatial vision (i.e., the ability of an animal to locate visual cues). The visual stimuli to which *C. tuberculatus* oriented had angular sizes as small as 10°, which likely requires spatial resolution finer than the individual eyespots these animals can provide (Kingston et al. 2018). The eyespots of *C. tuberculatus* are separated by angles as narrow as ~0.5°, so these chitons may acquire spatial information by comparing visual input between adjacent eyespots (Kingston et al. 2018). If so, the distributed visual system of *C. tuberculatus* may function like a compound eye in which the eyespots fill the functional roles of ommatidia.

Field observations support the possibility that chitons with eyespots engage in shelter-seeking behaviors informed by spatial information about light. Populations of *C. tuberculatus* forage at night and return to shaded rock refuges at sunrise. Individual *C. tuberculatus* have been observed traveling orthogonally to the rays of the rising sun instead of crawling directly away from them, as would be predicted if these animals were engaged in negative phototaxis (Crozier 1921). Further, *C. tuberculatus* relocated from their home rocks were found to travel consistently to the dark refuges nearest to them, supporting the importance of visual cues to the shelter-seeking behaviors of this species (Crozier 1921).

Spatial information about light does not appear to influence the defensive clamping responses of chitons with eyespots. In behavioral trials, two species of *Chiton* with eyespots, *C. marmoratus* and *C. tuberculatus*, did not distinguish between the appearances of dark objects against light backgrounds and the uniform dimming of the overhead light field (Kingston et al. 2018). However, *C. tuberculatus* has been observed responding to a moving shadow cast by a fly 2 m away (Arey and Crozier 1919), so it remains possible that the light-influenced defensive reactions of at least some chitons with eyespots are informed by spatial cues.

6.3.3 Light-Influenced Behaviors in Chitons with Shell Eyes

Unlike any other chitons examined thus far, those with shell eyes use spatial information about light to inform their defensive clamping responses. The eye-bearing chiton *A. granulata*, for example, is more likely to respond defensively to sudden appearances of dark overhead objects against light backgrounds than to equivalent, uniform decreases in downwelling irradiance (Speiser et al. 2011). The angular resolution demonstrated by *A. granulata* in these behavioral trials (~9°) is consistent with the angular resolution predicted by ray-tracing simulations for individual shell eyes (Speiser et al. 2011; Li et al. 2015). Other species with shell eyes, such as *O. neglectus*, display defensive responses to the dimming of light, but it is not known if these responses are informed by spatial information (Boyle 1969b). We also have yet to learn if chitons with shell eyes use spatial information to influence their light-guided locomotory behaviors.

Intriguingly, *A. granulata* demonstrates similar angular resolution in behavioral trials in which they are submerged in water and those in which they are surrounded by air (Speiser et al. 2011). Being able to see equally well in water and air would benefit an intertidal species like *A. granulata*, but it is challenging for animals to build eyes that work well in both conditions. Optical systems well-suited for water tend to be overfocused for air and those well-suited for air tend to be under-focused for water (Land and Nilsson 2012). The birefringent lenses of *A. granulata* may address this problem by simultaneously forming two images. If one of the images falls on the retina in water and the other image falls on the retina in air, it may explain why the distributed visual system of *A. granulata* appears to function equally well in both media (Speiser et al. 2011; Li et al. 2015). Birefringent lenses may be useful for intertidal animals, but they have drawbacks: if a retina receives both an in-focus image and an out-of-focus image, the scene being viewed by the eye will be perceived with lower clarity and contrast. We hope future studies will address whether or not birefringence is an adaptive feature of the aragonite lenses of intertidal chitons.

6.4 Neuroanatomy of Chitons

Most animals process spatial information about light using distinct, specialized structures in their central nervous systems. Cephalopods, for example, have large optic lobes associated with each of their eyes that can account for up to two-thirds of the total mass of their central nervous system (Kerbl et al. 2013; Shigeno et al. 2018). Likewise, arthropods process spatial information from each of their eyes using a series of optic neuropils including the lamina and medulla and, if present, additional neuropils like the lobula and lobula plate. These optic neuropils can account for up to three quarters of the neurons in an arthropod's brain (Strausfeld 1976; Homberg 2020).

Given their neuroanatomy, chitons may process visual information in dissimilar ways from any animal with a well-characterized visual system. Unlike vertebrates, cephalopods, and arthropods, chitons have nervous systems that lack prominent ganglia and thus lack the canonical optic lobes usually associated with eyes (Hubrecht 1882; Sumner-Rooney and Sigwart 2018). Instead of having the majority of neurons in their central nervous system condensed into ganglia, chitons have nervous systems composed primarily of medullary cords (Faller et al. 2012; Sumner-Rooney and Sigwart 2018). Like ganglia, medullary cords have outer layers of neuronal somata and inner cores of neuropil, but cord-based nervous systems differ structurally from those based on ganglia. In ganglia-based nervous systems, neuronal somata are concentrated within discrete ganglia, whereas in cord-based nervous systems, neuronal somata are less concentrated because they are distributed lengthwise along medullary cords (Richter et al. 2010).

Three concentric loops of neuropil are the most prominent components of the cord-based nervous systems of chitons (Fig. 6.6). These loops include the lateral (= pallial), ventral (= pedal), and cerebral neuropil. The lateral and ventral neuropil loops are present as medullary cords through most of the bodies of chitons. When they are not in contact with other neuropil loops, they are referred to as the lateral and ventral nerve cords, respectively. The lateral neuropil, the outermost loop, travels along the roof of the pallial cavity for most of its longitudinal path. The ventral neuropil, the middle loop, follows the lateral margins of the foot. The cerebral neuropil, the innermost loop, is part of the circumesophageal nerve ring that circles the mouth. The lateral and ventral neuropils join with the cerebral neuropil at the lateral margins of the circumesophageal nerve ring, such that the anterior half of the circumesophageal nerve ring is comprised of three layers of neuropil divided by layers of neuronal somata (Sumner-Rooney and Sigwart 2018).

Along with the prominent medullary cords, the nervous systems of chitons include small ganglia and numerous peripheral nerves (Fig. 6.6). The circumesophageal nerve ring connects to two small pairs of ganglia associated with the radula: the subradular ganglia and the buccal ganglia (Eernisse and Reynolds 1994). The ventral nerve cords are linked to each other and to the lateral nerve cords by numerous lateral connections (i.e., commissures). These periodic lateral connections between longitudinal medullary cords give the chiton nervous system a lattice-like

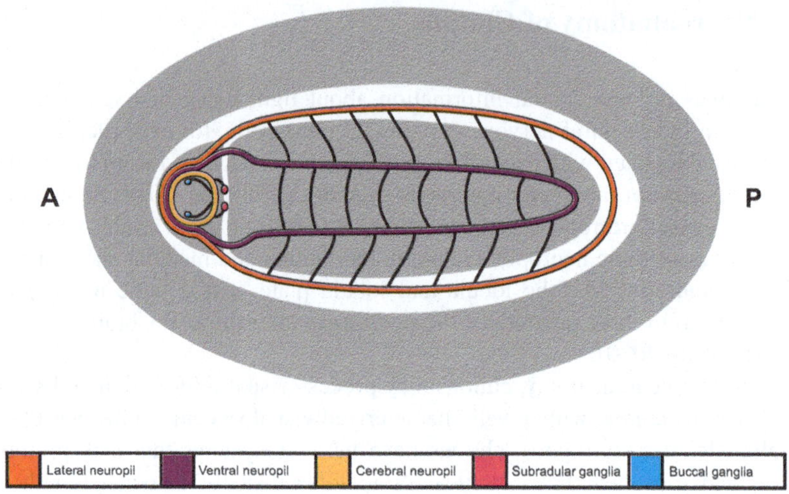

Fig. 6.6 A schematic of the chiton nervous system, shown from a ventral view (adapted from Sumner-Rooney and Sigwart 2018). The girdle, foot, and mouth are displayed as gray silhouettes. The commissures connecting the lateral and ventral nerve cords are shown in black. Otherwise, the structures are color-coded as indicated in the figure legend. Body axis is indicated as anterior-posterior (A-P)

appearance (Faller et al. 2012; Sumner-Rooney and Sigwart 2018). After exiting the shell plates, nerves associated with aesthetes join the lateral neuropil. Beyond this point, we have yet to learn where these nerves travel or to which structures they make synaptic connections. Until we solve this mystery, we will not know where or how chitons process sensory information from their aesthetes, eyespots, and shell eyes (Moseley 1885; von Knorre 1925; Boyle 1974; Fischer 1988).

6.5 Function and Evolution of Distributed Visual Systems in Chitons

6.5.1 How Do Light-Sensing Structures Relate to Light-Influenced Behaviors in Chitons?

Chitons with eyespots or shell eyes demonstrate visual abilities that are not observed in species that only have aesthetes. By orienting to spatial cues (Kingston et al. 2018), *C. tuberculatus*, a species with eyespots, demonstrates both spatial resolution (i.e., the ability to detect spatial cues) and spatial vision (i.e., the ability to locate spatial cues). Species that only have aesthetes are light-responsive, but they have yet to demonstrate either spatial resolution or spatial vision (Boyle 1972; Speiser et al. 2011). Enhancing phototactic behavior with spatial input may be useful to *C. tuberculatus* and other chitons with eyespots because it helps them locate dark refuges in structurally complex environments in which ambient light gradients are unreliable.

By distinguishing between the appearances of dark overhead objects and equivalent, uniform decreases in light levels, *A. granulata*, a chiton with shell eyes, demonstrates behaviors consistent with spatial resolution, but not necessarily spatial vision. Spatial resolution in the absence of spatial vision is thought to be a property of "burglar alarm" visual systems (Nilsson 1994). These sorts of visual systems help animals detect threats by making it possible for them to perceive the movements of small objects and distinguish shadows cast by objects from uniform changes in light conditions. They do not, however, make it possible for an animal to locate an object relative to its own position.

It is curious that *C. tuberculatus*, a chiton with eyespots, demonstrates spatial vision in its shelter-seeking behaviors, but does not distinguish between the appearances of overhead objects and uniform changes in light levels (Kingston et al. 2018). If defensive responses influenced by spatial information are indeed absent in chitons with eyespots, it suggests distributed visual systems incorporating eyespots and shell eyes may be associated with modifications to different types of light-influenced behaviors. Given available information, eyespots and shell eyes appear to be associated with modifications to shelter-seeking behaviors and defensive clamping responses, respectively. Further behavioral experiments will be necessary to understand how differences between the distributed light-sensing systems of chitons relate to differences between their light-influenced behaviors.

The evolution of eyespots and eyes in chitons, as in other animals, may be associated with functional trade-offs. Light-detecting organs with more numerous photoreceptors tend to provide finer-grained spatial resolution, but this may come at the cost of lower sensitivity because photoreceptors will tend to gather fewer photons as their receptive fields shrink (Nilsson 2013). Consequently, chitons that only have aesthetes may be more sensitive to small changes in light conditions than chitons with eyespots or eyes. The results of behavioral experiments on chitons have supported this prediction thus far. For example, the eyeless chitons *Ischnochiton maorianus* and *Ischnochiton rissoi* (as *Chiton rissoi*) were found to crawl away from a light source more rapidly and with greater precision than *O. neglectus*, a chiton with shell eyes (Boyle 1972). Further, the defensive responses of eyeless chitons have been found to be more sensitive than those with eyespots or eyes: *Chaetopleura apiculata*, an eyeless chiton, responds to decreases in irradiance as small as 1%, whereas chitons with eyespots, such as *C. marmoratus* and *C. tuberculatus*, and chitons with eyes, such as *A. granulata*, appear to only respond to decreases in irradiance of 5% or greater (Speiser et al. 2011; Kingston et al. 2018). These observations imply that the presence of eyespots or shell eyes in chitons may be associated with diminished photosensory function for the smaller yet potentially more sensitive aesthete organs.

6.5.2 Do Ecological Factors Help Explain Why Some Chitons Have Eyes When Many Do Not?

The intersection between ecological factors and material constraints may help explain the geographic distributions and habitat preferences of chiton species with eyespots and shell eyes. Like the surrounding shell material, the lenses of *A. granulata* are made of polycrystalline aragonite (Li et al. 2015). To minimize the scattering of light by the interfaces between them, the aragonite crystals in the lenses of *A. granulata* are larger and more highly aligned than those in the surrounding shell material, which tends to have a cross-lamellar microstructure (Li et al. 2015). These material properties make the lenses more transparent than the rest of the shell but also make them easier to fracture (Li et al. 2015). It is also apparent from observation that the lenses of chitons erode: in the shell plates of *A. granulata*, for example, it is common to see pigmented pits where eyes used to be located (Speiser et al. 2011). Because it is challenging for chitons (or any animal) to make macroscale, shell-based structures that are simultaneously transparent and resistant to fracture, we predict chitons with shell lenses may be uncommon in erosion-prone environments and show preference for microhabitats low in erosion-causing factors such as sand scour.

6.5.3 How Do Chitons Process Visual Information?

Given the computational demands of visual processing, one might expect to find neuroanatomical differences between chitons with eyespots or shell eyes and those that only have aesthetes. Despite this expectation, neuroanatomy appears to be relatively consistent across chitons (Sumner-Rooney and Sigwart 2018). If all chitons lack prominent ganglia, species with eyespots or shell eyes likely process information from their distributed visual systems using neuropil in their medullary cords. This presents an opportunity for discovery because we do not know how animals with cord-based nervous systems process visual information. Most organisms process visual information in ordered and layered central structures (e.g., optic lobes) in which an initial layer of neuropil integrates visual information topographically and subsequent layers extract higher-order visual information (e.g., shape recognition) that is then used to inform behavior. If chitons integrate visual information topographically and then extract higher-order visual information, we predict they do so using distributed visual-motor circuits. If we are correct, it may help explain why morphological differences between the light-sensing organs of chitons correspond to differences in light-influenced behaviors, but not obvious differences in neuroanatomy.

6.6 Future Directions

Along with ongoing morphological, physiological, and behavioral studies, molecular investigations will help us learn how the sensory systems of chitons function and how they have evolved. A species-level molecular phylogeny for Chitonidae, for example, will help us determine if the shell eyes of species within Acanthopleurinae and Toniciinae evolved from structures like the eyespots seen in species of Chitoninae or, alternately, if eyespots and eyes in Chitoninae evolved separately from aesthetes like those found across fossil and extant chitons. A better understanding of relationships among chiton families and genera will help us ask questions at broader taxonomic scales, such as did the shell eyes of *Schizochiton* evolve separately from those found in species within Acanthopleurinae and Toniciinae?

New genetic resources will help us predict the functions of the shell-embedded sensory organs of chitons by characterizing their molecular components. Transcriptome sequencing, for example, suggests the components of several different phototransduction pathways are expressed by cells within the shell plates of chitons: r-opsin and xenopsin are expressed in shell plate tissue from *A. granulata*, and a RGR/retinochrome-type opsin is expressed in corresponding tissue from both *Chiton virgulatus*, a species with aesthetes and eyespots, and *Stenoplax conspicua*, a species that only has aesthetes (Ramirez et al. 2017). Targeted studies of gene and protein expression will also be beneficial, such as recent work demonstrating the larval eyespots of *L. asellus* include photoreceptors that co-express r-opsin and xenopsin (Vöcking et al. 2015; Vöcking et al. 2017). Forthcoming molecular studies in chitons will be aided by new genomic resources, including a sequenced genome for *A. granulata* (Varney et al. 2021). Past and current work points toward chitons as a promising group in which to pursue an integrative understanding of the structure, function, and evolution of distributed light-sensing systems.

References

Arey LB, Crozier WJ (1919) The sensory responses of chiton. J Exp Zool 29:157–260. https://doi.org/10.1002/jez.1400290203

Baxter JM, Jones AM (1984) The valve morphology of *Callochiton achatinus* (Mollusca: Polyplacophora: Ischnochitonidae). J Zool 202:549–560. https://doi.org/10.1111/j.1469-7998.1984.tb05051.x

Baxter JM, Jones AM, Sturrock MG (1987) The ultrastructure of aesthetes in *Tonicella marmorea* (Polyplacophora; Ischnochitonina) and a new functional hypothesis. J Zool 211:589–604. https://doi.org/10.1111/j.1469-7998.1987.tb04473.x

Baxter JM, Sturrock MG, Jones AM (1990) The structure of the intrapigmented aesthetes and the properiostracum layer in *Callochiton achatinus* (Mollusca: Polyplacophora). J Zool 220:447–468. https://doi.org/10.1111/j.1469-7998.1990.tb04318.x

Blumrich J (1891) Das Integument der Chitonen. Z Wiss Zool 52:404–476

Boyle PR (1969a) Rhabdomeric ocellus in a chiton. Nature 222:895–896. https://doi.org/10.1038/222895a0

Boyle PR (1969b) Fine structure of the eyes of *Onithochiton neglectus* (Mollusca: Polyplacophora). Z Zellforsch Mikrosk Anat 102:313–332. https://doi.org/10.1007/BF00335443

Boyle PR (1972) The aesthetes of chitons. I. Role in the light response of whole animals. Mar Freshw Behav Physiol 1:171–184. https://doi.org/10.1080/10236247209386896

Boyle PR (1974) The aesthetes of chitons. II. Fine structure in *Lepidochitona cinereus* (L.). Cell Tissue Res 153:383–398. https://doi.org/10.1007/BF00229166

Boyle PR (1976) The aesthetes of chitons. Shell surface observations. Cell Tissue Res 172:379–388. https://doi.org/10.1007/bf00399520

Brooker LR, Shaw JA (2012) The chiton radula: a unique model for biomineralization studies. Adv Top Biomineralization 1:65–84. https://doi.org/10.5772/31766

Bullock RC (1988) The genus *Chiton* in the new world (Polyplacophora: Chitonidae). The Veliger 31:141–191

Carter JG, Hall RM (1990) Polyplacophora, Scaphopoda, Archaeogastropoda and Paragastropoda (Mollusca). In: Carter JG (ed) Skeletal biomineralization: patterns, processes and evolutionary trends. Van Nostrand Reinhold, New York. https://doi.org/10.1007/978-1-4899-5391-9

Checa AG, Vendrasco MJ, Salas C (2017) Cuticle of Polyplacophora: structure, secretion, and homology with the periostracum of conchiferans. Mar Biol 164:64. https://doi.org/10.1007/s00227-017-3100-6

Connors MJ, Ehrlich H, Hog M et al (2012) Three-dimensional structure of the shell plate assembly of the chiton *Tonicella marmorea* and its biomechanical consequences. J Struct Biol 177:314–328. https://doi.org/10.1016/j.jsb.2011.12.019

Connors M, Yang T, Hosny A et al (2019) Bioinspired design of flexible armor based on chiton scales. Nat Commun 10:5413. https://doi.org/10.1038/s41467-019-13215-0

Crozier WJ (1921) "Homing" behavior in *chiton*. Am Nat 55:276–281. https://doi.org/10.1086/279812

Eernisse DJ (1988) Reproductive patterns in six species of Lepidochitona (Mollusca: Polyplacophora) from the pacific coast of north america. Biol Bull 174:287–302. https://doi.org/10.2307/1541955

Eernisse DJ (1998) Class Polyplacophora. In: Scott PV, Blake JA (eds) Taxonomic atlas of the benthic fauna of the Santa Maria Basin and the Western Santa Barbara channel, vol 8. The Mollusca, part 1: Aplacophora, Polyplacophora, Scaphopoda, Bivalvia and Cephalopoda. Santa Barbara Museum of Natural History, pp 49–73

Eernisse DJ (2007) Chitons. In: Denny MW, Gaines SD (eds) Encyclopedia of Tidepools and rocky shores. University of California Press, Berkeley

Eernisse DJ, Reynolds PD (1994) Polyplacophora. In: Harrison FW, Kohn AJ (eds) Microscopic anatomy of invertebrates, Mollusca I, vol 5. Wiley, New York

Faller S, Faller S, Rothe BH et al (2012) Comparative neuroanatomy of Caudofoveata, Solenogastres, Polyplacophora, and Scaphopoda (Mollusca) and its phylogenetic implications. Zoomorphology 131:149–170. https://doi.org/10.1007/s00435-012-0150-7

Fernandez CZ, Vendrasco MJ, Runnegar B (2007) Aesthete canal morphology in twelve species of chiton (Polyplacophora). The Veliger 49:51–69

Fischer FP (1979) Die Ästheten von *Acanthochiton fascicularis* (Mollusca, Polyplacophora). Zoomorphologie 92:95–106. https://doi.org/10.1007/BF00999837

Fischer FP (1980) Fine structure of the larval eye of *Lepidochitona cinerea* L. (Mollusca: Polyplacophora). Spixiana 3:53–57

Fischer FP (1988) The ultrastructure of the aesthetes in *Lepidopleurus cajetanus* (Polyplacophora: Lepidopleurina). Am Malacol Bull 6:153–159. https://doi.org/10.1111/j.1469-7998.1993.tb02671.x

Fischer FP, Renner M (1978) Die Feinstruktur der Ästheten von *Chiton olivaceus* (Mollusca, Polyplacophora). Helgolander Wiss Meeresun 31:425–443. https://doi.org/10.1007/BF02189492

Haas W, Kriesten K (1978) Die Ästheten mit intrapigmentärem Schalenauge von *Chiton marmoratus* L. (Mollusca, Polyplacophora). Zoomorphologie 90:253–268. https://doi.org/10.1007/BF01007694
Heath H (1904) The larval eye of chitons. Proc Acad Natl Sci Phila 56:257–259
Henry JQ, Okusu A, Martindale MQ (2004) The cell lineage of the polyplacophoran, *Chaetopleura apiculata*: variation in the spiralian program and implications for molluscan evolution. Dev Biol 272:145–160. https://doi.org/10.1016/j.ydbio.2004.04.027
Homberg U (2020) Visual circuits in arthropod brains. J Comp Physiol A 206:105–107. https://doi.org/10.1007/s00359-020-01407-9
Hubrecht AAW (1882) A contribution to the morphology of the Amphineura. J Cell Sci 2:212–228. https://doi.org/10.1242/jcs.s2-22.86.212
Irisarri I, Uribe JE, Eernisse DJ, Zardoya R (2020) A mitogenomic phylogeny of chitons (Mollusca: Polyplacophora). BMC Evol Biol 20:22. https://doi.org/10.1186/s12862-019-1573-2
Kerbl A, Handschuh S, Nödl M-T et al (2013) Micro-CT in cephalopod research: investigating the internal anatomy of a sepiolid squid using a non-destructive technique with special focus on the ganglionic system. J Exp Mar Biol Ecol 447:140–148. https://doi.org/10.1016/j.jembe.2013.02.022
Kingston ACN, Chappell DR, Speiser DI (2018) Evidence for spatial vision in *Chiton tuberculatus*, a chiton with eyespots. J Exp Biol 221:jeb183632. https://doi.org/10.1242/jeb.183632
Kirschvink JL, Lowenstam HA (1979) Mineralization and magnetization of chiton teeth: Paleomagnetic, sedimentologic, and biologic implications of organic magnetite. Earth Planet Sci Lett 44:193–204. https://doi.org/10.1016/0012-821X(79)90168-7
Kocot KM, Cannon JT, Todt C et al (2011) Phylogenomics reveals deep molluscan relationships. Nature 477:452–456. https://doi.org/10.1038/nature10382
Kowalevsky MA (1883) Embryogénie du *Chiton polii* (Philippi) avec quelques remarques sur le développement des autres Chitons. Anaales du Museum d'Histoire Naturelle de Marseille 1:1–46. https://doi.org/10.5962/bhl.title.46987
Land MF, Nilsson D-E (2012) Animal eyes, 2nd edn. Oxford University Press, Oxford. https://doi.org/10.1093/acprof:oso/9780199581139.001.0001
Leise EM (1986) Chiton integument: development of sensory organs in juvenile *Mopalia muscosa*. J Morphol 189:71–87. https://doi.org/10.1002/jmor.1051890107
Leise EM (1988) Sensory organs in the hairy girdles of some Mopaliid chitons. Am Malacol Bull 6:141–151
Li L, Connors MJ, Kolle M et al (2015) Multifunctionality of chiton biomineralized armor with an integrated visual system. Science 350:952–956. https://doi.org/10.1126/science.aad1246
Lowenstam HA (1962) Magnetite in denticle capping in recent chitons (Polyplacophora). Geol Soc Am 73:435–438. https://doi.org/10.1130/0016-7606(1962)73[435:MIDCIR]2.0.CO;2
Moseley HN (1885) On the presence of eyes in the shells of certain Chitonidae and on the structures of these organs. Q J Microsc Sci 25:37–60. https://doi.org/10.1242/jcs.s2-25.97.37
Nilsson D-E (1994) Eyes as optical alarm systems in fan worms and ark clams. Philos Trans R Soc B 346:195–212. https://doi.org/10.1098/rstb.1994.0141
Nilsson D-E (2013) Eye evolution and its functional basis. Vis Neurosci 30:5–20. https://doi.org/10.1017/S0952523813000035
Nowikoff M (1907) Über die Rückensinnesorgane der Placophoren nebst einigen Bemerkungen über die Schale derselben. Z Wiss Zool 88:154–186
Nowikoff M (1909) Über die intrapigmentären Augen der Placophoren. Z Wiss Zool 93:668–680
Omelich P (1967) The behavioral role and the structure of the aesthetes of chitons. The Veliger 10:77–82
Puchalski SS, Eernisse DJ, Johnson CC (2008) The effect of sampling bias on the fossil record of chitons (Mollusca, Polyplacophora). Am Malacol Bull 25:87–95. https://doi.org/10.4003/0740-2783-25.1.87

Ramirez MD, Pairett AN, Pankey MS et al (2017) The last common ancestor of most bilaterian animals possessed at least nine opsins. Genome Biol Evol 8:3640–3652. https://doi.org/10.1093/gbe/evw248

Richter S, Loesel R, Purschke G et al (2010) Invertebrate neurophylogeny: suggested terms and definitions for a neuroanatomical glossary. Front Zool 7:29. https://doi.org/10.1186/1742-9994-7-29

Rosen MD, Stasek CR, Hermans CO (1979) The ultrastructure and evolutionary significance of the ocelli in the larva of *Katharina tunicata* (Mollusca: Polyplacophora). The Veliger 22:173–178

Schwabe E (2010) Illustrated summary of chiton terminology. Spixiana 33:171–194

Shaw JA, Macey DJ, Brooker LR, Clode PL (2010) Tooth use and wear in three iron-biomineralizing mollusc species. Biol Bull 218:132–144. https://doi.org/10.1086/BBLv218n2p132

Shigeno S, Andrews PLR, Ponte G, Fiorito G (2018) Cephalopod brains: an overview of current knowledge to facilitate comparison with vertebrates. Front Physiol 9:952. https://doi.org/10.3389/fphys.2018.00952

Sigwart JD, Sumner-Rooney LH (2021) Continuous and regular expansion of a distributed visual system in the eyed chiton *Tonicia lebruni*. Biol Bull 240:1–11. https://doi.org/10.1086/712114

Sigwart JD, Sutton MD (2007) Deep molluscan phylogeny: synthesis of palaeontological and neontological data. Proc R Soc B 274:2413–2419. https://doi.org/10.1098/rspb.2007.0701

Sigwart JD, Sumner-Rooney LH, Schwabe E et al (2014) A new sensory organ in "primitive" molluscs (Polyplacophora: Lepidopleurida), and its context in the nervous system of chitons. Front Zool 11:7. https://doi.org/10.1186/1742-9994-11-7

Sigwart JD, Vermeij GJ, Hoyer P (2019) Why do chitons curl into a ball? Biol Lett 15:20190429. https://doi.org/10.1098/rsbl.2019.0429

Sirenko B (2006) New outlook on the system of chitons (Mollusca: Polyplacophora). Venus 65:27–49. https://doi.org/10.18941/venus.65.1-2_27

Smith SA, Wilson NG, Goetz FE et al (2011) Resolving the evolutionary relationships of molluscs with phylogenomic tools. Nature 480:364–367. https://doi.org/10.1038/nature10526

Speiser DI, Eernisse DJ, Johnsen S (2011) A chiton uses aragonite lenses to form images. Curr Biol 21:665–670. https://doi.org/10.1016/j.cub.2011.03.033

Speiser DI, DeMartini DG, Oakley TH (2014) The shell-eyes of the chiton *Acanthopleura granulata* (Mollusca, Polyplacophora) use pheomelanin as a screening pigment. J Nat Hist 48:2899–2911. https://doi.org/10.1080/00222933.2014.959572

Strausfeld NJ (1976) Atlas of an insect brain. Springer-Verlag, New York. https://doi.org/10.1007/978-3-642-66179-2

Sturrock MG, Baxter JM (1993) The ultrastructure of the aesthetes of *Leptochiton asellus* (Polyplacophora: Lepidopleurina). J Zool 230:49–61. https://doi.org/10.1111/j.1469-7998.1993.tb02671.x

Sturrock MG, Baxter JM (1995) The fine structure of the pigment body complex in the intrapigmented aesthetes of *Callochiton achatinus* (Mollusca: Polyplacophora). J Zool 235:127–141. https://doi.org/10.1111/j.1469-7998.1995.tb05133.x

Sumner-Rooney LH, Sigwart JD (2015) Is the Schwabe organ a retained larval eye? Anatomical and behavioural studies of a novel sense organ in adult *Leptochiton asellus* (Mollusca, Polyplacophora) indicate links to larval photoreceptors. PLoS One 10:e0137119. https://doi.org/10.1371/journal.pone.0137119

Sumner-Rooney L, Sigwart JD (2018) Do chitons have a brain? New evidence for diversity and complexity in the polyplacophoran central nervous system. J Morphol 29:751. https://doi.org/10.1002/jmor.20823

Treves K, Traub W, Weiner S, Addadi L (2003) Aragonite formation in the chiton (Mollusca) girdle. Helv Chim Acta 86:1101–1112. https://doi.org/10.1002/hlca.200390096

Varney RM, Speiser DI, McDougall C et al (2021) The iron-responsive genome of the chiton *Acanthopleura granulata*. Genome Biol Evol 13:evaa263. https://doi.org/10.1093/gbe/evaa263

Vendrasco MJ, Fernandez CZ, Eernisse DJ, Runnegar B (2008) Aesthete canal morphology in the Mopaliidae (Polyplacophora). Am Malacol Bull 25:51–69. https://doi.org/10.4003/0740-2783-25.1.51

Vinther J, Sperling EA, Briggs DEG, Peterson KJ (2012) A molecular palaeobiological hypothesis for the origin of aplacophoran molluscs and their derivation from chiton-like ancestors. Proc R Soc B 279:1259–1268. https://doi.org/10.1098/rspb.2011.1773

Vöcking O, Kourtesis I, Hausen H (2015) Posterior eyespots in larval chitons have a molecular identity similar to anterior cerebral eyes in other bilaterians. EvoDevo 6:40. https://doi.org/10.1186/s13227-015-0036-0

Vöcking O, Kourtesis I, Tumu SC et al (2017) Co-expression of xenopsin and rhabdomeric opsin in photoreceptors bearing microvilli and cilia. eLife 6:e23435. https://doi.org/10.7554/eLife.23435

von Knorre H (1925) Die Schale und die Rückensinnesorgane von *Trachydermon* (*Chiton*) *cinereus* L. und die ceylonischen Chitonen der Sammlung Plate (Fauna et Anatomia ceylanica, III, Nr 3). Jenaische Zeitschrift für Naturwissenschaft herausgegeben von der medizinisch-naturwissenschaftlicken Gessellschaft zu Jena 61:469–632

Yonge CM (1939) On the mantle cavity and its contained organs in the Loricata (Placophora). J Cell Sci 2:367–390. https://doi.org/10.1242/jcs.s2-81.323.367

Chapter 7
The Visual System of Myriapoda

Andy Sombke and Carsten H. G. Müller

Abstract Eyes are manifold in the animal kingdom. Most arthropod groups, possess eyes, which can be found in various structural and functional specifications at either side of the head (lateral eyes) and/or at the head's frontal face (median eyes). Arthropods are not only famous for their highly ordered, often modular cellular organization, as exemplified by compound eyes, but are also considered fascinating objects in evolutionary research. In this respect, it was myriapod eyes and the specific structure of their visual neuropils that disclosed possible evolutionary pathways of lateral eyes in major arthropod groups and also hold an enormous resolving potential to evaluate the still uncertain phylogenetic interrelationships among euarthropods. Myriapods are mandibulate arthropods and the presumed sister group to Pancrustacea. While Symphyla and Pauropoda are blind, Chilopoda (centipedes) and Diplopoda (millipedes) possess lateral eyes composed of genuine or evolutionarily transformed ommatidia. The partly constant cell patterns in the ommatidia of scutigeromorph centipedes and in the miniaturized ommatidia of penicillate millipedes, along with crystalline cone cells found therein, are thought to be close to the ground pattern of Mandibulata. The cone-less eyes in pleurostigmophoran centipedes and chilognathan millipedes may have evolved convergently along both lineages. A feature shared by all myriapod eyes is the dual-type retinula (bilayered retinula consisting of distal and proximal retinula cells). Eye-bearing centipedes and millipedes possess two distinct visual neuropils in the lateral protocerebrum, commonly termed lamina and medulla, which are innervated by short and long retinula cell axons, which is probably close to the ground pattern of mandibulate arthropods. Small photoreceptor organs, so-called accessory lateral eyes, may be

A. Sombke (✉)
Center for Anatomy and Cell Biology, Cell and Developmental Biology,
Medical University of Vienna, Vienna, Austria

C. H. G. Müller
Zoological Institute and Museum, General and Systematic Zoology, University of Greifswald,
Greifswald, Germany
e-mail: carstmue@uni-greifswald.de

E. Buschbeck, M. Bok (eds.), *Distributed Vision*, Springer Series in Vision
Research, https://doi.org/10.1007/978-3-031-23216-9_7

present within the brain. Myriapods exhibit negative phototaxis to a certain degree, and even some blind species show flight behavior when illuminated.

Keywords Chilopoda · Diplopoda · Anatomy · Evolution · Development · Ecology

7.1 Introduction

Myriapods are exclusively terrestrial arthropods and mostly edaphic organisms that live in leaf litter, under bark, and on or inside the soil. Most myriapods are millipedes (Diplopoda) with approximately 12,000 species described. Centipedes (Chilopoda) comprise about 3500 species, while Symphyla and Pauropoda comprise 200 and 835 species, respectively (Minelli 2011; Brewer et al. 2012). Centipedes are predators possessing a pair of venomous claws, called forcipules, which are evolutionarily transformed appendages of the first trunk segment. Millipedes are detritivores and possess fused trunk segments (diplosegments) with two pairs of legs starting from the fifth trunk segment toward the posterior end of the trunk. The blind and unpigmented Symphyla usually possess 12 leg pairs and a pair of spinnerets. The blind and minute Pauropoda possess eight to 11 pairs of legs and a pair of three-branched antennae. While in Symphyla and Pauropoda eyes are absent, centipedes and millipedes possess lateral eyes (e.g., Paulus 2000; Müller et al. 2011; Müller and Sombke 2015) (Fig. 7.1). Median eyes are always absent in Myriapoda (Müller et al. 2011; Müller and Sombke 2015). Small photoreceptor organs – so-called accessory lateral eyes – may be present within the brain in some centipedes and millipedes.

Research on myriapod eyes looks back on a tradition that started in the pioneer era of dissection- and histology-based zootomy at the end of the nineteenth century (e.g., Graber 1880; Grenacher 1880; Hesse 1901; Heymons 1901) and was much later on further elaborated by the use of transmission electron microscopy (TEM) (Bedini 1968, 1970; Joly 1969; Bähr 1971, 1974; Paulus 1979; Spies 1981; Paulus 2000; Müller et al. 2003b, 2007; Heithier and Melzer 2005; Meyer-Rochow et al. 2006; Müller and Meyer-Rochow 2006a, b; Müller and Rosenberg 2006). However, until now, our knowledge of the structural diversity of myriapod eyes is anything but complete on the ultrastructural level. Among those centipede subgroups having eyes, the state of knowledge is often based on the description of a single or few species (Müller and Rosenberg 2009). In millipedes, the situation is even less resolved, with basic anatomical descriptions still missing for a number of subgroups (Müller and Sombke 2015).

Like crustaceans and hexapods, scutigeromorph centipedes possess real compound eyes composed of discrete optical units, called ommatidia (Paulus 1979; Müller et al. 2003b). All other centipedes – the Pleurostigmophora – possess cup-shaped eyes without crystalline cones, which, because of their aberrant cellular architecture, have been termed "spaced eyelets" (Strausfeld et al. 2016), "simple,

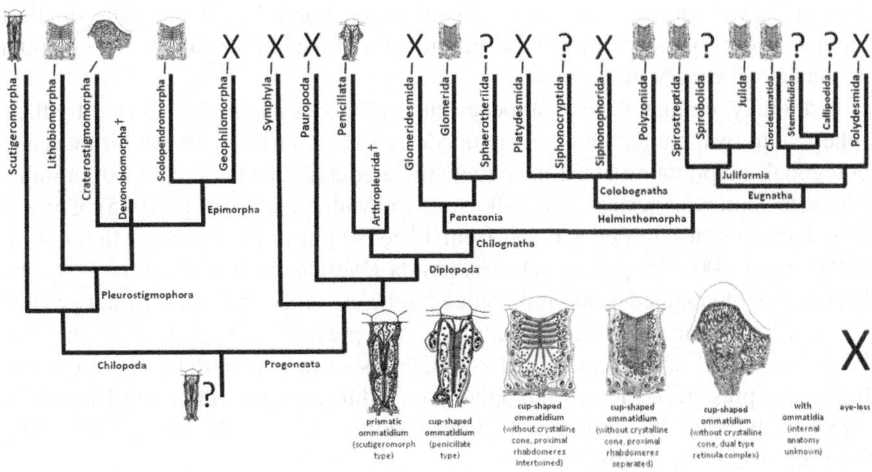

Fig. 7.1 Myriapod phylogeny and distribution of ommatidial types (combined from Shear and Edgecombe (2010) and Blanke and Wesener (2014)). Eyeless Siphoniulida whose phylogenetic affiliation is considered uncertain (Sierwald et al. 2003) were excluded. Ommatidial types are indicated by sketches adapted from Müller (2008). Based on the evolutionary scenario drawn by Harzsch et al. (2007), the last common ancestor of Myriapoda most likely possessed compound eyes, including ommatidia of the scutigeromorph type

stemma-like lateral eyes" (Bitsch and Bitsch 2005) or "lateral ocelli" (e.g., Paulus 2000; Harzsch et al. 2005). But, as of recently, these eyes have been considered as transformed ommatidia (Harzsch et al. 2007; Richter et al. 2010). Among Pleurostigmophora, ommatidia may occur as a single pair in some Lithobiomorpha and Craterostigmomorpha, as well as in the scolopendromorph family Mimopidae. In *Craterostigmus tasmanianus*, the single ellipsoid ommatidium is subdivided into a distinct anterior and posterior region (Müller and Meyer-Rochow 2006b). In other taxa, ommatidia are usually more numerous and range from constantly four (arranged in a 1 + 2 + 1 formation) in scolopendrid Scolopendromorpha (Müller and Meyer-Rochow 2006a) to a variable number of 1–49 in Lithobiomorpha (Zapparoli and Edgecombe 2011). Within the ommatidial field of Lithobiomorpha, there is a size gradient from a single "giant ocellus" to a very small ommatidium at the antero-ventral margin (Müller et al. 2011). Some henicopid and lithobiid Lithobiomorpha, as well as all species in Cryptopidae, Plutoniumidae, and Scolopocryptopidae (Scolopendromorpha), and all Geophilomorpha are blind. Most millipede species possess lateral eyes, with the exception of Glomeridesmida, Platydesmida, Siphonophorida, as well as some species in Juliformia and Polydesmida, which is most probably linked to cave life (Enghoff 1984; Blanke and Wesener 2014; Müller and Sombke 2015). In adult millipedes, the number of ommatidia often varies, e.g., 5–13 in Penicillata (e.g., Nguyen Duy-Jacquemin 1996; Short and Huynh 2006; Müller et al. 2007), but interspecific variations can be much higher, e.g., 30–57 in Pachybolidae (Enghoff 2011). If numerous ommatidia (="lateral ocelli") are present, they often are arranged in a triangular or rhomb-like cluster. The smallest ommatidium is usually found at the anterior end of the cluster, and new ommatidia are added with every molt.

Externally, ommatidia in centipedes and millipedes can be recognized by their spherical to oval corneal lens. In centipedes, the lens surface can exhibit the same polygonal sculpturation as on the cuticle (Müller et al. 2011). In contrast, in millipedes, the cuticular surface is usually smooth (Müller and Sombke 2015). Corneal lens diameters in millipedes range from 12 to 30 μm in Penicillata (Müller et al. 2007) up to 100–300 μm in Sphaerotheriida (Wesener and Sierwald 2005). The lateral eyes of centipedes and millipedes can be classified into two distinct groups: "genuine" ommatidia with crystalline cones present in Scutigeromorpha and Penicillata (Fig. 7.2a, c) and transformed, cup-shaped ommatidia without crystalline cones present in Pleurostigmophora and Chilognatha (Müller and Rosenberg 2006; Müller and Meyer-Rochow 2006b; Müller et al. 2007, 2011; Müller and Sombke 2015) (Fig. 7.2b).

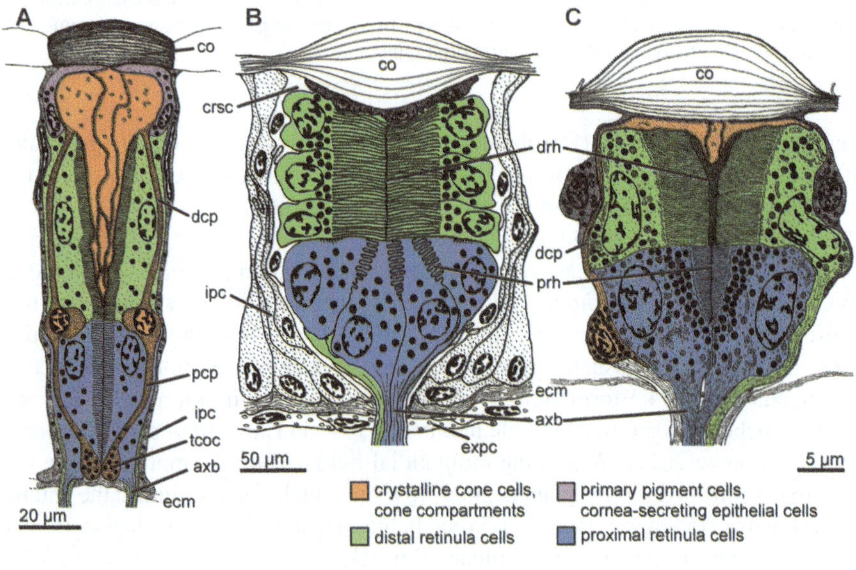

Fig. 7.2 Semischematic reconstruction of the ommatidial anatomy of centipedes and millipedes. (**a**) Ommatidium of *Scutigera coleoptrata* (Chilopoda, Scutigeromorpha) in mediolongitudinal section, modified after Müller et al. (2011). (**b**) Cup-shaped ommatidium of *Lithobius forficatus* (Chilopoda, Pleurostigmophora) in mediolongitudinal section, modified after Müller and Rosenberg (2006). (**c**) Cup-shaped ommatidium of *Phryssonotus platycephalus* (Diplopoda, Penicillata) in mediolongitudinal section, modified after Müller et al. (2007). Labels: *axb* axon bundle, *dcp* distal cone cell process, *co* corneal facet/lens, *crsc* circumretinular sheath cell, *expc* external pigment cell, *drh* distal rhabdom, *ecm* extracellular (basal) matrix of the ommatidia, *ipc* interommatidial pigment cell, *pcp* proximal cone cell process, *prh* proximal rhabdom, *tcoc* swollen termination of cone cell

7.2 Ommatidial Ground Pattern and Diverging Pathways

Specific ommatidial characters suggest that Myriapoda, Crustacea, and Hexapoda share a common origin (Mandibulata). Therefore, it is assumed that the last common ancestor of Mandibulata possessed ommatidia, most probably arrayed in a pair of proper compound eyes, constituted by a characteristic set of epithelial cells, among them eucone cells, primary pigment cells (=cornea-secreting or corneagenous cells), interommatidial pigment cells, and a double-layered formation of distal and proximal retinula cells. Additionally, the successive formation of the compound eye from an anterior proliferation zone can be considered a further constitutive character of Mandibulata (Harzsch et al. 2007; Müller et al. 2007, 2011; Edgecombe 2010; Müller and Sombke 2015).

The partly constant cell patterns in the ommatidia of scutigeromorph centipedes and in the miniaturized ommatidia of penicillate millipedes are thought to be close to the ground pattern of Mandibulata (Harzsch et al. 2005, 2007; Müller et al. 2007). Eucone cells with their nuclei placed outside (proximal of) the light-focusing cone compartments are most likely ancestral (Müller et al. 2003b; Harzsch et al. 2007; Müller et al. 2007, 2011). Strausfeld et al. (2016), however, favored a rather different evolutionary concept according to which apposition eyes composed of ommatidia with a dioptric apparatus, including a proper crystalline cone, were already present in the ground pattern of Euarthropoda. This concept implies that ommatidia of the scutigeromorph type (with only partly fixed cell numbers) derived from these ancestral tetraconate-like ommatidia. Nilsson and Kelber (2007) argued that the arrangement of scutigeromorph ommatidia is different from that of typical hexapod and crustacean ommatidia. These authors hypothesized that the multipartite scutigeromorph crystalline cone evolved to separate the distal rhabdomeres from the optical axis and, due to functional reasons, has an origin different from tetrapartite cones in Pancrustacea. Subsequently, the architecture of scutigeromorph ommatidia is merely considered to have shaped by virtue of functional requirements and clustered in an unusually large compound eye without having any phylogenetic significance at all. However, this functional modification has then to be assumed to have happened independently in the stem lineage of Penicillata as well (Müller et al. 2007). The multipartite crystalline cone of scutigeromorph centipedes (Figs. 7.2a and 7.3a–c) is mostly composed of four, rarely five, eucone cells, whereas in bristly millipedes (Penicillata), Müller et al. (2007) described tripartite crystalline cones composed of three eucone cells (Figs. 7.2c and 7.6a, b). Corresponding ultrastructural traits support the assumption that scutigeromorph and penicillate ommatidia are homologous, and thus, the lateral eyes of myriapods, despite appearing considerably different in structure, share a common ancestry (Müller et al. 2007).

A comprehensive phylogenetic analysis conducted by Müller (2008) suggested that cone-less eyes surrounded by a mantle of external (screening) pigment cells evolved independently in pleurostigmophoran Chilopoda and chilognathan Diplopoda. Consequently, a dual-type retinula, which is also present in scutigeromorph and penicillate ommatidia, is retained in the more or less spaced,

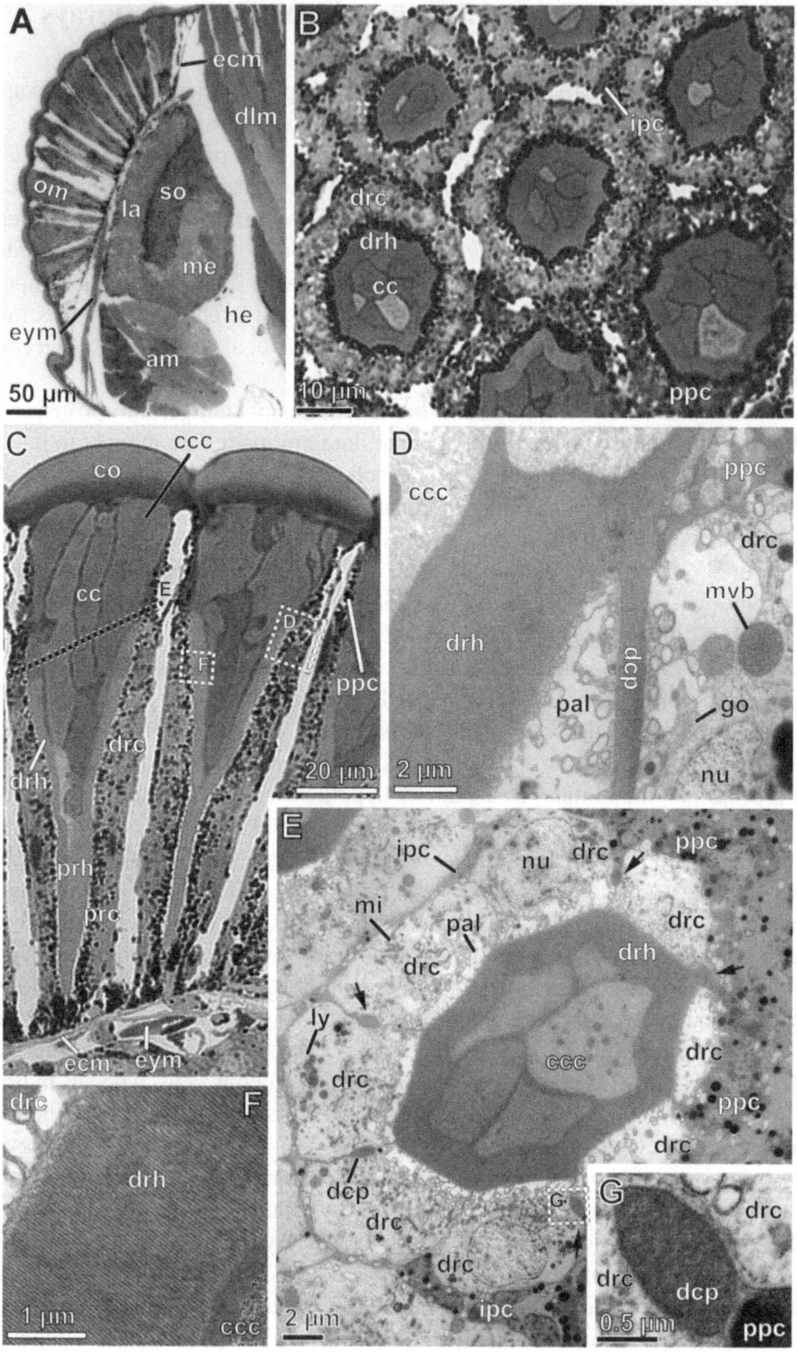
A
ecm
dlm
om
la
so
me
he
eym
am
50 µm
B
ipc
drc
drh
cc
10 µm
ppc
C
ccc
co
cc
E
D
F
ppc
drh
drc
prh
prc
20 µm
ecm
eym
D
ccc
ppc
drc
mvb
drh
dcp
pal
go
2 µm
nu
E
ipc
nu
drc
ppc
mi
pal
drc
drh
drc
ly
ccc
drc
drc
ppc
drc
dcp
drc
G
drc
ipc
2 µm
F
drc
drh
1 µm
ccc
G
drc
drc
dcp
0.5 µm
ppc

transformed, and cup-shaped ommatidia of Pleurostigmophora and Chilognatha (Figs. 7.1, 7.2b, 7.5a, e, and 7.7a, e). However, rhabdomeres of proximal retinula cells may join differently in pleurostigmophoran/chilognathan ommatidia (interdigitating vs. separated rhabdomeric microvilli). The following are the characteristics that these transformed pleurostigmophoran and chilognathan ommatidia have in common: (1) a cup-shaped profile, (2) loss of the crystalline cone, (3) increased and strongly variable cell numbers (particularly of retinula cells), (4) diversification of the strictly interommatidial pigment cells toward a system of glial-like sheath cells, and (5) deeply invaginated corneal lens secreted by an epithelium of flattened, unpigmented corneagenous cells. The above-listed set of traits characterizing pleurostigmophoran and chilognathan ommatidia may be considered a result of modification processes going on at an early stage of eye development, namely, after the formation of (myriapod) protommatidial precursors. Such modifications potentially involve fusion and/or secondary growth processes similar to developmental pathways known from fusion stemmata in insect larvae (compare review by Paulus 2000).

7.2.1 Lateral Eyes in Scutigeromorpha (Chilopoda)

The compound eye of Scutigeromorpha consists of 100 to 600 prismatic, closely adjoined ommatidia (Fig. 7.3a–c) (Hanström 1934; Müller et al. 2003b). In the apical region of the compound eye of *Scutigera coleoptrata*, ommatidia are of a hexagonal shape with a diameter of approximately 50 µm, while they possess a more

Fig. 7.3 Histology and ultrastructure of the ommatidia of *Scutigera coleoptrata* (Chilopoda, Scutigeromorpha). Overview and distal components. (**a–c**) Toluidine-blue histology: (**a**) cross-section through the dorsolateral region of the head with several ommatidia in longitudinal section. Note the lamina subjacent to the ommatidia and the arcuate medulla. (**b**) Oblique cross-section of several dorsal ommatidia showing a transitional gradient from the level of primary pigment cells (lower right) to the distal retinula (center and upper left). Note the highly variable number of compartments from distal to more proximal section levels of the crystalline cone. (**c**) Two ommatidia in mediolongitudinal section showing major constituents. (**d–g**) Ultrastructure of the distal region of the ommatidium (TEM). (**d**) Distal apical region of a distal retinula cell showing a distal cone cell process widening to form a compartment of the crystalline cone (sector indicated by a dashed box in C). Longitudinal section. (**e**) Oblique cross-section of the distal retinula of an ommatidium containing eight distal retinula cells (section plane indicated by a dashed line in C). Note the presence of five cone cell processes (dcp and black arrows) located in the infraretinular spaces. (**f**) Close-up of a distal rhabdomere demonstrating the highly ordered arrangement of thin rhabdomeric microvilli. (**g**) Close-up of a distal cone cell process in cross-section. Originals. Labels: *am* antenna-moving muscles, *cc* crystalline cone, *ccc* crystalline cone compartment, *co* corneal facet, *dlm* dorsomedian dilator muscles, *dcp* distal cone cell process, *drc* distal retinula cell(s), *drh* distal rhabdom, *ecm* fenestrated extracellular (basal) matrix of the compound eye, *eym* eye-associated muscles, *go* Golgi stacks, *he* hemolymphatic space, *ipc* interommatidial pigment cells, *la* lamina, *ly* lysosomes (late stage), *me* medulla, *mi* mitochondrion, *mvb* multivesicular body, *nu* nucleus, *om* ommatidium, *pal* perirhabdomeric ER (= "palisade ER"), *ppc* primary pigment cell(s) (=corneagenous cells), *prh* proximal rhabdom, *so* neuronal somata

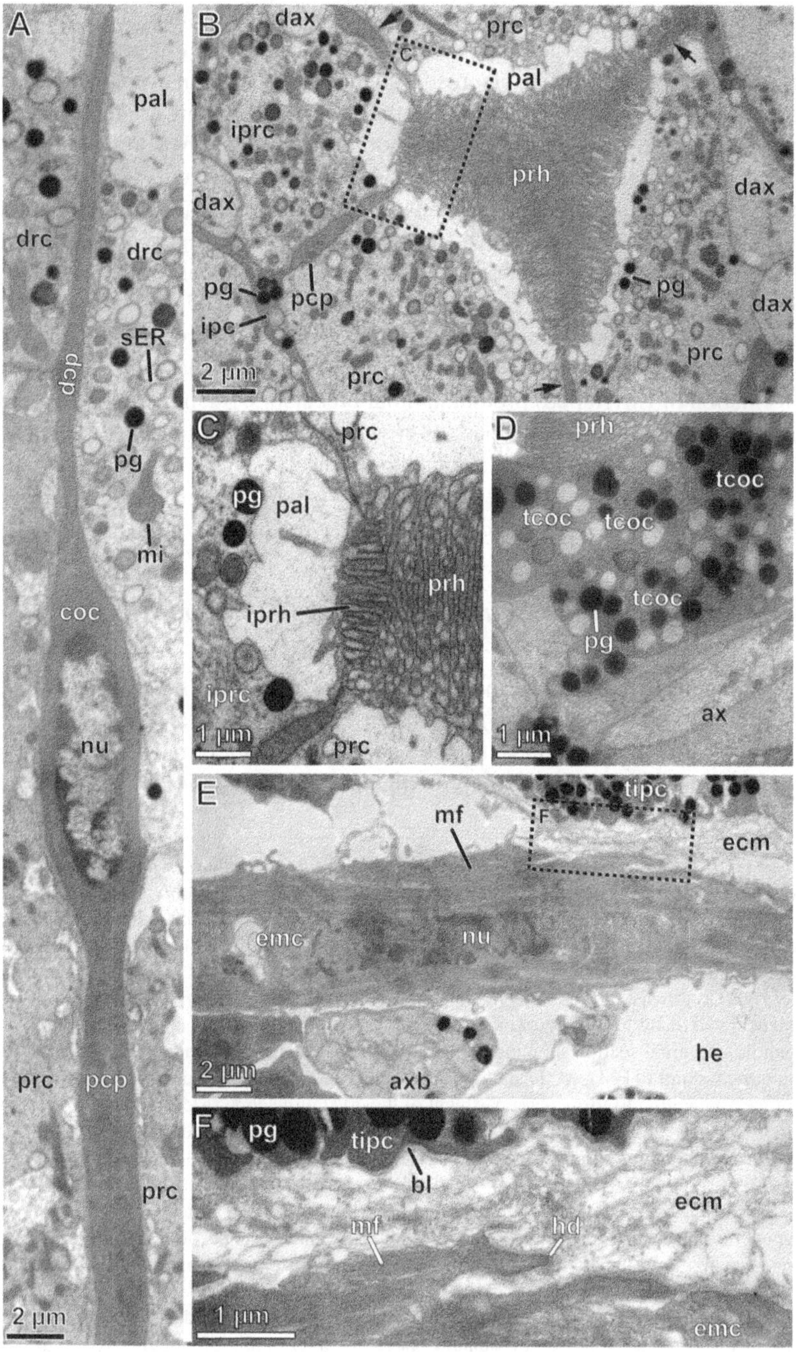
A
pal
drc
drc
sER
dcp
pg
mi
coc
nu
prc
pcp
prc
2 µm
B
dax
prc
C
pal
iprc
prh
dax
dax
pg
pcp
ipc
pg
dax
prc
2 µm
prc
C
prc
pg
pal
iprh
prh
iprc
1 µm
prc
D
prh
tcoc
tcoc
tcoc
tcoc
pg
ax
1 µm
E
tipc
mf
F
ecm
emc
nu
he
2 µm
axb
F
pg
tipc
bl
ecm
mf
hd
1 µm
emc

pentagonal shape at the margin of the eye (Müller et al. 2003b). At the corner points, where three ommatidia abut, small tricellular recto-canal epidermal glands are frequently found (Müller et al. 2003a). Each ommatidium contains a dioptric apparatus, a photoreceptive dual-type retinula, and a sheath of accessory (interommatidial) pigment cells shielding the ommatidium against scattered light (Figs. 7.2a and 7.3c, e). The number of constituting cells in a given ommatidium varies with cell type and the ommatidium's localization within the compound eye (Müller et al. 2003b; Harzsch et al. 2005, 2007).

The dioptric apparatus consists of a biconvex cornea, made of 8–10 corneagenous cells, and a multipartite crystalline cone. The corneagenous cells are arranged in a circle around the distal part of the crystalline cone (Fig. 7.3b). Their cytoplasm includes numerous screening pigment granules (ommochromes, 0.5–1.2 μm in diameter; see Figs. 7.2a and 7.3e). The crystalline cone strongly tapers to its proximal tip, which is located at the transition zone of the distal and proximal retinula (Figs. 7.2a and 7.3c). The crystalline cone mostly contains eight compartments that are unequally distributed, which are predominantly established by four eucone cells. Only peripheral ommatidia may contain five instead of four eucone cells. Their somata (nucleus-containing part of the cytoplasm) are placed in the proximal region of the ommatidium, coinciding with the proximal tip of the cone (Figs. 7.2a and 7.4a). Each cone cell soma projects two cytoplasmic, infraretinular processes strengthened by numerous microtubules (Figs. 7.2a, 7.3e–g, and 7.4a). The distal process ramifies into two substrands (illustrated in Müller et al. (2003b): see their, Figs. 1 and 4c), each of which makes a connection to the crystalline cone, the formation of which it contributes to with a single cone compartment (Figs. 7.2a and 7.3d).

Fig. 7.4 Ultrastructure of the ommatidia of *Scutigera coleoptrata* (Chilopoda, Scutigeromorpha). Proximal and subommatidial components. (**a**) Longitudinal section of a cone cell projecting through the distal and proximal retinula. A distal and a proximal cytoplasmatic process projects from the small, spindle-shaped soma. The distal process can be traced below the bifurcation level. (**b**) Cross-section of the proximal retinula below the soma level of the crystalline cone cells. Note the four proximal cone cell processes positioned between the four proximal retinula cells; the irregular (small) retinula cell is visible on the upper left. (**c**) Detail of the small rhabdomere contributed by the irregular proximal retinula cell (sector indicated by a dashed box in B). (**d**) Longitudinal view of the region immediately below the proximal rhabdom displaying the terminal swellings of the four cone cells. (**e**) Detail of eye-associated muscle cell/fiber abutting the extracellular (basal) matrix of the compound eye. Ommatidial axon bundles leaving the eye traverse the network of muscle fibers. (**f**) Attachment zone of an eye-associated muscle cell/fiber to the ecm (sector indicated by a dashed box in E). Adhesion and the presumed stretching power to be exerted on the ommatidia are ensured by hemidesmosomes. TEM. Originals. Labels: *ax* axons of retinula cells, *axb* axon bundle of an ommatidium, *bl* basal lamina (of interommatidial pigment cells), *coc* crystalline cone cell (soma region), *dax* axon of a distal retinula cell, *dcp* distal cone cell process, *drc* distal retinula cell, *ecm* fenestrated extracellular (basal) matrix of the compound eye, *emc* eye-associated muscle cell/fiber, *hd* hemidesmosome, *he* hemolymphatic space, *ipc* interommatidial pigment cells, *iprc* irregular proximal retinula cell, *iprh* irregular proximal rhabdomere, *mf* myofilaments, *mi* mitochondrion, *nu* nucleus, *om* ommatidium, *pal* perirhabdomeric ER (="palisade ER"), *pcp* proximal cone cell process, *pg* screening pigment granule, *prh* proximal rhabdom, *sER* smooth ER cisternae (highly active vesicular stage), *tcoc* swollen termination of cone cell, *tipc* swollen basal termination of interommatidial pigment cell

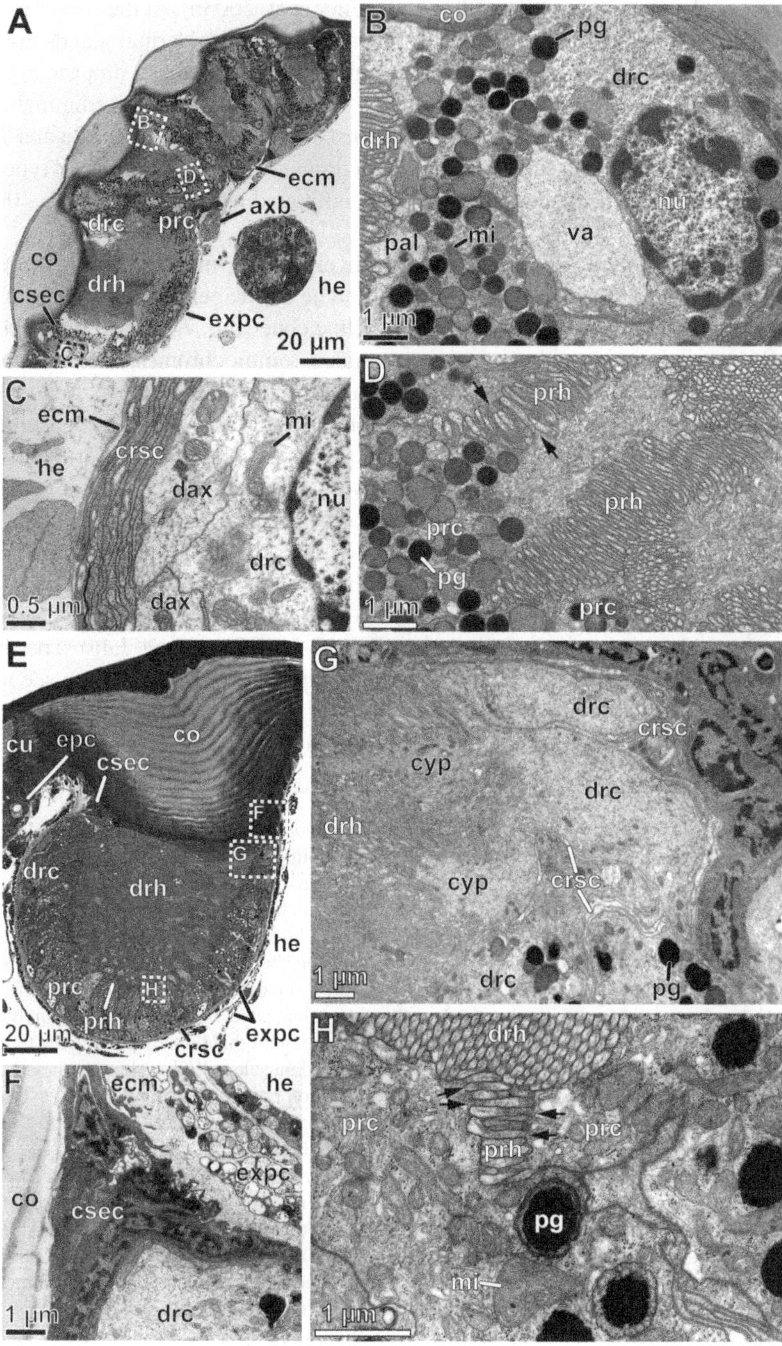
A
B
C
D
co
csec
drc
prc
drh
ecm
axb
he
expc
20 µm
B
pg
drc
drh
nu
pal
mi
va
1 µm
C
ecm
mi
he
crsc
dax
nu
drc
0.5 µm
dax
D
prh
prh
prc
pg
1 µm
prc
E
cu
epc
co
csec
F
G
drc
drh
H
he
prc
prh
20 µm
crsc
expc
G
drc
crsc
cyp
drc
drh
cyp
crsc
1 µm
drc
pg
F
ecm
he
expc
co
csec
1 µm
drc
H
drh
prc
prc
prh
pg
mi
1 µm

The proximal process projects down between two proximal retinula cells toward the basal matrix, to which it attaches with a terminal swelling filled with screening pigment granules (Figs. 7.2a and 7.4a, b, d). The retinula is stacked in two horizontal layers of retinula cells of distinct typologies (Figs. 7.2a and 7.3c). The distal retinula is composed of 8–13 cells forming a circular rhabdom surrounding the proximal part of the crystalline cone (see Fig. 7.3e for an example of a distal retinula equipped with only eight cells). The microvilli of the distal rhabdomeres are extraordinarily thin and strictly arrayed (Fig. 7.3d, f). The proximal retinula is always composed of four cells that form a closed, triangular rhabdom with broad, less ordered microvilli (Fig. 7.4b, c). One proximal retinula cell, called the irregular cell, contributes to the proximal rhabdom with a much smaller rhabdomere (Fig. 7.4b, c). Its specific optical role or innervation path is, however, unknown (but see discussion below). In contrast to Pleurostigmophora, rhabdomeric microvilli of the proximal retinula cells never interdigitate. The accessory (screening) pigment shield is composed of 14–16 cells, which are densely filled with pigment granules (0.4–0.8 µm in diameter). These extremely flattened cells are suspended from the innermost lamella of the surface cuticle to the basal matrix. Due to their spatial restriction to the

Fig. 7.5 Histology and ultrastructure of the cup-shaped ommatidia of lithobiomorph and scolopendromorph centipedes (Chilopoda, Pleurostigmophora). (**a–d**) Cup-shaped ommatidia of lithobiid Lithobiomorpha. (**a**) Horizontal section of the left head flank of *Lithobius dentatus* showing the size gradient in a longitudinally cut ommatidial cluster. Small-sized ommatidia are present anteriorly (upper right), whereas the biggest ommatidia are positioned posteriorly (lower left). Toluidine-blue histology. (**b–d**) Ultrastructure of cup-shaped ommatidia of *Lithobius mutabilis*. (**b**) Longitudinal section of a distal retinula cell with a straightly formed apex positioned in the distal-most circle of the multilayered distal retinula. Sector indicated exemplarily by a dashed box in A. (**c**) Cross-section through the periphery of the distal retinula wrapped by a multilayer of circumretinular sheath cells. Sector indicated exemplarily by a dashed box in A. (**d**) Longitudinal section of the apical region of several proximal retinula cells forming bidirectional rhabdomeres. Note that the microvilli of adjoined proximal rhabdomeres interdigitate (black arrows). Sector indicated exemplarily by a dashed box in A. (**e–h**) Cup-shaped ommatidium of *Scolopendra oraniensis* (Scolopendridae, Scolopendromorpha). Toluidine-blue histology (**e**) and ultrastructure based on TEM (**g–h**). (**e**) Posterior ommatidium in longitudinal section. (**f**) Transition zone of horizontally stacked cornea-secreting epithelial cells to the distal retinula, delimited to the ommatidium's periphery by the thick extracellular matrix and branches of external pigment cells. Longitudinal section. Sector indicated by a dashed box in E. (**g**) Longitudinal section of horizontally stacked distal retinula cells from the three distal-most circles. Note the finger-like projection of the apical membrane pointing toward the ommatidium's center and forming a circumapical rhabdomere. Sector indicated by a dashed box in E. (**h**) Longitudinal view of the region immediately below the distal rhabdom displaying a joint proximal rhabdomere with interdigitating microvilli (black arrows). Sector indicated by a dashed box in E. (**a–g**): Originals; H: reproduced from Müller and Meyer-Rochow 2006a. Labels: *axb* axon bundle of an ommatidium, *crsc* circumretinular sheath cells (somata and ramifying processes), *co* corneal lens, *csec* cornea-secreting epithelial cells, *cu* cuticle (surrounding corneal lenses), *cyp* thin cytoplasmic process of distal retinula cell, *dax* axon of a distal retinula cell, *drc* distal retinula cell, *drh* distal rhabdom, *ecm* extracellular (basal) matrix of ommatdium/ommatidial cluster, *epc* epidermal cells, *expc* external pigment cells, *he* hemolymphatic space, *mi* mitochondrion, *nu* nucleus, *pal* perirhabdomeric ER (="palisade ER"), *pg* screening pigment granule, *prh* proximal rhabdom, *va* vacuole interommatidial space, they are termed interommatidial pigment cells, thus providing an optical isolation sheath (Figs. 7.2a, 7.3b, e, 7.4b, e).

The entire compound eye is lined by a complex extracellular (basal) matrix produced by both cellular (processes of pigment cells, axons, glial cells, and musculature) and extracellular components (Müller et al. 2003b, 2011) (Fig. 7.2a). Two distinct bundles of numerous, slender, and striated muscle fibers project from the dorsal and frontal head cuticle and attach to the extracellular matrix along the bottom of the compound eye, ranging from the periphery to the center of the eye (Fig. 7.3a). This system was not disclosed in the descriptions by Müller et al. (2003a, b) and is revealed here for the first time. Both bundles of eye-associated muscle fibers meet and seem to intertwine below the central region of the compound eye. Each terminal of a muscle fiber is firmly attached to the basal matrix by hemidesmosomes (Fig. 7.4e, f). The fine-scale pattern of intertwining fibers, however, is yet unclear and needs to be explored by further studies. A three-dimensional analysis may show whether one or several muscle fibers target individual ommatidia. Whereas the position of the lamina somehow impedes the inward stretching of the eye toward the brain, a shifting of ommatidial constituents either firmly attached to (cone cells, interommatidial pigment cells) or passing through (retinula cells) the basal matrix may indeed occur. This may result in the shifting of the optical axes of distal and proximal rhabdomeres in these peripheral ommatidia to either an anterior or a posterior direction. Interestingly, a similar but less complex system of eye-associated muscles is found in dipteran flies (e.g., Burtt and Patterson 1970; Hengstenberg 1971, 1972). Here, a single muscle (musculus orbito-tentorialis), which is inserted at the tentorium at the posterior side of the head and attaches to the inner margin of the orbital ridge (see fig. 2 in Hengstenberg 1972), generates "clock spikes" of different frequencies in answer to sudden ambient light changes. The differential firing rate of these eye-associated muscles and the assumed retinal micro-movements support the detection of objects/patterns moving horizontally and crossing the visual field of the fly (Hengstenberg 1971, 1972; Viollet 2014). As to whether eye-associated musculature may enable or enhance motion vision in nocturnal scutigeromorphs remains unclear and should be the target of electrophysiological experiments and neuroethological studies.

7.2.2 Lateral Eyes (Cup-Shaped Ommatidia) in Pleurostigmophora (Chilopoda)

The lateral eyes of Pleurostigmophora are strongly transformed, more or less dispersed ommatidia with a cup-shaped profile. Each cup-shaped ommatidium includes an unequally biconvex (Figs. 7.1, 7.2b, 7.5a, e) to plane-convex (the latter only in Craterostigmomorpha) corneal lens but lacks a crystalline cone. In a cross-section, the corneal lens may appear circular or ovoid shaped, whereas the posterolateral ommatidium of the *Scolopendra* species shows an octagonal cross profile (Müller and Meyer-Rochow 2006a, b; Müller et al. 2011). In the cup-shaped ommatidia of lithobiomorph and scolopendromorph centipedes, the inner lens surface varies from slightly to strongly asymmetrical and is either moderately (*Lithobius* spp., *Eupolybothrus fasciatus*) or deeply invaginated (*Scolopendra cingulata* and *S. oraniensis*) into the cup (Müller and Meyer-Rochow 2006a; Müller and Rosenberg 2006). The corneal lens is produced by an epithelium of numerous (30 to over 2000) flattened corneagenous cells devoid of screening pigment granules, called cornea-secreting epithelial cells (e.g., Figs. 7.2b, 7.5a, e, f). In *Craterostigmus tasmanianus*, this layer consists of cubical cells, all displaying a homogenous cytoplasm (Müller and Meyer-Rochow 2006b). In all lithobiomorph and scolopendromorph species investigated, the layer of cornea-secreting epithelial cells is more heterogeneous since centromedian cells are much more elongated than the lateral ones (somata displaced to the periphery of the corneal lens). In *Scolopendra* spp., two types of cornea-secreting epithelial cells are evident: (1) short, plate-like cells that only surround the lateral margins of the corneal lens, which additionally are attached to protrusions of the cornea by microtubular bundles, and (2) cells that are limited to the proximal part of the corneal lens. In mediolongitudinal sections through each scolopendromorph ommatidium, four to six extremely flattened cornea-secreting epithelial cells can be found stacked horizontally to the side of the corneal lens (Fig. 7.5e, f). Each of these 230–2240 cells projects a thin, elongated cytoplasmic process toward the central subcorneal zone, where it attaches to a small central sector of the cornea (Müller and Rosenberg 2006; Müller and Meyer-Rochow 2006b; Müller et al. 2011).

A scolopendromorph ommatidium consists of 500 to over 1000 retinula cells, depending on the position of the ommatidium in the 1 + 2 + 1 cluster, as well as on the species and age of the animal. The highest yet counted number of retinula cells is found in the single ommatidium of *C. tasmanianus* (1000–1400). This high number might be indicative of dynamic neural summation (compare the discussion by Nilsson and Kelber (2007) and see also below). All yet examined pleurostigmophoran species possess a dual-type retinula comprising numerous distal and a much lower number of proximal retinula cells (e.g., Fig. 7.2b). Distal retinula cells are usually cubical or cylindrical, oriented perpendicular to the optical axis, and arranged in several rings stacked onto each other, giving the impression of a horizontal multilayer system (*Lithobius* spp., one to six layers (Fig. 7.2b); *E. fasciatus*, four to 12 layers; *Scolopendra* spp., 10–20 layers) (Müller and Meyer-Rochow 2006a; Müller and Rosenberg 2006). In *C. tasmanianus*, this pattern is different as the horizontally multilayered distal retinula is disintegrated and transformed into a much more complex cluster-like arrangement (Müller and Meyer-Rochow 2006b). Distal retinula cells with straight apex (*Lithobius* spp.) produce compact, rectangular rhabdomeres, which form a simple and fused rhabdom (Figs. 7.2b and 7.5b). In *E. fasciatus* and *Scolopendra* spp., each distal retinula cell projects a finger-like median process bearing a circumapical rhabdomere (Fig. 7.5g), hence forming a spatially more complex but still fused rhabdom (Müller and Meyer-Rochow 2006a; Müller and Rosenberg 2006) (see also Fig. 7.5e). In contrast, in *C. tasmanianus*, the clustered distal retinula cells possess knob-like or bilobed apices, resulting in the formation of cap-like, circumapical rhabdomeres, which generate a highly aberrant pattern of barely coherent, globular rhabdom subunits (Müller and Meyer-Rochow 2006b). Only 10% of all retinula cells constitute the proximal retinula, forming a homogenous layer at the bottom of the cup-shaped ommatidia in *Lithobius* spp., *E. fasciatus*, and *Scolopendra* spp. (Figs. 7.2b, 7.5a, e). Here, they are conical or club shaped and aligned parallel to the optical axis. Around the apex, each proximal retinula cell forms either a uni- or a bidirectional rhabdom, which can be very short (e.g., *Scolopendra oraniensis*: Fig. 7.5h) or elongated (e.g., *Lithobius mutabilis*: Fig. 7.5d). Microvilli are considerably separated from each other in a given proximal rhabdomere, allowing opposing rhabdomeres to interdigitate to form a common one (Figs. 7.2b and 7.5d, h), thus forming a fused, reticulated structure (Müller and Meyer-Rochow 2006a; Müller and Rosenberg 2006). Based on the disintegration of the multilayered system in *C. tasmanianus*, the proximal retinula is different from those described above. Here, the drop-shaped proximal retinula cells are arranged in scattered two-cell units. At the apex, each of these units produces a subrhabdom of two pectinate, interdigitating rhabdomeres (Müller and Meyer-Rochow 2006b).

Distal and proximal retinulae are tightly joined by 20–300 unpigmented cells, such as circumretinular sheath cells and, if present (only in clusters of closely adjoined ommatidia), also interommatidial sheath cells (e.g., Fig. 7.2b). The circumretinular sheath cell system is most elaborate in *C. tasmanianus* (Müller and Rosenberg 2006; Müller and Meyer-Rochow 2006b; Müller et al. 2011). Median processes may emerge to penetrate the distal retinula multilayer by projecting through the infraretinular spaces (Fig. 7.5g). Branching circumretinular (and interommatidial) sheath cell processes may multiple times envelop the somata of both distal and proximal retinula cells (Fig. 7.5c). These sheath cell processes also project proximally to the bottom of the cup, where they wrap the peripheral bundles of retinula cell axons and then resemble myelinated glial sheaths. Only in *Lithobius* spp. and *E. fasciatus* the 20–200 interommatidial (interocellar) sheath cells with electron-dense cytoplasm extend from the cornea down to the extracellular (basal) matrix (Fig. 7.2b). Comparable to the intraommatidial sheath cells, small vertical cytoplasmic processes branch with the processes of neighboring cells, thus, together with the basal matrix, forming a multilayered space between neighboring ommatidia (six to ten layers). In dark-adapted animals, the cytoplasm is heavily filled with polymorphic vacuoles that might provide a layer to reflect light, which may significantly increase sensitivity (Land 1972; Müller et al. 2011).

In Scolopendromorpha and Craterostigmomorpha, cup-shaped ommatidia are lined by a relatively thick extracellular (basal) matrix, which consists of both extracellular and cellular components. In *Scolopendra* spp., it contains a broad network of collagen fibers. In Lithobiomorpha, the basal matrix is relatively thin and hard to identify. In all pleurostigmophoran ommatidia examined so far, the basal matrix is only perforated near the base of each eye cup, where retinula cell axons (wrapped by proximal processes of circumretinular sheath cells) project into the visual nerve (Müller and Meyer-Rochow 2006a, b; Müller and Rosenberg 2006) (Fig. 7.2b). A sheath of multiple external pigment cells delimits the eye cup in *C. tasmanianus* or the entire ommatidial cluster in *Scolopendra* spp. (Fig. 7.5e, f), *Lithobius* spp. (Figs. 7.2b and 7.5a), and *E. fasciatus*. The cytoplasm of external pigment cells exhibits small electron-lucent vacuoles, polymorphic granules, and numerous electron-dense pigment granules (Fig. 7.5f). Only in *C. tasmanianus*, regions rich in electron-lucent granules alternate with regions rich in more osmiophilic granules. The main purpose of external pigment cells may be to shield the cup's retina from scattered light from below (Müller and Meyer-Rochow 2006a, b; Müller and Rosenberg 2006).

7.2.3 Lateral Eyes in Penicillata (Diplopoda)

Bristly millipedes (Penicillata) are only a few millimeters long and exhibit clusters of 4–6 (*Polyxenus lagurus*: Polyxenidae) up to 10–11 (*Phryssonotus platycephalus*: Synxenidae) spaced ommatidia located on small protuberances at the posterolateral head (Müller et al. 2007; Enghoff et al. 2015). Penicillate ommatidia are cup shaped, may include a rudimentary crystalline cone, and measure around 20 μm in diameter and 15–44 μm in length (Müller et al. 2007). Like Scutigeromorpha among centipedes, Penicillata represent a key taxon in myriapod phylogeny as they are the sister group to all remaining millipedes (see review by Edgecombe 2015). Their minute ommatidia became the subject of three TEM analyses, which, however, revealed conflicting results with respect to the cone cell apparatus. Spies (1981) described two cone compartments (vitreous bodies) present underneath the corneal lens, produced by two cone cells (vitreous body cells) located in the proximal part of each ommatidium in *P. lagurus*. Paulus (2000) rather observed a variable number of 2–4 cone compartments/cells. Müller et al. (2007) did not detect any cone cells in any ommatidium of *P. lagurus* but, instead, found three eucone cells to establish a minute tripartite crystalline cone in the synxenid *P. platycephalus* (Figs. 7.2c and 7.6b). Similar to the cone cell system in scutigeromorph ommatidia, cone cell somata are located outside of and proximally to their cone compartment, placed either at the medioproximal periphery (Fig. 7.2c) or at the bottom of the cup surrounding the retinular axon bundle and resting on the extracellular (basal) matrix (Fig. 7.6e). The three cone cell somata contain plenty of screening pigment granules (Fig. 7.6e). From each cone cell soma, a thin distal cytoplasmic process projects through the infraretinular space (Figs. 7.2c and 7.6c), which widens distally to form a single, flattened cone compartment wedged between the corneal lens and the distal rhabdom (Figs. 7.2c and 7.6a, b). The rudimentary nature of the tripartite crystalline cone as well as the absence of proximal cone cell processes are interpreted as a consequence of the miniaturization and cup-shaped appearance of the penicillate ommatidium (Müller et al. 2007).

Penicillate ommatidia possess flat, biconvex corneal lenses, which are secreted by four to five pigmented corneagenous cells. They are here termed primary pigment cells because of their circular arrangement and the presence of screening pigment granules (Figs. 7.2c and 7.6a) Based on a detailed description by Müller et al. (2007), these primary pigment cells may be considered a further argument in favor of the homology of scutigeromorph and penicillate ommatidia. The somata of the

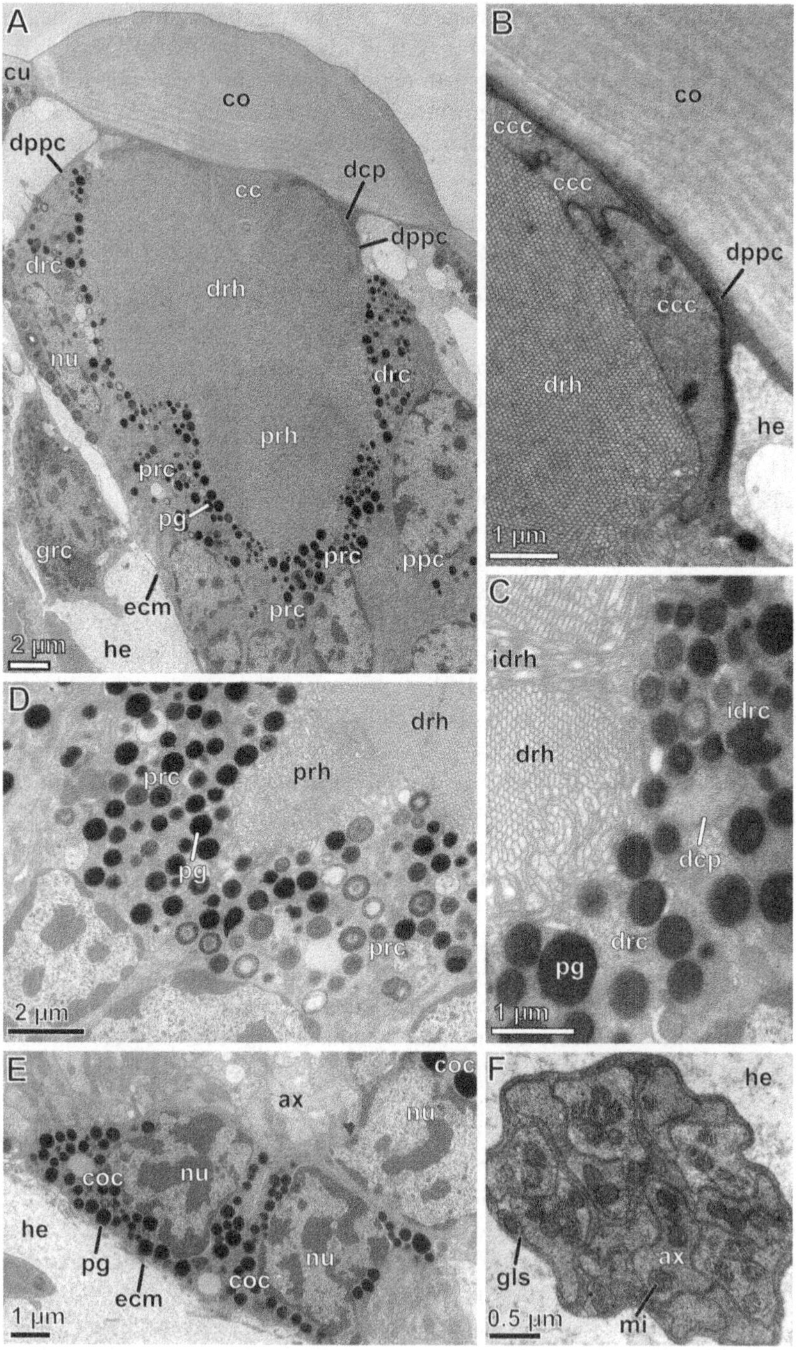
A
cu
co
dppc
dcp
cc
dppc
drc
drh
nu
drc
prh
prc
pg
grc
prc
ppc
ecm
prc
2 µm
he
B
co
ccc
ccc
dppc
ccc
drh
he
1 µm
C
idrh
idrc
drh
dcp
drc
pg
1 µm
D
drh
prc
prh
pg
prc
2 µm
E
coc
ax
nu
coc
nu
he
pg
ecm
nu
coc
1 µm
F
he
gls
ax
0.5 µm
mi

primary pigment cells are located far proximally from the corneal lens and shifted toward the interommatidial space (Figs. 7.2c and 7.6a). The dual-type retinula shows partial cell constancy: a variable number of 4–5 (*P. lagurus*) or 5–8 (*P. platycephalus*) voluminous distal retinula cells, whereas the number of proximal retinula cells is restricted to three. The distal retinula cells have straight apices and are arranged in a single circle (Figs. 7.2c and 7.6a). The cross-profile of the fused distal rhabdom is bilobed, appearing similar to an anvil or cloverleaf (compare fig. 4A and B in Müller et al. 2007). One of the distal retinula cells is irregular as it is considerably smaller and contributes to the distal rhabdom with only a minute rhabdomere (Fig. 7.6c). Similar to scutigeromorph ommatidia, the proximal rhabdom is fused and has a triangular cross-profile; opposing microvilli never interdigitate (Figs. 7.2c and 7.6d).

The thin and lamellated extracellular (basal) matrix lines the proximal periphery of the eye cup (Figs. 7.2c and 7.6a). It is established by various cellular constituents, such as the proximal retinula cells, ramifying processes of interstitial epidermal cells, and cone cell somata (only in *P. platycephalus*) (Müller et al. 2007). Ommatidial retinula axon bundles pierce the extracellular (basal) matrix (Fig. 7.2c). All penicillate ommatidia are devoid of external pigment cells (Spies 1981; Müller et al. 2007). Muscle fibers locally attach to the basal matrix (Müller et al. 2007). Like in scutigeromorphs (see Sect. 7.2.1 for discussion), these eye-associated muscles may alter the shape of the eye cup by local extension and thus initialize retinal micro-movements, resulting in the shift of the optical axes of distal and/or proximal rhabdomeres.

Fig. 7.6 Ultrastructure of the cup-shaped ommatidia of *Phryssonotus platycephalus* (Diplopoda, Penicillata). (**a**) Mediolongitudinal section of an ommatidium showing the dual-type retinula composed of distal and proximal retinula cells. The tripartite crystalline cone is wedged between the corneal lens and the subjacent distal rhabdom. A soma of a primary pigment cell is visible at the medioproximal rim of the ommatidial cup. (**b**) Paralongitudinal section showing the interspace of the corneal lens and distal rhabdom occupied by compartments of the three crystalline cone cells, as well as thin distal processes of the primary pigment cells. (**c**) Detail of the irregular distal rhabdomere formed by the tiny and irregular distal retinula cell in cross-section. Note the distal cone cell process projecting through the infraretinular space. (**d**) Detailed longitudinal view of the proximal retinula showing the proximal rhabdom being considerably smaller than the overlying distal rhabdom. (**e**) Bottom of the ommatidium with aligned intensively pigmented somata of the three cone cells. (**f**) Axon bundle of an ommatidium crosscut below the ommatidial cluster. Axons are encompassed by glial sheath. Originals. Labels: *ax* axon(s) of retinula cell(s), *ccc* cone cell compartment, *coc* crystalline cone cell (soma region), *cu* cuticle (surrounding corneal lenses), *dcp* distal cone cell process, *dppc* distal cytoplasmic process of primary pigment cell, *drc* distal retinula cell, *drh* distal rhabdom, *ecm* extracellular (basal) matrix of the ommatidia, *gls* glial sheath, *grc* granulocyte, *he* hemolymphatic space, *idrc* irregular distal retinula cell, *idrh* irregular distal rhabdomere, *mi* mitochondrion, *nu* nucleus, *pg* screening pigment granule, *prh* proximal rhabdom

7.2.4 Lateral Eyes in Chilognatha (Diplopoda)

Chilognatha with eyes show a great range of ommatidial numbers and arrangements. All ommatidia are cup shaped, lack a crystalline cone, and may be closely adjoined to form a functional compound eye (due to overlapping visual fields). For instance, a relatively low number of ommatidia is observed in *Glomeris* spp. (Glomerida); having five to nine ovoid ommatidia with a diameter of 50–150 μm present in one to two rows, the shape of the ommatidial cluster looks like a stretched drop (Bedini 1970; Spies 1981; Müller and Sombke 2015). A further example is *Polyzonium germanicum* (Colobognatha), exhibiting three to four ommatidia with a diameter of 30 μm arranged in a single row (Spies 1981). Other chilognaths possess clusters of several dozens of ommatidia. In *Ommatoiulus sabulosus* (Julida), 35–40 ommatidia are present in several rows in a triangular formation (Spies 1981). A comparable range is present in *Cylindroiulus punctatus*, which possesses 31–37 ommatidia with a diameter of approx. 50 μm (Kirwan and Nilsson 2019). In Craspedosomatidae, approximately 26 ommatidia are closely aggregated in a triangular field (Spies 1981; Müller and Sombke 2015).

The ommatidia of *Glomeris* spp. and *P. germanicum* include a biconvex, multilamellated corneal lens that is secreted by an unspecified number of flattened and strongly pigmented (with ommochromes) cornea-secreting epithelial cells (Fig. 7.7a). These cells project thin processes reaching between the rhabdom and the bottom side of the cornea (Bedini 1970; Müller and Sombke 2015). The ommatidia of julid millipedes are largely similar but slightly differ from glomerid ommatidia by having corneal lenses with a distinctly truncated base (Fig. 7.7e), secreted by weakly pigmented epithelial (corneagenous) cells. The retinulae of Chilognatha are composed of highly prismatic retinula cells in a cup-shaped arrangement (Fig. 7.7a, e). The ubiquitous presence of a dual-type retinula is likely, however, not definitely confirmed for all chilognathan eyes investigated (Müller and Sombke 2015). Usually, the distal retinula comprises multiple layers of cubical or highly prismatic cells in a circular formation, always with straight apices (Fig. 7.7b, g). Similar to most pleurostigmophoran ommatidia, the distal rhabdomeres are all tightly adjoined and form a fused rhabdom (Fig. 7.7c, f, g). Their microvilli are arrayed perpendicular to the optical axis. The total number of distal retinula cells is usually counted by hundreds and may vary considerably from 100 to more than 200 in Glomerida, 300–400 in Julida, and up to 450–500 in Craspedosomatidae (Bedini 1970; Spies 1981; Müller and Sombke 2015). The number of distal retinula cells encountered in each of the horizontally stacked circles is also high and may vary as well. For instance, a minimum number of 35 distal retinula cells was found in a given layer within the ommatidia of *C. punctatus* (Kirwan and Nilsson 2019). In glomerid, polyzoniid, craspedosomatid, and julid ommatidia, specialized retinula cells, equivalent to the proximal retinula cells in Pleurostigmophora, contribute to the multilayered retinula with a proximal portion. Compared to the distal retinula cells, these proximal retinula cells are often smaller, show a higher density of screening pigment granules, and are arranged in a cup-like epithelial formation (e.g., *G. marginata*: Fig. 7.7c, d; *P. germanicum*: Fig. 7.7f, g). The proximal rhabdomeres usually show a different arrangement than the overlying distal rhabdomeres (Fig. 7.7h). In contrast to pleurostigmophoran ommatidia, however, the microvilli of opposing proximal rhabdomeres do not interdigitate.

Ommatidia of Chilognatha are surrounded by different kinds of sheath cells. The lateral borders of adjacent eye cups are separated by pigmented epidermal cells (interommatidial sheath cells). Furthermore, multiple basal layers of spindle-shaped, pigmented sheath cells may be present (Spies 1981). These sheath cells surround the entire ommatidial cluster, subjacent to the extracellular (basal) matrix, hence called external pigment cells. In glomerid, craspedosomatid, and juliform millipedes, these external pigment cells surround the ommatidial cluster in several layers (Fig. 7.7c–f; see also figs. 9.4 and 9.5 in Müller and Sombke (2015); however, they are mislabeled therein as "interommatidial sheath cells"). Basally, the ommatidia are delimited by a fibrillous extracellular (basal) matrix, which is pierced by axon bundles (Fig. 7.7a, e). The extracellular (basal) matrix also envelops the optic nerve (Müller and Sombke 2015).

7.3 Eye Development

Eye development in Myriapoda has been studied in only a few species and generally is regulated by the *Pax6* gene family (e.g., Callaerts et al. 2006). In *Glomeris marginata* (Diplopoda), two different *Pax6* genes (*Pax6.1* and *Pax6.2*) were detected, which were interpreted as ancestral duplications in the arthropod stem lineage (Prpic 2005). The onset of expression starts in the procephalic region in largely overlapping domains, which temporally and spatially resembles the composite expressions of *ey* and *toy* in *Drosophila melanogaster* (Prpic 2005). Similarly to *G. marginata*, in the eyeless species *Strigamia maritima* (Chilopoda), *Pax6A* and *Pax6B* are expressed in the visual lobe as well as in the neuroectoderm of every segment along the trunk (Hunnekuhl 2013). Several other eye developmental transcription factors known from *D. melanogaster* are expressed in the visual lobes of *G. marginata* and/or *S. maritima*: *sloppy-pair* (*slp*) and *otx* (Steinmetz et al. 2010; Janssen et al. 2011; Hunnekuhl 2013), *dpp* and *hedgehog* (*HH*) (Prpic 2004; Janssen 2012; Hunnekuhl 2013), *dachshund* and *homothorax* (Prpic and Tautz 2003), and several *Wnt* genes (Janssen et al. 2004, 2010; Hayden and Arthur 2014; Janssen and Posnien 2014; Brena 2015). As discussed by Hunnekuhl (2013) and Brena (2015), this conservation in gene expression between myriapods and *D. melanogaster* may imply that the molecular signature is more associated with the early differentiation of the protocerebral area, presumably also involved in the formation of mushroom bodies (compare Sombke et al. 2011b; Sombke and Rosenberg 2015), than with the differentiation of eyes.

In *Scolopendra* spp. (Chilopoda), precursors can be found during early embryonic development. Before hatching, retinula cells and the visual nerve are already developed (Heymons 1901). After hatching, all four considerably spaced ommatidia

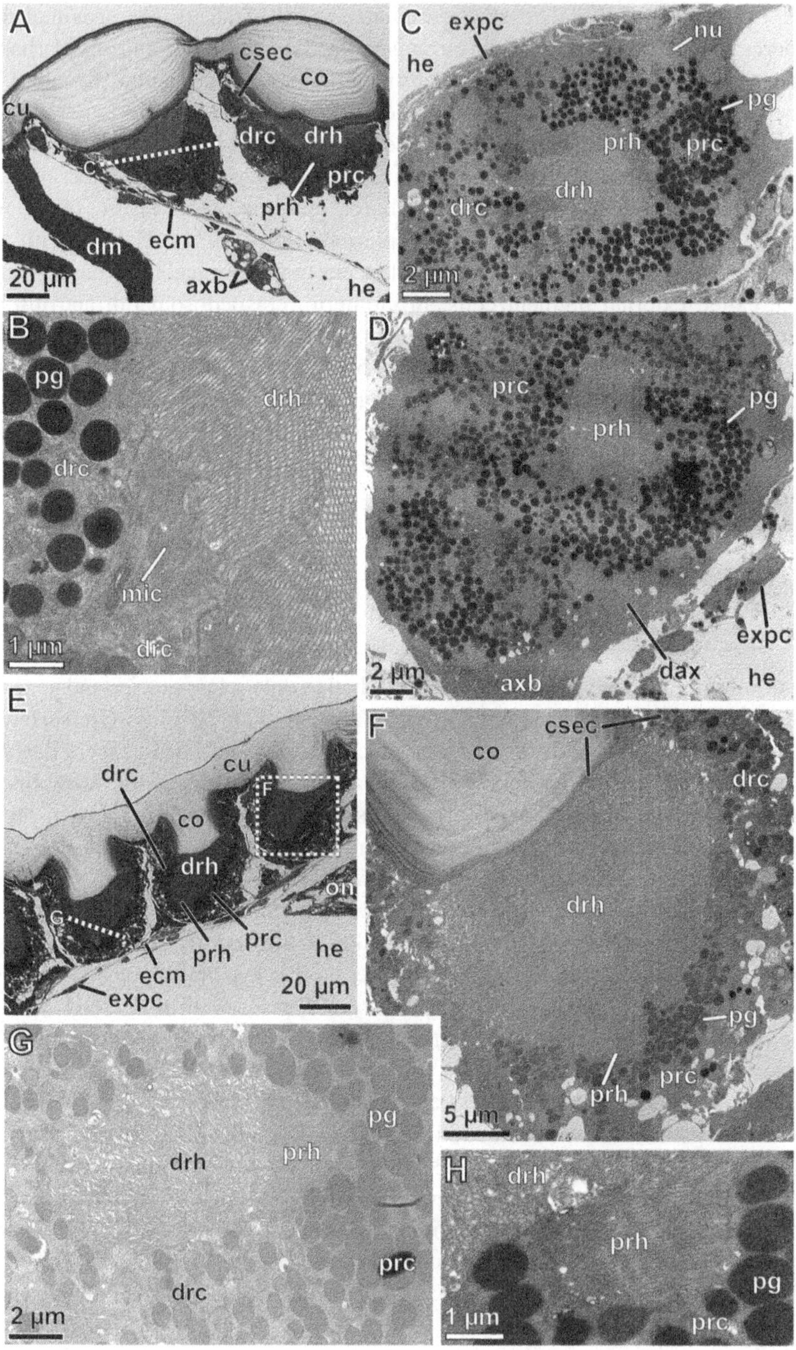
A
csec
co
cu
drc
drh
C
prc
prh
dm
ecm
20 µm
axb
he
C
expc
nu
he
pg
prh
prc
drh
drc
2 µm
B
pg
drh
drc
mic
1 µm
drc
D
prc
pg
prh
expc
2 µm
axb
dax
he
E
cu
drc
F
co
drh
on
G
prc
prh
he
ecm
20 µm
expc
F
csec
co
drc
drh
pg
prh
prc
5 µm
G
pg
prh
drh
prc
drc
2 µm
H
drh
prh
pg
1 µm
prc

are present and continuously grow in size with every molt, thus performing intercalary growth. The labeling of mitotic activity with BrdU revealed that a persistent proliferation takes place in scolopendromorph ommatidia, in particular pertaining to the corneagenous cells as well as, to an even greater extent, to the distal and proximal retinula cells (Harzsch et al. 2007). In contrast, the labeling of mitotic activity revealed a confined zone of mitotic cells in various postembryonic stages of *Scutigera coleoptrata*, including both amorphic juvenile and adult stages. This zone of active mitoses is restricted to the anterolateral rim of the growing compound eye, where lines of new ommatidia (protommatidia) are generated and subsequently added to the already differentiated union of ommatidia. The restriction of mitotic activity to this anterolateral growth zone implies that no cells of whatever type are added to scutigeromorph ommatidia once differentiated (Harzsch et al. 2007). A continuous proliferation across molts, similar to that detected in scolopendromorph eyes, was also suggested for the transformed ommatidia of Lithobiomorpha (Andersson 1976, 1981; Müller et al. 2011). The ablation of ommatidia in *Lithobius forficatus* is followed by regeneration. New ommatidia are added to the anterolateral

Fig. 7.7 Histology and ultrastructure of the cup-shaped ommatidia of glomerid and juliform millipedes (Diplopoda, Chilognatha). (**a–d**) Cup-shaped ommatidia of *Glomeris marginata* (Glomerida). (**a**) Cross-section of the left head flank showing two posteriorly located ommatidia in the longitudinal histological section (stained with toluidine-blue). The extracellular (basal) matrix and sheath of external pigment cells have loosened from the ommatidial bottom due to a fixation artifact. (**b–d**) Ultrastructural details revealed by TEM. (**b**) Straightly formed apices of several distal retinula cells in the longitudinal section. An extensive system of microtubules replaces the perirhabdomeric ER cisternae. (**c**) Oblique section through the medioproximal region of the dual-type retinula showing distal retinula cells to the left and proximal retinula cells to the right (section plane indicated by a dashed line in A). (**d**) Oblique section through the proximal retinula showing several heavily pigmented proximal retinula cells forming a fused proximal rhabdom. Axons of distal retinula cells project at the periphery of the eye cup and merge (along with the axons of the proximal retinula cells) into an axon bundle, seen at bottom of the image. Fixation artifacts affected the integrity of adjacent components, such as the extracellular matrix and external pigment cells, which appear disintegrated. (**e–h**) Cup-shaped ommatidia of an unidentified juliform millipede (Juliformia). (**e**) Cross-section through the left head flank and the triangular eye field with several closely adjoined ommatidia cut longitudinally. Toluidine-blue histology. (**f, g**) Ultrastructural details revealed by TEM. (**f**) Mediolongitudinal section of the ommatidium illustrating proportions of horizontally multistacked distal retinula cells versus the proximal retinula cells arranged in a flat cup (sector exemplarily indicated by a dashed box in E). Peripheral areas of the ommatidium partly appear disintegrated due to poor fixation of the axonal strands, extracellular (basal) matrix, and external pigment cells. (**g**) Obliquely cut transition of distal and proximal retinula cells. Distal retinula cells and related rhabdom, seen at the left, shading off into much more pigmented proximal retinula cells and rhabdom, at the right (section plane exemplarily indicated by a dashed line in E). (**h**) Close-up of proximal rhabdom in longitudinal section; microvilli of proximal rhabdomeres are cut longitudinally, perpendicular to the crosscut distal rhabdomeres on top. Originals. Labels: *axb* axon bundle of an ommatidium, *co* corneal lens, *csec* cornea-secreting epithelial cells (somata and subcorneal processes), *cu* cuticle (surrounding corneal lenses), *dax* axons of distal retinula cells, *dl* dilator muscle, *drc* distal retinula cell(s), *drh* distal rhabdom, *ecm* extracellular (basal) matrix of ommatdium/ommatidial cluster, *expc* external pigment cells, *he* hemolymphatic space, *mi* microtubules, *nu* nucleus, *on* optic nerve, *pg* screening pigment granule, *prh* proximal rhabdom, *va* vacuole

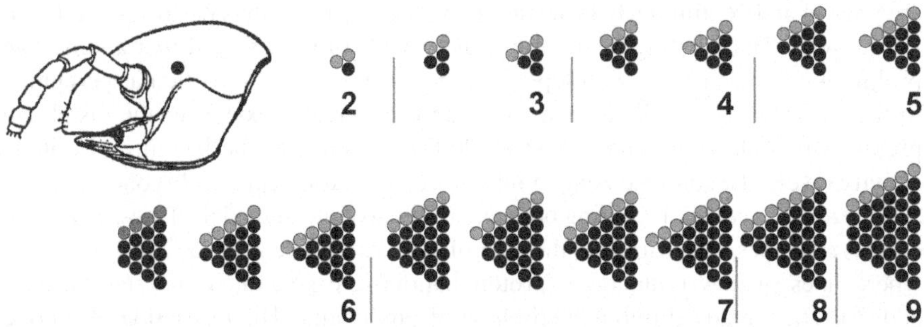

Fig. 7.8 Millipede eye development. Presence of ommatidia in several juvenile stages of *Cylindroiulus truncorum* (Diplopoda, Iulida). Ommatidia are not to scale with the schematic head. Grey circles indicate ommatidia that were added compared to the previous stage. Numbers indicate the quantity of ommatidial rows per eye field. Modified after Snodgrass (1952) and Harzsch et al. (2007)

tip of the ommatidial cluster (syn.: "ocellar field"). Their retinular axons become incorporated into the optic nerve (Joly and Herbaut 1968).

Likewise, in millipedes, new ommatidia are continuously added to the ommatidial cluster with every molt and, thereby, elongate the rows of earlier generated ommatidia (Peitsalmi and Pajunen 1991, 1992; Enghoff et al. 1993; Harzsch et al. 2006, 2007). This growth pattern results in the formation of a mostly triangular or bilobed cluster of ommatidia. Within this cluster, the ommatidia are closely adjoined and aligned in multiple rows reflecting their relative age and the taxon-specific pattern of eye development (Fig. 7.8). Therefore, counting eye rows is a generally accepted method to determine developmental stages in Julida and, to a certain degree, in Spirostreptida, Spirobolida, and Callipodida. However, Enghoff et al. (1993) stated that the traditional determination of developmental stages by counting the eye rows is inadequate in certain millipede species. Also, it was observed that some millipede species may lose entire eye rows in older stages or do not display a clear linear arrangement of their ommatidia anymore (Harzsch et al. 2007; Müller and Sombke 2015). Depending on the developmental stage, a given ommatidial cluster may contain anything from zero up to nine rows of ommatidia in *Cylindroiulus truncorum* (Fig. 7.8). Within each row, ommatidia display individual variations in age, shape, and size. The oldest and largest ommatidia are always found at the posterior or posteroventral edge of the cluster. A similar pattern is present in *Archispirostreptus gigas*, the ommatidia of which undergo successive intercalary growth. During postembryonic development, all ommatidia of *A. gigas* are surrounded by a distinct rim of mitotic cells (Harzsch et al. 2007). The intercalary growth in *S. oraniensis* (centipedes) resembles the continuing growth traced within the ommatidia of *A. gigas* (millipedes). The presence of a proliferation zone that generates new ommatidia at the anteroventral side of the eye field is a general feature in euarthropods, which is also found in Trilobita and Xiphosura, which likewise exhibit intercalary growth (Clarkson and Zhang 1991; Meadors et al. 2001; Smith et al. 2002; Harzsch et al. 2006, 2007). Scutigeromorpha and Pancrustacea differ from this pattern in that their proliferation zone generates ommatidia composed of a fixed array of cells with a restricted number. Thus, scutigeromorph ommatidia with partially fixed cell numbers (fixed number of four proximal retinula cells) are proposed to represent an intermediate stage in the evolution of mandibulate ommatidia (Harzsch et al. 2007).

7.4 Visual Neuropils Associated with Lateral Eyes

Very little is known about visual neuropils and information processing in myriapods, and many aspects of general architecture and neuronal connectivity remain unclear. Only five species have been investigated so far with regard to the neuronal architecture of visual neuropils (Chilopoda: *Scutigera coleoptrata*, *Thereuopoda clunifera*, *Lithobius forficatus, Scolopendra heros*; Diplopoda: *Orthoporus ornatus, Julus* sp.). Centipedes and millipedes carrying eyes composed of either regular or transformed and cup-shaped ommatidia possess two distinct visual neuropils in the lateral protocerebrum, commonly termed as lamina and medulla (e.g., Saint Remy 1887; Holmgren 1916; Hanström 1928; Hörberg 1931; Fahlander 1938; Melzer et al. 1996; Strausfeld 2005; Sombke et al. 2011a; Strausfeld 2012; Sombke and Harzsch 2015; Sombke and Rosenberg 2016). In Chilopoda and Diplopoda, the first-order integrating visual neuropil is the planar to convex lamina (="lamina ganglionaris") (Figs. 7.3a and 7.9a–e). In *S. coleoptrata* (Chilopoda), retinula cells are histaminergic, and long and short axons project from the ommatidia into the lamina and the medulla (Sombke and Harzsch 2015) (Fig. 7.9a, c). The majority of the axons of retinula cells (short axons) terminate in the lamina, where they form small club-shaped terminals in elongated, compact subdomains (Fig. 7.9c). Few axons (long, translaminar axons) pass through the lamina to innervate the medulla with larger club-shaped terminals (Fig. 7.9c). In *L. forficatus* (Chilopoda), likewise short and long retinula cell axons occur (Melzer et al. 1996) (Fig. 7.9d). Here, retinula cell axons have a constant diameter with regularly arranged bleb-like swellings, which are more pronounced near axon terminals. Short axons do not branch. Long axons exhibit intense branching (with two to five branches each) in the medulla with long collaterals that project to different regions of the neuropil (Fig. 7.9d). It is not known whether long axons possess synapses in the lamina as well (Melzer et al. 1996). In *S. coleoptrata*, within the area between lamina and medulla, long axons cross each other, forming a chiasm (Saint Remy 1887; Hörberg 1931; Sombke and Harzsch 2015) (Fig. 7.9c). Axons from the anteriormost domain in the lamina innervate the posteriormost domain of the medulla and vice versa (Sombke and Harzsch 2015). In the lamina of *S. coleoptrata*, each ommatidium is allocated to two relay neurons (monopolar cells), but no local interneurons were found (Strausfeld 2005). A cluster of neuronal somata is present between lamina and medulla. In *Scolopendra heros* (Chilopoda) and *Orthoporus ornatus* (Diplopoda), Strausfeld (2012) described

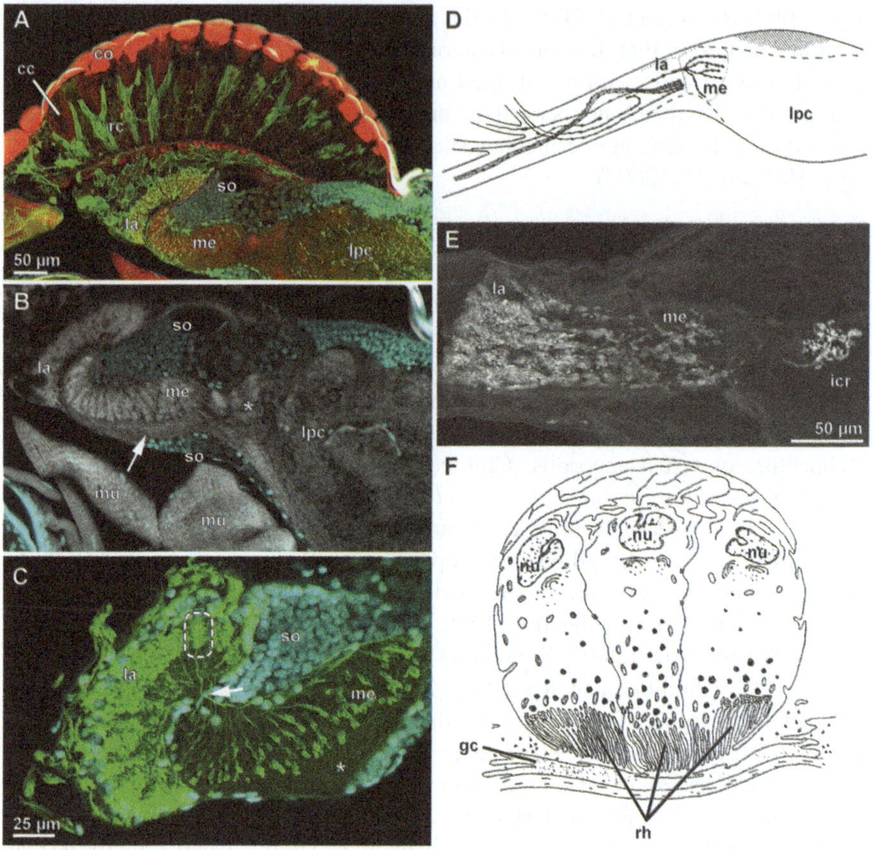

Fig. 7.9 Visual neuropils and intracerebral photoreceptors. (**a–c**) Immunohistochemical labeling of the visual neuropils of *Scutigera coleoptrata* (original and modified after Sombke and Harzsch 2015). (**a**) Labeling against synaptic proteins (red), histamine (green), and nuclei (cyan). Note that the red labeling of the ommatidial corneae is an artifact. Retinula cells are histaminergic. Terminations are evident in the lamina, the medulla, and sparsely in the lateral protocerebrum. (**b**) Immunohistochemical labeling against synaptic proteins (grey) and nuclei (cyan). The fibrillose subdomain (innervated by long retinula cell axons) is compartmentalized. The arrow marks the homogenous subdomain. Further medially, dense synaptic profiles (asterisk) are visible. (**c**) Immunohistochemical labeling against histamine (green) and nuclei (cyan). Short retinula cell axons terminate within distinct cartridges in the lamina (dashed rectangle). From/through each cartridge, a long, translaminar axon proceeds to the medulla. Long retinula cell axons cross each other in between the lamina and medulla, forming a chiasm (arrow). The homogenous subdomain of the medulla is not innervated by long retinula cell axons (asterisk). For more details, see Sombke and Harzsch (2015). (**d**) Retinula cell axons in *Lithobius forficatus* based on cobalt fillings, modified after Melzer et al. (1996). Short and long retinula cell axons terminate either in the lamina (short) or in the medulla (long). Short retinula cell axons do not branch, while long retinula cell axons exhibit intense branching in the medulla. (**e**) Immunohistochemical labeling against histamine in the lateral protocerebrum of *Lithobius forficatus*. The density of retinula cell terminals is much higher in the lamina than in the medulla. Additionally, several cells of the intracerebral photoreceptor organ are labeled. Axons of the spheroid organ might innervate the region of the medulla. Original. (**f**) Cellular architecture of the intracerebral photoreceptor organ in *L. forficatus*. Modified after Jamault-Navarro (1992). Labels: *cc* crystalline cone, *co* cornea, *gc* glia cell, *icr* intracerebral photoreceptor, *la* lamina, *lpc* lateral protocerebrum, *me* medulla, *mu* musculature, *nu* nucleus, *rh* rhabdomere, *rc* retinula cells, *so* somata

that photoreceptor axons project into the brain and segregate to different synaptic domains in the lateral protocerebrum.

In *S. coleoptrata*, the second-order visual neuropil is the spheroid to arcuate medulla, which is equipped with large tangential neurons that send their axons from their inner edge into the protocerebrum, where they terminate and make contact with the dendrites of descending neurons (Strausfeld 2005; Sombke and Harzsch 2015). A few somata are present at the distal anterior border of and proximal to the medulla (Fig. 7.9a–c) (Sombke and Harzsch 2015). The medulla appears subdivided into a fibrillose and a more homogenous subdomain, which was interpreted as an equivalent to the medulla externa (medulla) and medulla interna (lobula) of hexapods and crustaceans (Fig. 7.9b, c) (Hörberg 1931). The fibrillose subdomain (innervated by long retinula cell axons) is compartmentalized into parallel units (Fig. 7.9b). The homogeneous domain of the medulla is devoid of histaminergic retinula cell terminations. A slight layering or compartmentalization of the medulla was also described in millipedes (Holmgren 1916; Hanström 1928). Contrary to Hörberg's (1931) interpretation, the homogenous subdomain (in the second visual neuropil of *S. coleoptrata*) very likely does not correspond to a third visual neuropil present in Hexapoda and Malacostraca (compare Strausfeld 2012; Sombke and Harzsch 2015; Strausfeld and Olea-Rowe 2021). Besides lamina and medulla, dense synaptic areas are present in the lateral protocerebrum of *S. coleoptrata* (Fig. 7.9a) (Sombke and Harzsch 2015). A contralateral connection of the medullae was described in *S. coleoptrata*, *Lithobius* sp., *Scolopendra* sp., and *Julus* sp. (Saint Remy 1887; Haller 1904; Holmgren 1916; Hanström 1928). In *L. forficatus*, Golgi impregnations revealed large interneurons (presumably two neuron populations) that connect the medulla to the dorsal and ventrolateral protocerebrum (Melzer et al. 1996). In the proximal medulla, neurites form arborizations that are in close contact with terminals of long retinula cell axons. In the dorsal protocerebrum, large neurites form synaptic arborizations near the mushroom body pedunculus. In addition, collaterals project below the pedunculus, and large neurites possess further branches with synaptic arborizations that project to the ventrolateral protocerebrum (Melzer et al. 1996). According to Hanström (1928), the lateral protocerebrum (comprising the visual neuropils) is also connected ventrally to the central body via a commissure associated with a glomerular neuropil, which was termed "optic body." Originally, Holmgren (1916) introduced this term for isopods (Crustacea), where a small lentiform neuropil is a part of a commissure (optic commissure) that connects the second-order visual neuropils. Interestingly, these projections lead to a similar region where optic foci were described in Hexapoda (Strausfeld 1976; Melzer et al. 1996).

The morphological differentiation of short and long photoreceptor axons is correlated in general with the presence of different types of photoreceptive pigments (Melzer et al. 1996). In flies, axons of the retinula cells R7 and R8 (ultraviolet (UV) and blue-light receptive) project through the lamina to terminate in the medulla, crossing each other between the lamina and medulla (outer chiasm). In crayfish, only the irregular eighth retinula cell (blue/violet receptive) exhibits the same projection pattern (Strausfeld and Nässel 1981). The ommatidia of *S. coleoptrata* are highly UV receptive, and Meyer-Rochow et al. (2006) suggested that at least two visual pigments might be present in separate photoreceptive cells per ommatidium (but see also Sect. 7.6. below). Thus, a hypothetical UV-receptive retinula cell could be associated with long, translaminar axons, which might correlate with the irregular proximal retinula cell per ommatidium in *S. coleoptrata* (Müller et al. 2003b). At least in *S. coleoptrata* and partially in *L. forficatus*, both visual neuropils are organized in discrete termination sites, and long photoreceptor axons form a chiasm (a feature shared with hexapods and malacostracan crustaceans but not with branchiopod crustaceans). Consequently, lamina and medulla, short and long retinula cell axons, and the (outer) optic chiasms were hypothesized to be a part of the ground pattern of the visual system of Mandibulata (Melzer et al. 1996; Sombke and Harzsch 2015).

7.5 Intracerebral Photoreceptors

Within the protocerebrum of some centipedes and millipedes, intracerebral rhabdomeric photoreceptors were detected, the so-called accessory lateral eyes (Sahli 1966; Juberthie-Jupeau 1967; Jamault-Navarro 1992; Heithier and Melzer 2005; Nguyen Duy-Jacquemin 1974; Müller and Sombke 2015). In *Lithobius forficatus*, the organ is spheroid with a diameter of approx. 100 μm, located in the dorsolateral protocerebrum between the visual neuropils and the mushroom body (Jamault-Navarro 1992). It consists of approximately ten large cells that are histaminergic (Fig. 7.9e). Each of them possesses a deeply and regularly digitated internal membrane, similar to microvillar arrays in ommatidial rhabdoms (Fig 7.9f). The outer part is associated with a strongly granulated tissue layer and glial cells (Jamault-Navarro 1992). In *Cylindroiulus truncorum*, it is an ovoid structure with a dimension of about 20 × 5 μm, each positioned close to the ventral side of the medulla and embedded in a cluster of neuronal somata (Heithier and Melzer 2005). Each organ includes six retinula cells surrounding a central rhabdom-like structure that is star-shaped in cross-section. Retinula cell nuclei are arranged in two levels, and the microvilli of opposing cells may interdigitate locally. The retinula cells are overlaid by four cap cells and surrounded by glia-like cells (Heithier and Melzer 2005). Axons from all retinula cells presumably target the medulla. Based on their anatomy and similar structures in other arthropods (e.g., accessory eyes in Chelicerata and stemmata in Hexapoda; see also Spreitzer and Melzer 2003), Heither and Melzer (2005) suggested that these intracerebral photoreceptors are simple light receptors with no capability of delivering a resolved image. Simple nonocular receptors might serve a variety of roles, e.g., the monitoring of ambient light intensity, which is crucial for circadian rhythm. Likewise, in burrowing arthropods, nondirectional photoreception may trigger appropriate behaviors when breaking the substrate surface as well as finding shaded places (Nilsson 2009, 2013). The presence of lateral eyes and accessory eyes (intracerebral photoreceptors) thus suggests that there is a division of labor associated with different behaviors.

7.6 Visual Ecology, Physiology, and Behavior

In Scutigeromorpha, the distal rhabdom lines the crystalline cone, while the proximal rhabdom is located below its proximal tip (Fig. 7.2a). Thus, proximal retinula cells have narrow receptive fields centered on the ommatidial axis, whereas distal retinula cells can be expected to have wide receptive fields deviating strongly from the ommatidial axis (Müller et al. 2003b; Nilsson and Kelber 2007). Also, the crystalline cone separates distal receptor cells from each other, resulting in a pronounced off-axis sensitivity that increases the difference in receptive fields between distal and proximal receptors (Nilsson and Kelber 2007). In their detailed analysis, Nilsson and Kelber (2007) investigated the crystalline cone of *Scutigera coleoptrata*, which after dissection were optically homogeneous and had a low refractive index (<1.38). In measuring the focal length, they showed that the corneal lenses place a sharp image close to the proximal tip of the crystalline cone and that the cone has the function of a vitreous space. When the crystalline cone is missing (Pleurostigmophora and Chilognatha), the rhabdom reaches very close to the inner corneal surface (compare Fig. 7.1). Thus, distal retinula cells will have a wider angular sensitivity than proximal ones as they are also not shielded by other parts of the rhabdom (Nilsson and Kelber 2007). Another consequence of this rhabdom design is that the distal receptor cells of both sides of the eye have a different optic (visual) axis (as well as compared to proximal receptor cells). The optic axes of both distal and proximal rhabdomeres in the peripheral ommatidia of *S. coleoptrata* may be shifted dynamically due to the contracting action of associated muscle fibers (see Sect. 7.2.1). The architecture of the dual-type retinula (two or more layers) in combination with a high number of retinula cells might be indicative of dynamic neural summation (Nilsson and Kelber 2007). At high light intensities, proximal receptors may be used alone for high spatial resolution. With decreasing illumination, distal receptors could be added to make the spatial channel broader and more sensitive. Altogether, the myriapod ommatidium is thought to be built for maximizing the dynamic range of vision (Nilsson and Kelber 2007).

Only a few photopigments have been identified in myriapods. In *S. coleoptrata*, peropsin/RGR-like genes were identified (Henze and Oakley 2015), which implies that the last common ancestor of arthropods at least possessed one peropsin. Different (but single) r-opsins were identified in several centipede and millipede species (Fleming et al. 2018). Most myriapod visual opsins cluster in the arthropod LWS (long-wavelength-sensitive) clade, with the exception of the centipedes *S. coleoptrata* and *Craterostigmus tasmanianus*, which cluster in the MWS (middle-wavelength-sensitive) clade. Fleming et al. (2018) argued that the distribution of opsin paralogues across Myriapoda suggests that the molecular components of their visual system are degenerate in comparison to the base of their stem linage, which might be consistent with the highly modified ommatidia. In *S. coleoptrata*, electrophysiological recordings from their compound eyes revealed two sensitivity peaks, one in the vicinity to light at 448 nm (blue) and the second one in the ultraviolet region around 350 nm (Meyer-Rochow et al. 2006). At wavelengths longer than

549 nm, response curves fell linearly with a very weak response to red light. Although Meyer-Rochow et al. (2006) suggested that there could be two visual pigments in this species, Fleming et al. (2018) only found a single one, which points to the primary and secondary sensitivity/absorbance of a single blue-green sensitive opsin (compare also Cronin 2014). Similar electrophysiological results were obtained in *L. forficatus* (Meyer-Rochow et al. 2006). High UV sensitivity might help these species to discriminate secondarily illuminated patches from openings, crevices, and holes in the ground leading to the outside as most materials absorb some UV radiation, and open space can reliably be assessed by a determination of the UV amount present (Meyer-Rochow et al. 2006).

Centipedes exhibit negative phototaxis to a certain degree (but exceptions do occur), and even some blind pleurostigmophoran species (without transformed cup-shaped ommatidia) show flight behavior (Plateau 1887; Verhoeff 1902; Klein 1934; Scharmer 1935; Bauer 1955; Demange 1956; Görner 1959; Meske 1961). Bright light has no influence on the locomotion direction of *Scutigera coleoptrata* and *Scolopendra cingulata* while running, but both species eventually favor darker places (Görner 1959) (Fig. 7.10). Meske (1961) observed that specimens of *Lithobius forficatus* predominantly prefer shady places and actively avoid direct illumination. A dark-colored soil is preferred over a brighter one in *S. coleoptrata*, *L. forficatus*, *S. cingulata*, and *S. subspinipes* (positive scototaxis) (Klein 1934; Bauer 1955; Görner 1959; Meske 1961). Interestingly, the same preference is observed in blind species, such as Cryptopidae (Scolopendromorpha) and Geophilomorpha (Plateau 1886, 1887). Control experiments with eye-bearing species in total darkness or with two evenly dark-colored plates showed an even distribution, which points to the involvement of the visual system (Bauer 1955; Görner 1959; Meske 1961). It appears reasonable to assume that intracerebral

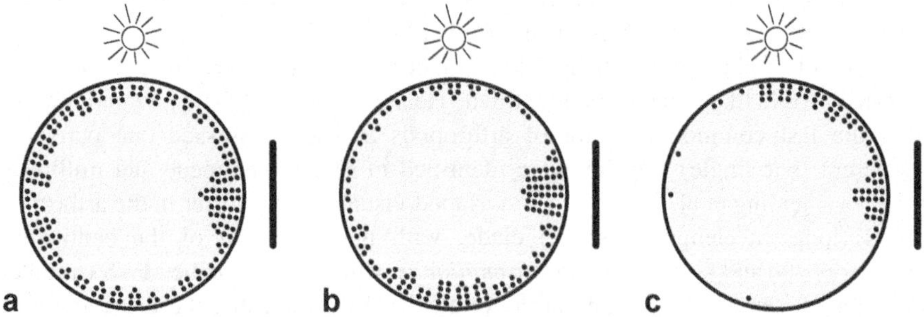

Fig. 7.10 Visual ecology in centipedes. Light and dark preferences of *Scutigera coleoptrata* (**a**), *Lithobius forficatus* (**b**), and *Scolopendra cingulata* (**c**). Experiments with artificial light source and a black plane. Specimens of *S. coleoptrata* (**a**) and *L. forficatus* (**b**) preferably running toward the black plane. The light source has no influence on the running direction. *L. forficatus* additionally prefers the light-averted side (**b**). In a and b, every point represents three trials. Specimens of *S. cingulata* (**c**; every point represents one trial) preferably running toward the light source and the black plane. Modified after Görner (1959)

photoreceptors (see above) are involved in simple light reception; however, they were not investigated in myriapods with regard to function.

Optomotoric experiments (behavior evoked by visual motion) revealed a weak but significant reaction in the direction of a rotating stimulus in *L. forficatus* (Meske 1961). This agrees with neuronal data and the presence of an optic chiasm. A reaction to polarized light has not been observed in centipedes (Görner 1959), even though at least the rhabdoms of most distal retinula cells in *S. coleoptrata* possess strictly parallel, unidirectional microvilli that should enable photoreceptors to detect linearly polarized light (Müller et al. 2003b). In contrast, Nilsson and Kelber (2007) argued that myriapod eyes are less suited for color and polarization vision at full spatial resolution and concluded that they are designed for monochrome vision with a unique mechanism for adaptation to changing light intensities. Reduced capabilities to detect colors were discovered experimentally by Meske (1961), who found *L. forficatus* to show no reaction to red light.

The cup-shaped ommatidia of *L. forficatus* exhibit fine structural changes in the light- and dark-adapted state (12:12 h cycle) (Bähr 1972; Müller and Rosenberg 2006). In light-adapted ommatidia, the distal retinula cells possess a ring of swollen cisternae of perirhabdomeric endoplasmatic reticulum around the axial rhabdom, which in the dark-adapted state are less voluminous (Müller and Rosenberg 2006). Additionally, screening pigment granules (ommochromes), highly abundant in both distal and proximal retinula cells, aggregate and establish a screening pigment shield around the entire rhabdom in the light-adapted phase, while rhabdomeric microvilli extend in the dark-adapted phase. The latter adaptation is typical for arthropod eyes so they can increase the diameter of rhabdoms and, thus, the sensitivity of the photoreceptors during the night (Meyer-Rochow 1999). Based on extracellular recordings in *L. forficatus*, Bähr (1965, 1967) showed that dark adaptation occurs at high stimulus intensities in a biphasic manner and seems to be completed after 20 min. Bähr (1967) postulated that this species orientates from bright to dark environments, and images or movements cannot be detected.

The exact role of vision in millipedes is difficult to assess and is poorly studied (Müller and Sombke 2015). Some species are known to be highly to extremely sensitive to light (Cloudsley-Thompson 1951b) or may possess a diurnal or nocturnal lifestyle (Cloudsley-Thompson 1951a; Banerjee 1967; Wongthamwanich et al. 2012) and/or exhibit both negative and positive phototaxis when illuminated at night (McKillup 1988). Studies on the visual ecology of eye-bearing millipedes may also be ambiguous as, for instance, in mass-migrating species such as *Ommatoiulus sabulosus*. Although this species was positively proved to be scototactic in lab experiments (Klein 1934), activity in both bright sunlight and at night has been observed (Dziadosz 1966; Fairhurst 1970; Demange 1960). Meyer-Rochow (2015) described the nocturnal mass migration in *Chamberlinius hualiensis*, a species that strongly avoids light. Presumably, intracerebral photoreceptors mediate light sensitivity. Other millipedes, e.g., *Coromus* sp., *Habrodesmus falx*, and *Ommatoiulus moreleti*, are highly attracted to artificial light at night and under laboratory conditions (Toye 1966; McKillup 1988), which was interpreted as a misled orientation behavior in the process of navigation. The visual system of *Cylindroiulus punctatus* presumably functions for seeking dark places and shelter, and it was concluded that predator avoidance or the visual recognition of conspecifics are impossible (Kirwan and Nilsson 2019). Investigations on the spatial resolution and visual performance of their eyes revealed that they could resolve a stimulus of 56° period. Thus, *C. punctatus* can only resolve a 1 m wide log from a distance of 2.9 m away (Kirwan and Nilsson 2019).

Acknowledgments We thank Ludwig Jonas (Electron Microscopic Center, University of Rostock, Germany) as well as Rabea Schlüter (Imaging Center of the Department of Biology, University of Greifswald, Germany) and Gabriele Uhl (Zoological Institute and Museum, University of Greifswald, Germany) for their technical assistance and permissions to use transmission electron microscopes. We are grateful to Caroline Viertel (Zoological Institute and Museum, University of Greifswald, Germany) for the immunohistochemical experiments in *Lithobius forficatus* and Iulia Barutia (Department of Evolutionary Biology, University of Vienna, Austria) for linguistic comments. Stefan Fischer (Structural Microscopy Core Facility, University of Tübingen, Germany) provided expertise and references regarding the probable impact of eye-associated muscles on optics in scutigeromorph ommatidia.

References

Andersson G (1976) Post-embryonic development of *Lithobius forficatus* (L.) (Chilopoda: Lithobiidae). Entomol Scand 7:161–168

Andersson G (1981) Taxonomical studies on the post-embryonic development in Swedish Lithobiomorpha (Chilopoda). Entomol Scand Suppl 16:105–124

Bähr RR (1965) Ableitung lichtinduzierter Potentiale von den Augen von *Lithobius forficatus* L. Naturwissenschaften 52:459

Bähr RR (1967) Elektrophysiologische Untersuchungen an den Ocellen von *Lithobius forficatus* L. Z Vgl Physiol 55:70–102

Bähr RR (1971) Die Ultrastruktur der Photorezeptoren von *Lithobius forficatus* L. (Chilopoda: Lithobiidae). Z Zellforsch Mikrosk Anat 116:70–93

Bähr RR (1972) Licht- und dunkeladaptive Änderungen der Sehzellen von *Lithobius forficatus* L. (Chilopoda: Lithobiidae). Cytobiologie 6:214–233

Bähr RR (1974) Contribution to the morphology of chilopod eyes. Symp Zool Soc Lond 32:383–404

Banerjee B (1967) Diurnal and seasonal variations in the activity of the millipedes *Cylindroiulus punctatus* (Leach), *Tachypodoiulus niger* (Leach) and *Polydesmus angustus* Latzel. Oikos 18:141–144

Bauer K (1955) Sinnesökologische Untersuchungen an *Lithobius forficatus*. Zoologische Jahrbücher Allgemeine Zoologie und Physiologie der Tiere 65:267–300

Bedini C (1968) The ultrastructure of the eye of a centipede *Polybothrus fasciatus* (Newport). Monitore Zoologico Italiano – Ital J Zool 2:31–47

Bedini C (1970) The fine structure of the eye in *Glomeris* (Diplopoda). Monitore Zoologico Italiano (NS) 4:201–219

Bitsch C, Bitsch J (2005) Evolution of eye structure and arthropod phylogeny. In: Koenemann S (ed) Crustaceans & Arthropod Relationships. CRC Press, Taylor & Francis Book, New York, pp 185–214

Blanke A, Wesener T (2014) Revival of forgotten characters and modern imaging techniques help to produce a robust phylogeny of the Diplopoda (Arthropoda, Myriapoda). Arthropod Struct Dev 43:63–75

Brena C (2015) Myriapoda. In: Wanninger A (ed) Evolutionary developmental biology of invertebrates 3. Springer, Vienna, pp 141–189
Brewer MS, Sierwald P, Bond JE (2012) Millipede taxonomy after 250 years: classification and taxonomic practices in a mega-diverse yet understudied arthropod group. PLoS One 7:e37240
Burtt ET, Patterson JA (1970) Internal muscle in the eye of an insect. Nature 228:183–184.
Callaerts P, Clements J, Francis C, Hens K (2006) Pax6 and eye development in Arthropoda. Arthropod Struct Dev 35:379–391
Clarkson ENK, Zhang X-G (1991) Ontogeny of the carboniferous trilobite *Paladin eichwaldi shunnerensis* (King, 1914). Trans R Soc Edinb Earth Sci 82:277–295
Cloudsley-Thompson JL (1951a) Studies in diurnal rhythms I. rhythmic behaviour in millipedes. J Exp Biol 28:165–172
Cloudsley-Thompson JL (1951b) On the responses to environmental stimuli, and the sensory responses of millipedes (Diplopoda). Proc Zool Soc London 121:253–277
Cronin TW (2014) Visual ecology. Princeton University Press, Princeton
Demange J-M (1956) Contribution a l'étude de la biologie, en captivité de *Lithobius piceus gracilitarsis* Bröl. (Myriapode-Chilopode) Bulletin du Muséum national d'Histoire naturelle (Paris) Sér 2, 28:388–393
Demange J-M (1960) Sur un important rassemblement de *Schizophyllum sabulosum* L. (Myriapode-Diplopode). Cahiers des Naturalistes: Bulletin des Naturalistes Parisiens 16:89–91
Dziadosz C (1966) Materialy do znajomosci rozmieszczejnia krocinogów (Diplopoda) w Polsce. Fragmenta Faunistica 13:1–31
Edgecombe GD (2010) Arthropod phylogeny: an overview from the perspectives of morphology, molecular data and the fossil record. Arthropod Struct Dev 39:74–87
Edgecombe GD (2015) Diplopoda – phylogenetic relationships. In: Minelli A (ed) Treatise on zoology – anatomy, taxonomy, biology. The Myriapoda, vol 2. Brill, Leiden, pp 353–362
Enghoff H (1984) Phylogeny of millipedes – a cladistic analysis. J Zool Syst Evol Res 22:8–26
Enghoff H (2011) East African giant millipedes of the tribe Pachybolini (Diplopoda, Spirobolida, Pachybolidae). Zootaxa 2753:1–41
Enghoff H, Dohle W, Blower JG (1993) Anamorphosis in millipedes (Diplopoda)—the present state of knowledge with some developmental and phylogenetic considerations. Zool J Linnean Soc 109:103–234
Enghoff H, Golovatch SI, Short M, Stoev P, Wesener T (2015) Diplopoda – taxonomic overview. In: Minelli A (ed) Treatise on zoology – anatomy, taxonomy, biology, vol 2. The Myriapoda, Brill, Leiden, pp 363–453
Fahlander K (1938) Beiträge zur Anatomie und systematischen Einteilung der Chilopoden. Zoologiska Bidrag fran Uppsala 17:1–148
Fairhurst CP (1970) Activity and wandering in *Tachypodoiulus niger* (Leach) and *Schizophyllum sabulosum* (L.). Bulletin du Muséum National d'Histoire Naturelle (Paris) 41:61–66
Fleming JF, Kristensen RM, Sørensen MV, Park T-YS, Arakawa K, Blaxter M, Rebecchi L, Guidetti R, Williams TA, Roberts NW, Vinther J, Pisani D (2018) Molecular palaeontology illuminates the evolution of ecdysozoan vision. Proc R Soc B Biol Sci 285:20182180
Görner P (1959) Optische Orientierungsreaktionen bei Chilopoden. Z Vgl Physiol 42:1–5
Graber V (1880) Über das unicorneale Tracheaten-und speciell das Arachnoideen-und Myriopoden-Auge. Arch Mikrosk Anat 17:58–93
Grenacher H (1880) Über die Augen einiger Myriapoden. Zugleich eine Entgegnung an V. Graber. Archiv für mikroskopische Anatomie 18:415–467
Haller B (1904) Über den allgemeinen Bau des Tracheatensyncerebrums. Arch Mikrosk Anat 65:181–279
Hanström B (1928) Vergleichende Anatomie des Nervensystems der Wirbellosen Tiere unter Berücksichtigung seiner Funktion. Julius Springer Verlag, Berlin
Hanström B (1934) Bemerkungen über das Komplexauge der Scutigeriden. Lunds Universitets Arskrift 30:1–14

Harzsch S, Müller CHG, Wolf H (2005) From variable to constant cell numbers: cellular characteristics of the arthropod nervous system argue against a sister-group relationship of Chelicerata and Myriapoda but favour the Mandibulata concept. Dev Genes Evol 215:53–68

Harzsch S, Melzer RR, Müller CHG (2006) Eye development in Myriapoda: implications for arthropod phylogeny. Nor J Entomol 53:187–190

Harzsch S, Melzer RR, Müller CHG (2007) Mechanisms of eye development and evolution of the arthropod visual system: the lateral eyes of myriapoda are not modified insect ommatidia. Org Divers Evol 7:20–32

Hayden L, Arthur W (2014) The centipede *Strigamia maritima* possesses a large complement of Wnt genes with diverse expression patterns. Evol Dev 16:127–138

Heithier N, Melzer RR (2005) The accessory lateral eye of a diplopod, *Cylindroiulus truncorum* (Silvestri, 1896) (Diplopoda: Julidae). Zoologischer Anzeiger – A J Comp Zool 244:73–78

Hengstenberg R (1971) Das Augenmuskelsystem der Stubenfliege *Musca domestica*. Kybernetik 9:56–77

Hengstenberg R (1972) Eye movements in the housefly *Musca domestica*. In: Wehner R (ed) Information processing in the visual Systems of Anthropods: symposium held at the Department of Zoology, University of Zurich, march 6–9, 1972. Springer, Berlin, Heidelberg, pp 93–96

Henze MJ, Oakley TH (2015) The dynamic evolutionary history of Pancrustacean eyes and opsins. Integr Comp Biol 55:830–842

Hesse R (1901) Untersuchungen über die Organe der Lichtempfindung bei niederen Thieren. VII. Von den Arthropoden Augen. Zeitschrift für wissenschaftliche Zoologie (Leipzig) 70:347–473

Heymons R (1901) Die Entwicklungsgeschichte der Scolopender. Zoologica 33:1–244

Holmgren N (1916) Zur vergleichenden Anatomie des Gehirns: Von Polychaeten, Onychophoren, Xiphosuren, Arachniden, Crustaceen, Myriapoden und Insekten. Vorstudien zu einer Phylogenie der Arthropoden. Kungliga Svenska Vetenskapsakademiens Handligar 56:1–315

Hörberg T (1931) Studien über den komparativen Bau des Gehirns von *Scutigera coleoptrata* L. Lunds Universitets Arsskrift NF Avd 2(27):1–24

Hunnekuhl VS (2013) Molecular subdivision and cell type specification in centipede head and brain development. Ph.D. Thesis. University of Cambridge

Jamault-Navarro C (1992) Sur la Présence d'une Structure Rhabdomérique Localisée Intracérébralement dans le Protocérébron de *Lithobius forficatus* L (Myriapode, Chilopode). Ber.nat-medVerein Innsbruck Suppl 10:81–86

Janssen R (2012) Segment polarity gene expression in a myriapod reveals conserved and diverged aspects of early head patterning in arthropods. Dev Genes Evol 222:299–309

Janssen R, Posnien N (2014) Identification and embryonic expression of *Wnt2*, *Wnt4*, *Wnt5* and *Wnt9* in the millipede *Glomeris marginata* (Myriapoda: Diplopoda). Gene Expr Patterns 14:55–61

Janssen R, Prpic N-M, Damen WGM (2004) Gene expression suggests decoupled dorsal and ventral segmentation in the millipede *Glomeris marginata* (Myriapoda: Diplopoda). Dev Biol 268:89–104

Janssen R, Le Gouar M, Pechmann M, Poulin F, Bolognesi R, Schwager EE, Hopfen C, Colbourne JK, Budd GE, Brown SJ, Prpic N-M, Kosiol C, Vervoort M, Damen WG, Balavoine G, McGregor AP (2010) Conservation, loss, and redeployment of Wnt ligands in protostomes: implications for understanding the evolution of segment formation. BMC Evol Biol 10:374

Janssen R, Budd GE, Damen WGM (2011) Gene expression suggests conserved mechanisms patterning the heads of insects and myriapods. Dev Biol 357:64–72

Joly R (1969) Sur l'ultrastructure de l'oeil de *Lithobius forficatus* L. (Myriapode Chilopode). Comptes rendus hebdomadaires des Séances de l'Académie des Sciences, Série D 268:3180–3182

Joly R, Herbaut C (1968) Sur la régénération oculaire chez *Lithobius forficatus* L. (Myriapode Chilopode). Archives de zoologie expérimentale et générale 109:591–613

Juberthie-Jupeau L (1967) Existence d'organes neuraux intracérébraux chez les Glomeridia (Diplopodes) épigés et carvernicoles. Comptes rendus hebdomadaires des Séances de l'Académie des Sciences, Série D 264:89–92

Kirwan JD, Nilsson D-E (2019) A millipede compound eye mediating low-resolution vision. Vis Res 165:36–44

Klein K (1934) Über die Helligkeitsreaktionen einiger Arthropoden. Zeitschrift für wissenschaftliche Zoologie (Leipzig) 145:1–38

Land MF (1972) Mechanisms of orientation and pattern recognition by jumping spiders (Salticidae). In: Wehner R (ed) Information processing in the visual Systems of Arthropods. Springer, Berlin, pp 231–247

McKillup SC (1988) Behaviour of the millipedes *Ommatoiulus moreletii*, *Ophyiulus verruculiger* and *Oncocladosoma castaneum* in response to visible light; an explanation for the invasion of houses by *Ommatoiulus moreletii*. J Zool 215:35–46

Meadors S, McGuiness C, Dodge FA, Barlow RB (2001) Growth, visual field, and resolution in the juvenile *Limulus* lateral eye. Biol Bull 201:272–274

Melzer RR, Petykó Z, Smola U (1996) Photoreceptor axons and optic neuropils in *Lithobius forficatus* (Linnaeus, 1758) (Chilopoda, Lithobiidae). Zool Anz 235:177–182

Meske C (1961) Untersuchungen zur Sinnesphysiologie von Diplopoden und Chilopoden. J Comp Physiol A 45:61–77

Meyer-Rochow VB (1999) Compound eye: circadian rhythmicity, illumination, and obscurity. In: Eguchi E, Tominaga Y (eds) Atlas of arthropod sensory receptors. Springer, Tokyo, pp 97–124

Meyer-Rochow VB (2015) Compound eyes of insects and crustaceans: some examples that show there is still a lot of work left to be done: compound eyes of insects and crustaceans. Insect Science 22:461–481

Meyer-Rochow VB, Müller CHG, Lindström M (2006) Spectral sensitivity of the eye of *Scutigera coleoptrata* (Linnaeus, 1758) (Chilopoda: Scutigeromorpha: Scutigeridae). Appl Entomol Zool 41:117–122

Minelli A (ed) (2011) The Myriapoda. Brill, Leiden

Müller CHG (2008) Vergleichend-ultrastrukturelle Untersuchungen an Augen ausgewählter Hundertfüsser (Mandibulata: Chilopoda) und zur Bedeutung von Augenmerkmalen für die phylogenetische Rekonstruktion der Euarthropoda. Cuvillier Verlag

Müller CHG, Meyer-Rochow VB (2006a) Fine structural organization of the lateral ocelli in two species of *Scolopendra* (Chilopoda: Pleurostigmophora): an evolutionary evaluation. Zoomorphology 125:13–26

Müller CHG, Meyer-Rochow VB (2006b) Fine structural description of the lateral ocellus of *Craterostigmus tasmanianus* Pocock, 1902 (Chilopoda: Craterostigmomorpha) and phylogenetic considerations. J Morphol 267:850–865

Müller CHG, Rosenberg J (2006) Homology of lateral ocelli in the Pleurostigmophora? New evidence from the retinal fine structure in some lithobiomorph species (Chilopoda: Lithobiidae). Nor J Entomol 53:165–186

Müller CHG, Rosenberg J (2009) Morphology is still an indispensable discipline in zoology: facts and gaps from Chilopoda. Soil Organisms 81:387–398

Müller CHG, Sombke A (2015) Diplopoda – Sense organs. In: Minelli A (ed) Treatise on zoology-anatomy, taxonomy, biology. The Myriapoda, Brill, Leiden, pp 181–235

Müller CHG, Rosenberg J, Meyer-Rochow VB (2003a) Hitherto undescribed interommatidial exocrine glands in Chilopoda. Afr Invertebr 44:185–197

Müller CHG, Rosenberg J, Richter S, Meyer-Rochow VB (2003b) The compound eye of *Scutigera coleoptrata* (Linnaeus, 1758) (Chilopoda: Notostigmophora): an ultrastructural reinvestigation that adds support to the Mandibulata concept. Zoomorphology 122:191–209

Müller CHG, Sombke A, Rosenberg J (2007) The fine structure of the eyes of some bristly millipedes (Penicillata, Diplopoda): additional support for the homology of mandibulate ommatidia. Arthropod Struct Dev 36:463–476

Müller CHG, Sombke A, Hilken G, Rosenberg J (2011) Chilopoda – Sense Organs. In: Minelli A (ed) Treatise on zoology-anatomy, taxonomy, biology. The Myriapoda, Brill, Leiden, pp 235–278

Nguyen Duy-Jacquemin M (1974) Les organes intracérébraux de *Polyxenus lagurus* et comparaison avec les organes neuraux d'autres diplopodes. Symp Zool Soc Lond 32:211–216

Nguyen Duy-Jacquemin M (1996) Systématique et biogéographie des diplopodes pénicillates des Îles Canaries et du Cap Vert. Mémoires du Muséum National d'Histoire Naturelle 169:113126

Nilsson D-E (2009) The evolution of eyes and visually guided behaviour. Philos Trans R Soc Lond B Biol Sci 364:2833–2847

Nilsson D-E (2013) Eye evolution and its functional basis. Vis Neurosci 30:5–20

Nilsson D-E, Kelber A (2007) A functional analysis of compound eye evolution. Arthropod Struct Dev 36:373–385

Paulus HF (1979) Eye structure and the monophyly of the Arthropoda. In: Gupta AP (ed) Arthropod phylogeny. Van Nostrand Reinhold, New York, pp 299–383

Paulus HF (2000) Phylogeny of the Myriapoda–Crustacea–Insecta: a new attempt using photoreceptor structure. J Zool Syst Evol Res 38:189–208

Peitsalmi M, Pajunen VI (1991) Mapping the ocular field in *Proteroiulus fuscus* (Am Stein) (Diplopoda, Blaniulidae). Ann Zool Fenn 28:31–39

Peitsalmi M, Pajunen VI (1992) Eye growth in *Choneiulus palmatus* (Nemec) and *Nopoiulus kochii* (Gervais) (Diplopoda, Blaniulidae). Ann Zool Fenn 29:39–46

Plateau F (1886) Recherches sur la perception de la lumière par les Myriapodes aveugles. Journal de l'anatomie et de la physiologie normales et pathologiques de l'hommes et des animaux 22:431–457

Plateau F (1887) Recherches expérimentales sur la vision chez les Arthropodes (première partie). – a. Résumé des travaux effectués jusqu'en 1887 sur la structure et le fonctionnement des yeux simples. b. Vision chez les Myriopodes. Bulletins de l'Académie royale des sciences, des lettres et des beaux-arts de Belgique 3. Sér.: 407–448

Prpic N-M (2004) Homologs of *wingless* and *decapentaplegic* display a complex and dynamic expression profile during appendage development in the millipede *Glomeris marginata* (Myriapoda: Diplopoda). Front Zool 1:1–12

Prpic N-M (2005) Duplicated *Pax6* genes in *Glomeris marginata* (Myriapoda: Diplopoda), an arthropod with simple lateral eyes. Zoology 108:47–53

Prpic N-M, Tautz D (2003) The expression of the proximodistal axis patterning genes *distal-less* and *dachshund* in the appendages of *Glomeris marginata* (Myriapoda: Diplopoda) suggests a special role of these genes in patterning the head appendages. Dev Biol 260:97–112

Richter S, Loesel R, Purschke G, Schmidt-Rhaesa A, Scholtz G, Stach T, Vogt L, Wanninger A, Brenneis G, Doring C, Faller S, Fritsch M, Grobe P, Heuer C, Kaul S, Moller O, Muller C, Rieger V, Rothe B, Stegner M, Harzsch S (2010) Invertebrate neurophylogeny: suggested terms and definitions for a neuroanatomical glossary. Front Zool 7:29

Sahli F (1966) Contribution a l'étude de la péridomorphose et du système neurosécréteur des Diplopodes Iulides. PhD, Université de Dijon

Saint Remy G (1887) Contribution a l'étude du cerveau chez les arthropodes trachéates. Archives de zoologie experimentale et generale 2:1–274

Scharmer J (1935) Die Bedeutung der Rechts-Links-Struktur und die Orientierung bei *Lithobius forficatus* L. Zoologische Jahrbücher Allgemeine Zoologie und Physiologie der Tiere 54:459–506

Shear WA, Edgecombe GD (2010) The geological record and phylogeny of the Myriapoda. Arthropod Struct Dev 39:174–190

Short M, Huynh C (2006) Redescription of *Phryssonotus novaehollandiae* (Silvestri, 1923) with details of post-embryonic stadia. Nor J Entomol 53:211–222

Sierwald P, Shear WA, Shelley RM, Bond JE (2003) Millipede phylogeny revisited in the light of the enigmatic order Siphoniulida. J Zool Syst Evol Res 41:87–99

Smith K, Ridings C, Dodge FA, Barlow RB (2002) Development of the lateral eyes of juvenile *limulus*. Biol Bull 203:222–223

Snodgrass RE (1952) A textbook of arthropod anatomy. Comstock, New York

Sombke A, Harzsch S (2015) Immunolocalization of histamine in the optic neuropils of *Scutigera coleoptrata* (Myriapoda: Chilopoda) reveals the basal organization of visual systems in Mandibulata. Neurosci Lett 594:111–116

Sombke A, Rosenberg J (2015) Diplopoda – nervous and neuroendocrine systems. In: Minelli A (ed) Treatise on zoology-anatomy, taxonomy, biology. The Myriapoda, Brill, Leiden, pp 161–179

Sombke A, Rosenberg J (2016) Myriapoda. In: Schmidt-Rhaesa A, Harzsch S, Purschke G (eds) Structure and evolution of invertebrate nervous systems. Oxford University Press, Oxford, pp 478–491

Sombke A, Harzsch S, Hansson BS (2011a) Organization of Deutocerebral Neuropils and Olfactory Behavior in the centipede *Scutigera coleoptrata* (Linnaeus, 1758) (Myriapoda: Chilopoda). Chem Senses 36:43–61

Sombke A, Rosenberg J, Hilken G (2011b) Chilopoda-the nervous system. In: Minelli A (ed) Treatise on zoology-anatomy, taxonomy, biology. The Myriapoda, Brill, Leiden, pp 217–234

Spies T (1981) Structure and phylogenetic interpretation of diplopod eyes (Diplopoda). Zoomorphology 98:241–260

Spreitzer A, Melzer RR (2003) The nymphal eyes of *Parabuthus transvaalicus* Purcell 1899 (Buthidae): an accessory lateral eye in a scorpion. Zool Anz 242:137–143

Steinmetz PR, Urbach R, Posnien N, Eriksson J, Kostyuchenko RP, Brena C, Guy K, Akam M, Bucher G, Arendt D (2010) *Six3* demarcates the anterior-most developing brain region in bilaterian animals. EvoDevo 1:14

Strausfeld NJ (1976) Atlas of an insect brain. Springer, Berlin, Heidelberg

Strausfeld NJ (2005) The evolution of crustacean and insect optic lobes and the origins of chiasmata. Arthropod Struct Dev 34:235–256

Strausfeld NJ (2012) Arthropod brains. Evolution, functional elegance, and historical significance. Belknap, Cambridge

Strausfeld NJ, Nässel DR (1981) Neuroarchitecture of brain regions that subserve the compound eyes of Crustacea and insects. In: Autrum H (ed) Handbook of sensory physiology. Springer, Berlin, pp 1–132

Strausfeld NJ, Olea-Rowe B (2021) Convergent evolution of optic lobe neuropil in Pancrustacea. Arthropod Struct Dev 61:101040

Strausfeld NJ, Ma X, Edgecombe GD, Fortey RA, Land MF, Liu Y, Cong P, Hou X (2016) Arthropod eyes: the early Cambrian fossil record and divergent evolution of visual systems. Arthropod Struct Dev 45:152–172

Toye SA (1966) The reactions of three species of Nigerian millipedes (*Spirostreptus assiniensis*, *Oxydesmus* sp., and *Habrodesmus falx*) to light, humidity and temperature. Entomol Exp Appl 9:468–483

Verhoeff KW (1902) Abteilung Gliederfüssler: Arthropoda Klasse Chilopoda. In: Bronn's Klassen und Ordnungen des Tier-Reichs. Akademische Verlagsgesellschaft, Leipzig, p 725

Viollet S (2014) Vibrating makes for better seeing: from the Fly's micro-eye movements to Hyperacute visual sensors. Front Bioeng Biotechnol 2:9

Wesener T, Sierwald P (2005) New giant pill-millipede species from the littoral forest of Madagascar (Diplopoda, Sphaerotheriida, Zoosphaerium). Zootaxa 1097:1–60

Wongthamwanich N, Panha S, Sitthicharoenchai D, Seelanan T, Enghoff H, Thirakhupt K (2012) Daily activities of the Giant pill-millipede *Zephronia* cf. *viridescens* Attems, 1936 (Diplopoda: Sphaerotheriida: Zephroniidae) in a deciduous Forest in northern Thailand. Zool Stud 51:913–926

Zapparoli M, Edgecombe GD (2011) Order Lithobiomorha. In: Minelli A (ed) Treatise on zoology-anatomy, taxonomy, biology. The Myriapoda, Brill, Leiden, pp 371–389

Chapter 8
Insect Dorsal Ocelli: A Brief Overview

Emily Baird and Ayse Yilmaz

Abstract The dorsal ocelli of adult insects are simple eyes that can complement or act independently from the compound eye visual system. While notably absent in some species, they are present across the insect phylogeny but can vary dramatically in their anatomy, neuronal organisation, physiology and function. In some species, they appear to only detect changes in light intensity, while in others, they modulate compound eye responses, mediate orientation using polarised light or function as a *zeitgeber* in the timing of daily activity. Their variability across, and even within, species suggests that insect dorsal ocelli are relatively malleable organs, and there is still much that remains to be learned about the ecological and phylogenetic factors that have shaped them. In this chapter, we provide a brief overview of the exciting and remarkable findings of the past century of research into the dorsal ocelli of insects. Our aim is to inspire future studies into these dynamic sensory organs, not only to improve our understanding of this enigmatic sensory system but also to bring deeper insights into the fascinating world of insect sensory biology.

Keywords Dorsal ocelli · Insect · Vision · Eyes · Behavior

8.1 Introduction

In addition to a pair of compound eyes, many adult insects have a visual system comprising one to three single-lens eyes located on the dorsal surface of the head, known as dorsal ocelli (or small eyes, Figs. 8.1 and 8.2). Dorsal ocelli are present even in primitive insect lineages and are likely to be older than insects themselves ((Buschbeck and Friedrich 2008), Fig. 8.3). Although the optical characteristics of

E. Baird (✉)
Department of Zoology, Stockholm University, Stockholm, Sweden
e-mail: emily.baird@zoologi.su.se

A. Yilmaz
Department of Biology, Lund University, Lund, Sweden

E. Buschbeck, M. Bok (eds.), *Distributed Vision*, Springer Series in Vision Research, https://doi.org/10.1007/978-3-031-23216-9_8

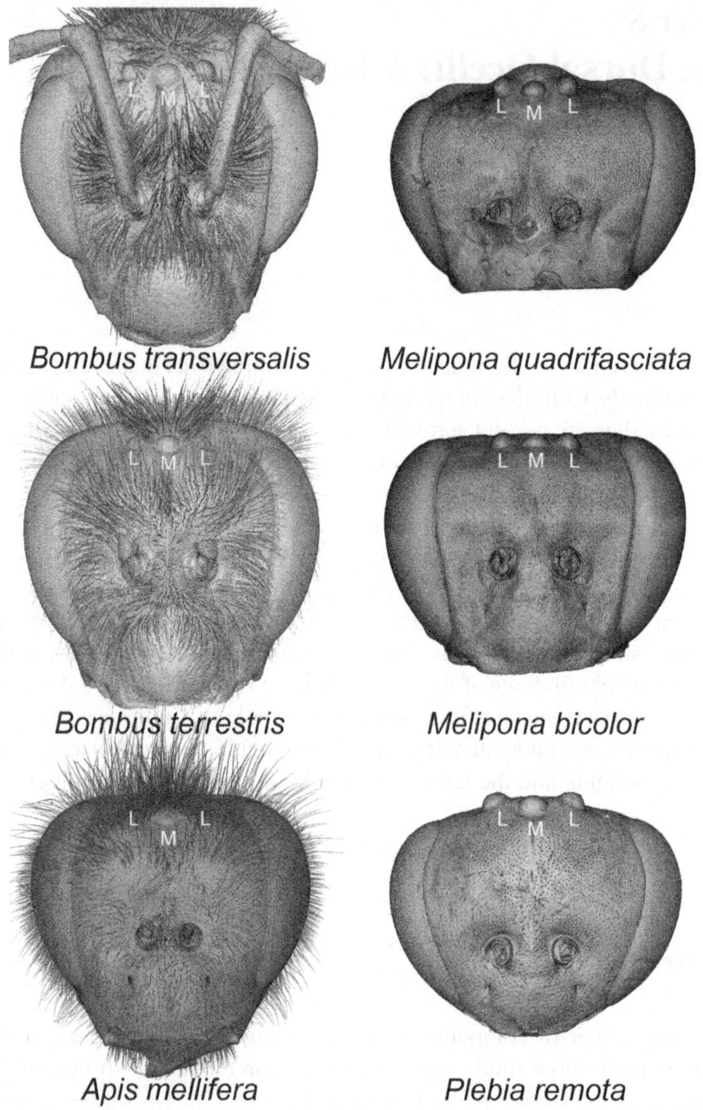

Fig. 8.1 The ocelli of different bee species. Volume renderings from X-ray microtomographic scans of the heads of workers of six bee species showing how the placement, arrangement and relative size of the ocelli (marked with L for lateral ocellus and M for median ocellus) vary between species from the same genus (*Bombus* and *Melipona*). All species live in forested habitats, with the exception of *Bombus terrestris* and *Apis mellifera*, which typically forage in open environments such as meadows. Note that the heads have been scaled to similar widths and the hair has been removed from the *Melipona* and *Plebia* species

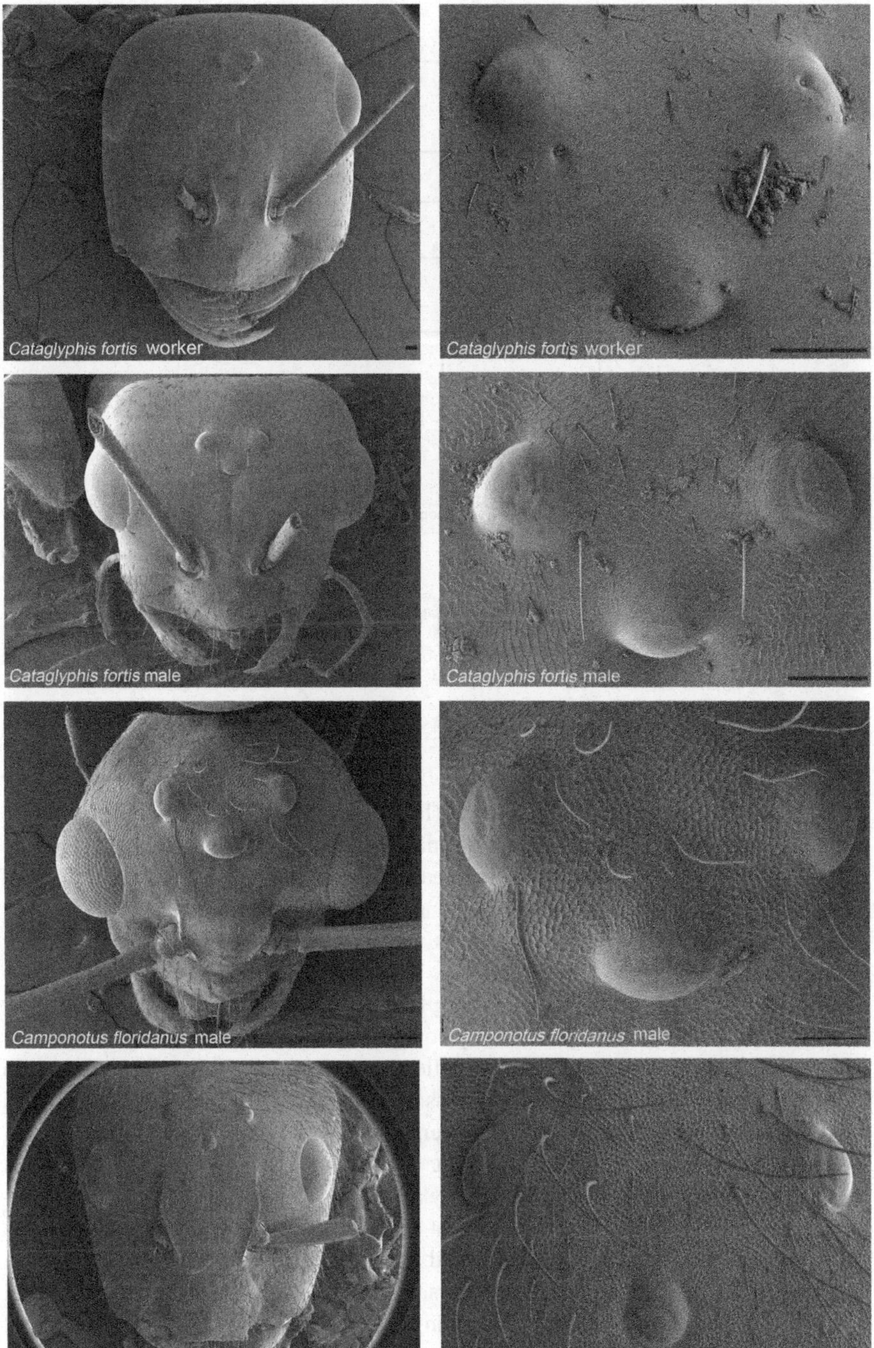

Fig. 8.2 The ocelli of different ant species. Scanning electron microscope (SEM) images of the heads and ocelli of *Cataglyphis fortis*, *Camponotus floridanus* and *Camponotus rufipes* ants, demonstrating how their relative size and position vary between species and sexes. Males and queens are capable of flight, while workers are not. Scale bars = 100 μm

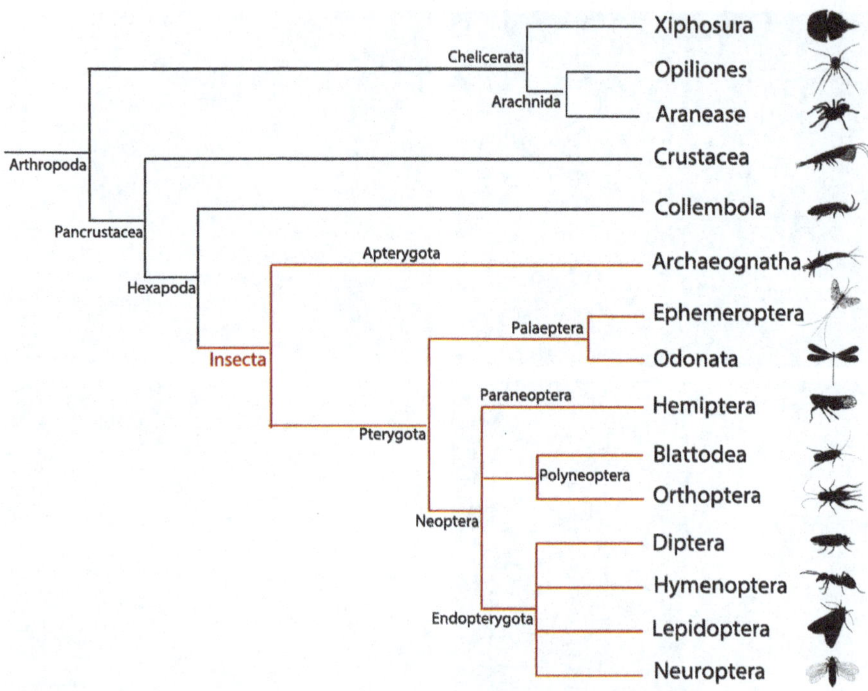

Fig. 8.3 A phylogenetic tree of the arthropoda showing the lineages within Insecta that have developed ocelli (marked in red)

external ocelli vary across insect species (Figs. 8.1 and 8.2), they nonetheless share the same basic design principles. External dorsal ocelli (ocelli with lenses that extend beyond the outer surface of the head) are typically covered with high aperture dioptric lenses and possess an iris, a vitreous body, a retina and an ocellar neuropil in which photoreceptor axons make a connection with interneurons (Goodman 1981). To early investigators, dorsal ocelli appeared to be ubiquitous among flying insects, strongly suggesting that they play an important role in flight (Kalmus 1945). This idea is particularly compelling in the case of some ant species, where winged castes have ocelli and flightless castes do not, suggesting that they do indeed provide information that is necessary for flight. While this explanation is appealing in its simplicity – that the challenges of flight require an additional visual system – it is unfortunately not satisfying once a more detailed survey of ocelli is carried out. For example, ocelli are largely absent in many taxa where flight is common, including the Coleoptera, the largest order of insects (although they are present in several families (Hatch 1926)), and some subfamilies of Diptera (otherwise generally known for their prominent dorsal ocelli). Moreover, they are present in many flightless insect species, suggesting that their functionality must extend beyond flight control.

Despite having been the focus of numerous detailed studies for well over a century, the role of ocelli in most insects remains unclear. Early behavioural studies

Table 8.1 Comparison between typical visual qualities in the dorsal ocelli and compound eyes of insects

Visual quality	Dorsal ocelli compared to compound eyes
Sensitivity	Higher
Resolution	Lower
Processing speed	Higher
Field of view	Larger
Wavelength sensitivity	Mono- or dichromatic (UV and green), compound eyes typically trichromatic (UV, blue, green)
Focus	Unfocussed generally

assumed that, like the compound eyes, the ocelli supported form vision, but with subsequent anatomical studies suggesting that they are unfocussed, investigators began to explore other potential functions (see historical overview in Cassier (1962)). With recent detailed anatomical analyses across a broad range of insect orders suggesting that they may indeed receive focussed images on the retina, these ideas have now come full circle, although whether they truly provide form vision and what function(s) they might have remain largely unclear. The more we learn about their structure, physiology and connectivity, the more complex and enigmatic the ocelli become. At the very least, they appear to be malleable structures that, over the evolutionary history of insects, have been co-opted to perform multiple functions both within and between species, genera and orders (Mizunami 1994). Understanding ocellar structure and function across a range of insects might therefore not only provide important insights into their evolutionary history but also reveal the types of visual information that ocelli provide for survival and reproduction within a particular ecological niche.

The aim of this chapter is to provide a brief overview of what is known about the dorsal ocelli of insects and, where possible, to compare their properties with those of the compound eyes (Table 8.1). We focus primarily on the external ocelli as they have been subject to the most rigorous investigations, but it is important to note that internal ocelli – photoreceptive structures lying under the cuticle and often not visible to the naked eye – continue to be discovered in a range of insects previously thought to be anocellate. For readers interested in more detail of what is known about dorsal ocelli, we suggest beginning with the extensive and insightful reviews of Goodman (1981) and Mizunami (1994).

8.2 Ocellar Structure and Neuronal Organisation

8.2.1 Ocellar Lenses

Ocellar lenses show considerable variations in shape and form, even within the same insect groups (Figs. 8.1 and 8.2). For instance, the ocelli of honeybees, bumblebees and orchid bees have thick (Fig. 8.4a), biconvex lenses with a cup-shaped

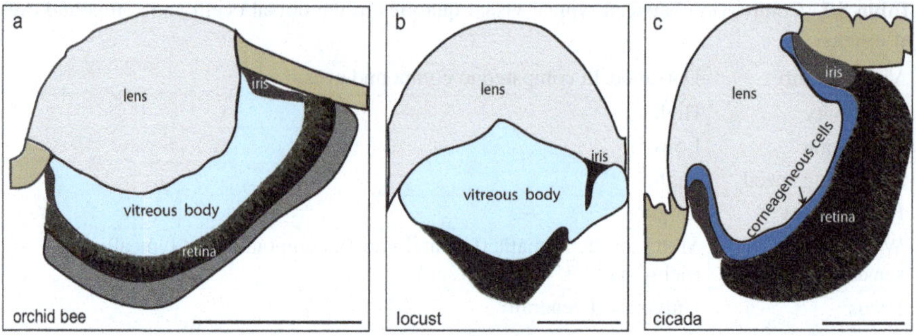

Fig. 8.4 Schematic drawing of external ocellar lens structure showing considerable variations in shape in (**a**) the orchid bee *Euglossa imperialis*, (**b**) the locust *Locusta migratoria* and (**c**) the cicada *Psaltoda moerens*. Schematic structures were traced from figures from Taylor et al. (2016), Berry et al. (2007b) and Ribi and Zeil (2015), respectively (from **a** to **c**). Scale bars = 200 μm

outer surface and an asymmetrical inner surface (Hung and Ibbotson 2014; Ribi et al. 2011; Taylor et al. 2016; Wilby et al. 2019). Similarly, the corneal lens of the nocturnal hemipteran *Triatoma infestans* is thick, biconvex around the center and asymmetrical on the inner surface (Insausti and Lazzari 2002), while both lateral and median ocelli of locusts have a large, thick and uniformly curved lenses across the vertical and horizontal planes and have a relatively flat inner surface (Berry et al. 2007b; Wilson 1978) (Fig. 8.4b). In dragonflies, the median ocellus possesses a thick lens that has a strongly concave inner lens surface near its centre (Stange et al. 2002), while the lenses of the lateral ocelli are strongly curved on the outer surface and flat along the inner surface (Berry et al. 2007b). Considering the flat ocellar lenses of cockroaches (considered to be the largest among known insects), the champagne-cork shape lenses of cicada ((Ribi and Zeil 2015), Fig. 8.4c) and the irregular 'nipple' pattern reported on the outer lenses of the blowfly, external ocellar lenses do not appear to have any common feature apart from the fact that they have wide visual fields that extend beyond those of the compound eyes (Goodman 1981; Insausti and Lazzari 2002; Mizunami 1994; Taylor et al. 2016; Wilby et al. 2019). One relatively consistent feature, however, is that nocturnal insects tend to have relatively larger ocelli than their diurnal counterparts (although the ocelli of hornets appear to be an exception (Kelber et al. 2011)), providing strong support for the hypothesis that they play an important role in sensing small changes in light intensity in dim light (Berry et al. 2011; Kerfoot 1967; Narendra and Ribi 2017; Warrant et al. 2006).

8.2.2 *Field of View*

Despite its importance for understanding the functional role of ocelli, the ocellar field of view has only been quantified in a few species: locusts (Cornwell 1955), dragonflies (Stange et al. 2002), flies (Schuppe and Hengstenberg 1993), bumblebees (Wilby et al. 2019) and orchid bees (Taylor et al. 2016). The field of view of the median ocellus – typically directed dorso-frontally – together with the fields of view of the lateral ocelli – typically directed dorso-laterally – covers almost the entire dorsal hemisphere, extending below the horizon (Cornwell 1955; Schuppe and Hengstenberg 1993; Stange et al. 2002; Taylor et al. 2016; Wilby et al. 2019). The fields of view of the lateral and median ocelli of all studied species have a binocular overlap, while in the orchid bee *Euglossa imperialis*, the visual fields of all three ocelli overlap, creating a trinocular region (Taylor et al. 2016). While the function of the overlapping fields of view remains unclear, Wilson (1978) hypothesised that the wide, dorsally oriented field of view of ocelli would be well suited for flight stabilisation by monitoring global changes in light intensity, a hypothesis that is supported by behavioural data from locusts (Goodman 1965; Taylor 1981a, b) and dragonflies (Stange 1981; Stange and Howard 1979) and electrophysiological data from flies (Parsons et al. 2006, 2010).

8.2.3 Focal Plane

Generally, the ocelli of most insect species are under-focussed (Berry et al. 2007a; Ribi et al. 2011; Schuppe and Hengstenberg 1993) (Fig. 8.5a, b), suggesting that, rather than generating form vision, they provide information about light intensity integrated over a large visual field. Interestingly, in several species, some regions of the ocellar retina have been discovered to receive focussed light. For example, the dragonfly median ocellus and the ocelli of paper wasps and orchid bees have lenses having a focal plane that lies within the retina (Stange et al. 2002; Taylor et al. 2016; Warrant et al. 2006) (Fig. 8.5c). It is unclear, however, what information these focussed regions convey or why, although modelling in dragonflies suggests that it may reflect a matched filter for the horizon to aid in flight stabilisation (Berry et al. 2006; Stange et al. 2002).

8.2.4 Internal Organisation

Beneath the corneal lens, there is typically a thin layer of corneageneous cells that secrete the corneal lens during development (Goodman 1970; Insausti and Lazzari 2002; Ribi et al. 2011; Toh et al. 1983). In some insects, a small number of elongated cells form an irregularly shaped vitreous body or clear zone that separates the corneagenous and retinular cells from the inner lens surface (Berry et al. 2007a; Ribi et al. 2011). Retinal cells, which are arranged perpendicularly to the internal surface of the corneal lens, lie directly behind the corneagenous cells and contain several thousand rhabdomeric photoreceptors (Goodman 1981). Ocellar photoreceptors can be highly sensitive to ultraviolet (UV) light (e.g. *Locusta* (Wilson 1978), *Drosophila* (Sabat et al. 2016)), green light (e.g. *Periplaneta* (Goldsmith and Ruck

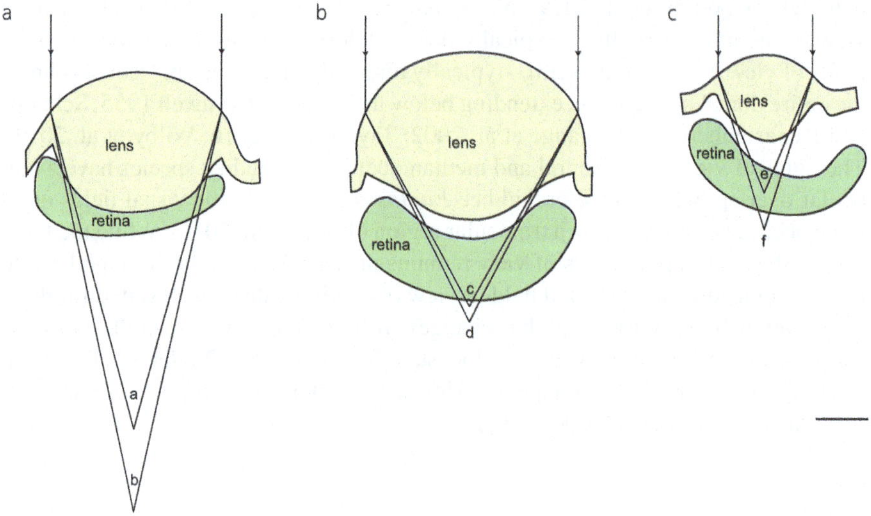

Fig. 8.5 Optical properties of the median ocelli of hymenopterans showing the positions of astigmatic focal planes for the minor (**a**, **c**, **e**) and major axes (**b**, **d**, **f**) in (**a**) the nocturnal halictid bee *Megalopta genalis*, (**b**) the nocturnal paper wasp *Apoica pallens*, and (**c**) the diurnal paper wasp *Polistes occidentalis*, as measured using the hanging drop method (Modified from Warrant et al. (2006)). Scale bar = 100 μm

1958), *Megalopta* (Berry et al. 2011)) or both (e.g. *Apis* (Goldsmith 1960; Goldsmith and Ruck 1958; Ogawa et al. 2017)) and thus may be functional in detecting a strong contrast between the sky and the ground (Chappell and DeVoe 1975; Wilson 1978) and aid in colour constancy (Garcia et al. 2017). In many species, the ocellar retinae appear to be divided into dorsal and ventral sections with different lengths of rhabdomeres and viewing regions, suggesting functional differences (Hung et al. 2013; Narendra et al. 2016; Ribi and Zeil 2018; Taylor et al. 2016). For instance, being directed skyward, the shorter cells (less sensitive) in the ventral retina may detect intensity changes in the sky where it is brightest, while the longer photoreceptors (more sensitive) in the dorsal retina may detect intensity changes at the horizon (Berry et al. 2006, 2007a, b; Hung et al. 2013; Ribi et al. 2011; van Kleef et al. 2013).

In some insects, the specific arrangements of rhabdoms – with a more proximal placement near the centre of the ocellar retina and with elongated lengths – suggest the existence of foveal regions with enhanced sensitivity and resolution (Berry et al. 2007b). However, the typical arrangement of rhabdoms seems disorganised and is potentially optimised for light absorption rather than form vision (Goodman 1981) as discrimination of small objects with such an arrangement is unlikely (Berry et al. 2007b). This is further supported by the observation that rhabdoms are often wide, fused and continuous (Berry et al. 2007b; Goodman 1981; Narendra et al. 2016; Taylor et al. 2016). However, the number of retinular cells that contribute to the formation of the rhabdom varies across species and even within an ocellus. For instance, diurnal insects such as the honeybee *Apis mellifera* (Ribi and Zeil 2018) and the desert ant *Cataglyphis bicolor* (Penmetcha et al. 2019) have two retinular cells, while in the nocturnal cockroach *Periplaneta americana*, the number of retinula cells that form the rhabdom may vary between two and six (Toh et al. 1983). In the nocturnal moth *Trichoplusia ni* (Lepidoptera, Noctuidae), there may even be six or seven retinula cells (Dow and Eaton 1976). However, in nocturnal *Myrmecia* ants, the diurnal ant *Melophorus bagoti* and the diurnal hoverfly *Eristalis tenax*, each retinula cell contributes microvilli to multiple rhabdoms (Narendra and Ribi 2017; Penmetcha et al. 2019; Ribi and Zeil 2018). The variation in ocellar rhabdom morphology across insect orders suggests that it might be quite plastic both in both a developmental and evolutionary sense. Indeed, variations in rhabdom morphology may be related to a trade-off between sampling and sensitivity, although whether this is the case remains unclear.

Another adaptation for sensitivity observed in insect dorsal ocelli is the presence of screening pigments within the retinular cells. These pigments modify sensitivity by adjusting the amount of incident light reaching the retina by migrating distally from the proximal areas of the ocelli in bright light conditions (screening the rhabdoms from one another) and migrating in the opposite direction in low light conditions. Densely but irregularly packed layered pigment cells between the lens and the receptor layer – sometimes directly behind the peripheral edge of the inner lens (i.e., locust (Berry et al. 2007b), orchid bee (Taylor et al. 2016)) – form a pigmentary iris, a mobile pigment sheath that can open and close the aperture of the lens (Goodman 1981; Stavenga et al. 1979; Wilson 1975). In many insects, these pigment cells dynamically change position during different states of light adaptations to control the photon flux in the photoreceptors (Berry et al. 2007a; Stavenga et al. 1979; Wilson 1975), although in some insects, these changes vary not with light intensity but rather with age (Insausti and Lazzari 2000). Additional light capture in some species – such as honeybees, cockroaches, dragonflies and flies – is also achieved using a tapetal layer placed behind the ocellar retinae that maximises the chance of absorbing photons by reflecting them back through the receptor cells (Mizunami 1994).

Investigations into the shape and arrangement of ocellar rhabdoms have also provided good evidence that they are likely to be sensitive to polarised light in many insect species (Mote and Wehner 1980; Ribi and Zeil 2018; Taylor et al. 2016; Warrant et al. 2006; Zeil et al. 2014). The ability of the ocelli to detect and use polarised light has also been suggested for several insects by both behavioural analyses (Fent 1986; Fent and Wehner 1985; Wellington 1953, 1974) and electrophysiological analyses (Geiser and Labhart 1982; Mote and Wehner 1980). Interestingly, polarisation sensitivity appears to be traded off against absolute sensitivity and is likely a strong indicator of a species' ecology – while diurnal honeybees (Geiser and Labhart 1982), bumblebees (Zeil et al. 2014), orchid bees (Taylor et al. 2016) and ants (Fent and Wehner 1985; Narendra and Ribi 2017) all appear to have polarisation-sensitive ocellar rhabdoms, nocturnal sweat bees (Berry et al. 2011) and nocturnal *Myrmecia* ants (Narendra and Ribi 2017) do not.

8.2.5 Neuronal Connectivity

The outputs of photoreceptor neurons in the ocellar retina synapse onto only a few tens of second-order neurons in the ocellar plexus (at least for the insects studied to date, like locusts, cockroaches and honeybees (Mizunami 1994)). These second-order neurons are divided into two categories: a small number of large L-neurons with large axon diameters and a larger number of small S-neurons with 5 μm or narrower axon diameters (Goodman 1981; Ribi et al. 2011). Of these two, L-neurons (which reach up to 30 μm in diameter in some insect species (Chappell et al. 1978)) are particularly suitable for detecting rapid changes in light intensity due to their high signal speed and low latency (Wilson 1978).

The target neuropils of the ocellar system are similar across insects: the visual (optic lobes), olfactory (antennal lobes) and mechanosensory systems (dorsal deutocerebrum and tritocerebrum), as well as higher associative brain regions (mushroom bodies and central complex) and the premotor and thoracic motor centres. However, some features of the organisation do vary. Mizunami (1994) described three neuronal synapsing patterns that most likely also reflect the functional roles of the ocelli in the visual behaviours of different species: bisynaptic, trisynaptic and the intermediate system. By conveying the ocellar signal from the photoreceptors directly to the target neuropils (Mobbs 1985; Pan and Goodman 1977) using a large number of second-order neurons, the bisynaptic system is possibly an adaptation to fast signal transmission at the cost of sensitivity (increased numbers of interneurons increase the sensitivity but reduce the transmission rate) and exists in all tested holometabolous insects investigated (Mizunami 1994). Most hemimetabolous insects, on the other hand, have the slower but more sensitive trisynaptic system, where photoreceptor signals first converge onto second-order neurons in the ocellar plexus and then convey information to the target neuropils through a large number of third-order neurons in the ocellar tract neuropil or ocellar nerve (Mizunami 1995). The intermediate system has both bi- and trisynaptic pathways, where both second-order and third-order neurons transmit their signals to their target neuropils.

8.3 Function

Research to date has suggested that ocelli have a range of different functions (Table 8.2) that fall into three general groupings: firstly, they work in parallel with the compound eyes by providing similar visual information to the brain; secondly, they regulate behaviours mediated by the compound eyes; and, finally, they modulate neuronal signals across different brain regions. Below, we present an overview of the functions that the ocelli have most convincingly been shown to fulfil.

Table 8.2 Functions attributed to ocelli

Function	Taxa	References	Comments
Phototaxis	Locusts, flies, crickets, bugs, honeybees	Kastberger (1990), Kastberger and Kranner (2000), Kastberger and Schuhmann (1992), and Lazzari et al. (1998)	Modulates phototactic orientation mediated by the compound eyes, mediates negative phototaxis in bugs
Timing of activity	Honeybees, moths	Eaton et al. (1983), Schricker (1965), Sprint and Eaton (1987), and Wunderer and Jan De Kramer (1989)	
Circadian rhythmicity	Crickets	Rence et al. (1988)	Circadian control of compound eye sensitivity
Flight stabilisation	Locusts, dragonflies, flies	Parsons et al. (2006), Parsons et al. (2010), Schuppe and Hengstenberg (1993), Stange (1981), Stange and Howard (1979), and Taylor (1981a, b)	Most likely occurs through integration with inputs from the compound eyes
Orientation using celestial cues	Flies, ants, bumblebees	Fent and Wehner (1985), Schwarz et al. (2011a, b), Wellington (1953), and Wellington (1974)	In *Cataglyphis fortis*, ocelli mediate the orientation using polarised light alone; in *Melophorus bagoti*, the celestial cues they use have not been established
Colour constancy	Honeybees	Garcia et al. (2017)	Theoretically plausible but not demonstrated behaviourally

8.3.1 Phototactic Organs

The first investigations into the function of the ocelli focussed on their ability to support form vision, with this work ultimately failing to provide any clear conclusions (Cassier 1962). With subsequent studies revealing that the ocelli of many taxa would unlikely support form vision, investigators turned to the potential role of ocelli in phototactic behaviour, that is, orientation towards or away from the direction of light. To date, only one study has provided evidence that the ocelli alone could mediate (negative) phototactic behaviour (in the bug *Triatoma infestans* (Lazzari et al. 1998)), while others instead found that phototactic responses are primarily mediated by the compound eyes but that input from the ocelli can modify their strength, speed and accuracy (Goodman 1970; Mizunami 1994). In addition to modifying phototactic responses, ocelli have also been shown to mediate the strength of the compound-eye-mediated startle response in locusts – which is initiated by rapid changes in light intensity (Goodman 1968) – and to modulate motion vision processing in the compound eyes at different light intensities in cockroaches (Honkanen et al. 2018).

8.3.2 Timing of Activity

The indication that ocelli modulate compound eye responses at different light levels in combination with the physiological evidence that they encode information about absolute light intensity inspired investigators to explore their role in mediating the timing of daily activity. Schricker (1965) observed that the ocelli of honeybees mediated the initiation and cessation of foraging activity at sunrise and sunset. Similarly, inputs from the ocelli were found to regulate the timing of the onset of nocturnal activity in cabbage looper moths (Eaton et al. 1983; Sprint and Eaton 1987; Wunderer and Jan De Kramer 1989). In combination with the general observation that ocelli in nocturnal insects are relatively larger than in their diurnal relatives – making them more sensitive to dim-light conditions (e.g., (Berry et al. 2011; Kerfoot 1967; Narendra and Ribi 2017; Warrant et al. 2006)) – these findings suggest that ocelli provide information about absolute light intensity and changes in light intensity that then appear to modulate behavioural state. Indeed, in crickets, the ocelli have been found to play a role in setting the circadian singing rhythm (Rence et al. 1988). It is interesting to note that while ocelli are involved in the timing of activity in both honeybees and cabbage looper moths, ocellar morphology in these species is quite different. Most notably, cabbage looper moths have only two ocelli (in many other moth species, they are even subcuticular), while honeybees have three ocelli, which are relatively larger and optically more elaborate. Thus, while ocelli may share a function in the timing of daily activity across insect orders, the large differences in their structure suggest that, through evolution, they have likely also been co-opted for other tasks (as noted by Wilson (1978)).

8.3.3 Flight Stabilisation

Ocelli have long been thought to be important for flight due to their prevalence in a broad range of flying insects (Kalmus 1945), where they often lie in a triangular arrangement high up on the head. Considering their large visual fields, their high sensitivity and their ability to detect rapid changes in light intensity (especially with respect to the compound eyes, Table 8.1), Wilson (1978) proposed that ocelli might function to rapidly detect instability in flight by monitoring the position of the horizon, particularly at low light intensities. This hypothesis was later confirmed in tethered preparations of locusts (Goodman 1965; Taylor 1981a) and dragonflies (Stange 1981) and has been supported by electrophysiological recordings of motion-sensitive neurons in the fly (Parsons et al. 2006, 2010). Interestingly, studies on freely flying bees suggest that a lack of ocellar input does not inhibit flight but that ocelli may improve the precision and speed of the stabilisation information provided by the compound eyes (Kastberger 1990; Kastberger and Kranner 2000;

Kastberger and Schuhmann 1992; Schricker 1965). Given the evidence to date, there seems to be little doubt that flight stabilisation is a central ocellar function in flying insects.

8.3.4 *Orientation*

The discovery that fly larvae appeared to use their lateral ocelli to orient using polarised light prompted Wellington (1953) to test if the dorsal ocelli of the adults of the flesh fly *Sarcophaga aldrichi* were also capable of this. By rotating a polarising filter over walking flies, Wellington was able to show that they would orient to the plane of polarised light when the compound eyes were fully covered. Considering this result, and the finding that ocelli are more sensitive than compound eyes, Wellington (1974) suggested that the ocelli might also enable diurnal insects to extend their activity periods into twilight, such as bumblebees. By combining observations on the activity period of bumblebees with occluded ocelli and observations on the differences in free-flight behaviour during the day and at dusk, Wellington (1974) concluded that the compound eyes mediate landmark navigation in bright light (causing homing bees to zigzag between landmarks) and that the ocelli mediate orientation using polarised light at twilight (permitting bees to make straight flights home). The possibility that the ocelli may play a role in orientation behaviour has also been explored in ants, where their presence in the flightless castes of some species remained (and largely still remains) a mystery. Fent and Wehner (Fent and Wehner 1985) discovered that the desert ant *Cataglyphis fortis* was capable of orienting homeward when only polarised light was visible in the sky and when the compound eyes were occluded, providing further evidence that hymenopteran ocelli can detect and use polarised light for orientation. Another ant species, *Melophorus bagoti*, has also been shown to use its ocelli for compass orientation (Schwarz et al. 2011a, b), although their ocellar rhabdom structure suggests that they are unlikely to be able to use polarised light for this (Penmetcha et al. 2019). Moreover, in *Cataglyphis nodus* ants, a close overlap between ocellar interneurons and afferents from the Johnston's organ in the posterior slope region suggests that they integrate multiple compass cues (i.e. mechanosensory and visual information) for robust navigation (Grob et al. 2020). Overall, there is good evidence to suggest that the ocelli can play a functional role in orientation, at least in hymenopteran and dipteran insects. Whether this is also the case for other insects remains to be investigated.

8.4 Conclusions

While much research effort has been dedicated to understanding the function of ocelli, it is clear that there is still much more to learn. Due partly to their diminutive size, these sensory organs are challenging to investigate anatomically, physiologically and behaviourally. This not only means that it is difficult to assess their function in ecologically relevant ways but also that it is difficult to perform informative comparative analyses across closely related species.

Our understanding of the factors that shape the morphology and function of ocelli has been limited, in part, by available methods. Firstly, the commonly used method of covering or ablating the ocelli in tethered insects and then observing the resulting behavioural response to visual stimuli is limited because it generates a highly unnatural situation for the animal and may therefore not provide a complete reflection of their function. Studies on freely moving insects have been fruitful in providing insights into the function of ocelli in a natural context, but much more comparative work is required to understand their role in different species. The second issue that makes elucidating the functional role of ocelli problematic is the lack of systematic comparative studies within taxa that have ocellate and anocellate members. For this, ants are a wonderful model system as many species have ocelli in flying castes, while they are absent in the non-flying castes (with the exception of a few families). Another interesting model system is the tabanid flies, where some members have highly reduced or potentially absent ocelli, while others have the well-developed ocelli typical of most flies. Investigations into these groups of insects, which explore differences in the brains, sensory systems and behaviour of ocellate and anocellate members, could prove important for fully understanding the function of ocelli and the factors that have shaped them.

As is hopefully clear from this overview, there is much that remains to be learned about the dorsal ocelli of insects. In particular, many questions remain about how their optics and neural anatomy vary with ecology and phylogeny. Answering these questions not only would provide interesting insights into the factors that shape sensory systems but may also lead to lightweight bioinspired technologies for autonomous flight control and navigation. Similarly, comparative investigations into the neuronal and synaptic organisation of neurons in the ocellar systems within and between species, between diurnal and nocturnal species and between ocellate and anocellate members would be particularly helpful for understanding the plasticity of sensory systems and insect brains, as well as the drivers that form them. Another interesting avenue for future research would be to perform comparative studies across the developmental stages to better understand the evolutionary origins of ocelli and to answer, for example, why some species have three ocelli while others have only one or two or one and why in some species they are subcuticular. It would be also very interesting to determine the relative contribution of the compound eyes and the ocelli may make to spatial orientation via the optomotor response. Future investigations into insect ocelli will undoubtedly provide exciting and remarkable findings that would help us learn more about not only this enigmatic sensory system but also the endlessly fascinating world of insects and their sensory biology.

References

Berry R, Stange G, Olberg R, van Kleef J (2006) The mapping of visual space by identified large second-order neurons in the dragonfly median ocellus. J Comp Physiol A 192:1105–1123

Berry R, Kleef J, Stange G (2007a) The mapping of visual space by dragonfly lateral ocelli. J Comp Physiol A 193:495–513

Berry RP, Warrant EJ, Stange G (2007b) Form vision in the insect dorsal ocelli: an anatomical and optical analysis of the locust ocelli. Vis Res 47:1382–1393

Berry RP, Wcislo WT, Warrant EJ (2011) Ocellar adaptations for dim light vision in a nocturnal bee. J Exp Biol 214:1283–1293

Buschbeck EK, Friedrich M (2008) Evolution of insect eyes: tales of ancient heritage, deconstruction, reconstruction, remodeling, and recycling. Evol Edu Outreach 1:448–462

Cassier P (1962) Role des ocelles frontaux chez *Locusta migratoria migratorioides* (R. et F.). Insects Sociaux 9:213–230

Chappell RL, DeVoe RD (1975) Action spectra and chromatic mechanisms of cells in the median ocelli of dragonflies. J Gen Physiol 65:399–419

Chappell RL, Goodman LJ, Kirkham JB (1978) Lateral ocellar nerve projections in the dragonfly brain. Cell Tissue Res 190:99–114

Cornwell PB (1955) The functions of the ocelli of *Calliphora* (Diptera) and *Locusta* (Orthoptera). J Exp Biol 32:217–237

Dow MA, Eaton JL (1976) Fine structure of the ocellus of the cabbage looper moth (*Trichoplusia ni*). Cell Tissue Res 171:523–533

Eaton JL, Tignor KR, Holtzman GI (1983) Role of moth ocelli in timing flight initiation at dusk. Physiol Entomol 8:371–375

Fent K (1986) Polarized skylight orientation in the desert ant *Cataglyphis*. J Comp Physiol A 158:145–150

Fent K, Wehner R (1985) Oceili: a celestial compass in the desert ant *Cataglyphis*. Science 228:192–194

Garcia JE, Hung Y-S, Greentree AD, Rosa MGP, Endler JA, Dyer AG (2017) Improved color constancy in honey bees enabled by parallel visual projections from dorsal ocelli. PNAS 114:7713–7718

Geiser FX, Labhart T (1982) Electrophysiological investigation on theocellar retina of the honeybee (*Apis mellifera*). Verh Dtsch Zool Ges 75:307

Goldsmith TH (1960) The nature of the retinal action potential, and the spectral sensitivities of ultraviolet and green receptor systems of the compound eye of the worker honeybee. J Gen Physiol 43:775–799

Goldsmith TH, Ruck PR (1958) The spectral sensitivities of the dorsal ocelli of cockroaches and honeybees; an electrophysiological study. J Gen Physiol 41:1171–1185

Goodman LJ (1965) The role of certain optomotor reactions in regulating stability in the rolling plane during flight in the desert locust *Schistocera gregaria*. J Exp Biol 42:382–407

Goodman LJ (1968) Interaction between ocellar and compound eye visual inputs in *Schistocerca gregoria*. In: 13th international congress of entomology. Moscow

Goodman LJ (1970) The structure and function of the insect dorsal ocellus. In: Beament JWL, Treherne JE, Wigglesworth VB (eds) Advances in insect physiology, vol 7. Academic, Cambridge, MA

Goodman LJ (1981) Organisation and physiology of the insect dorsal ocellar system. In: Autrum H (ed) Handbook of sensory physiology, vol VII/6C. Springer, Berlin

Grob R, Tritscher C, Grubel K, Stigloher C, Groh C, Fleischmann PN, Rossler W (2020) Johnston's organ and its central projections in *Cataglyphis* desert ants. J Comp Neurol 529:2138–2155

Hatch MH (1926) Notes on the morphology of the eyes of Coleoptera. J N Y Entomol Soc 34:343–349
Honkanen A, Saari P, Takalo J, Heimonen K, Weckström M (2018) The role of ocelli in cockroach optomotor performance. J Comp Physiol A 204:231–243
Hung YS, Ibbotson MR (2014) Ocellar structure and neural innervation in the honeybee. Front Neuroanat 8:6
Hung YS, van Kleef JP, Stange G, Ibbotson MR (2013) Spectral inputs and ocellar contributions to a pitch-sensitive descending neuron in the honeybee. J Neurophysiol 109:1202–1213
Insausti TC, Lazzari CR (2000) An ocellar "pupil" that does not change with light intensity, but with the insect age in Triatoma infestans. Mem Inst Oswaldo Cruz 95:743–746
Insausti TC, Lazzari CR (2002) The fine structure of the ocelli of *Triatoma infestans* (Hemiptera: Reduviidae). Tissue Cell 34:437–449
Kalmus H (1945) Correlations between flight and vision, and particularly between wings and ocelli, in insects. Proc R Entmol Soc Lond 20:84–96
Kastberger G (1990) The ocelli control the flight course in honeybees. Physiol Entomol 15:337–346
Kastberger G, Kranner G (2000) Visualization of multiple influences on ocellar flight control in giant honeybees with the data-mining tool Viscovery SOMine. Behav Res Methods Instrum Comput 32:157–168
Kastberger G, Schuhmann K (1992) Ocellar occlusion effect on the flight behaviour of homing honeybees. J Insect Physiol 39:589–600
Kelber A, Jonsson F, Wallen R, Warrant E, Kornfeldt T, Baird E (2011) Hornets can fly at night without obvious adaptations of eyes and ocelli. PLoS One 6:e21892
Kerfoot WB (1967) Correlation between ocellar size and foraging activities of bees (Hymenoptera Apoidea). Am Nat 101:65–70
Lazzari CR, Reiseman CE, Insausti TC (1998) The role of the ocelli in the phototactic behaviour of the haematophagous bug *Triatoma infestans*. J Insect Physiol 44:1159–1162
Mizunami M (1994) Information processing in the insect ocellar system: comparative approaches to the evolution of visual processing and neural circuits. In: Evans PD (ed) Advances in insect physiology, vol 25. Academic, Cambridge, MA
Mizunami M (1995) Functional diversity of neural organization in insect ocellar systems. Vis Res 35:443–452
Mobbs PG (1985) Brain structure. In: Kerkut GA (ed) Comprehensive insect physiology, biochemistry and pharmacology, vol 5. Pergamon Press, Oxford
Mote MI, Wehner R (1980) Functional characteristics of photoreceptors in the compound eye and ocellus of the desert ant, *Cataglyphis bicolor*. J Comp Physiol A 137:63–71
Narendra A, Ribi WA (2017) Ocellar structure is driven by the mode of locomotion and activity time in Myrmecia ants. J Exp Biol 220:4383–4390
Narendra A, Ramirez-Esquivel F, Ribi WA (2016) Compound eye and ocellar structure for walking and flying modes of locomotion in the Australian ant, *Camponotus consobrinus*. Sci Rep 6:22331
Ogawa Y, Ribi W, Zeil J, Hemmi JM (2017) Regional differences in the preferred e-vector orientation of honeybee ocellar photoreceptors. J Exp Biol 220:1701–1708
Pan KC, Goodman LJ (1977) Ocellar projections within the central nervous system of the worker honey bee, *Apis mellifera*. Cell Tissue Res 176:505–527
Parsons MM, Krapp HG, Laughlin SB (2006) A motion-sensitive neurone responds to signals from the two visual systems of the blowfly, the compound eyes and ocelli. J Exp Biol 209:4464–4474
Parsons MM, Krapp HG, Laughlin SB (2010) Sensor fusion in identified visual interneurons. Curr Biol 20:624–628
Penmetcha B, Ogawa Y, Ribi WA, Narendra A (2019) Ocellar structure of African and Australian desert ants. J Comp Physiol A Neuroethol Sens Neural Behav Physiol 205:699–706
Rence BG, Lisy MT, Garves BR, Quinlan BJ (1988) The role of ocelli in circadian singing rhythms of crickets. Physiol Entomol 13:201–212

Ribi WA, Zeil J (2015) The visual system of the Australian 'Redeye' cicada (*Psaltoda moerens*). Arthropod Struct Dev 44:574–586

Ribi W, Zeil J (2018) Diversity and common themes in the organization of ocelli in Hymenoptera, Odonata and Diptera. J Comp Physiol A 204:505–517

Ribi W, Warrant EJ, Zeil J (2011) The organization of honeybee ocelli: regional specializations and rhabdom arrangements. Arthropod Struct Dev 40:509–520

Sabat D, Priyadarsini S, Mishra M (2016) Understanding the structural and developmental aspect of simple eye of Drosophila: the ocelli. J Cell Signal 1:109

Schricker B (1965) Die orientierung der honigbiene in der dämmerung. Zugleich ein beitrag zur frage der ocellen funktion bei bienen zeitschrift fur vergleichende. Physiologie 49:420–458

Schuppe H, Hengstenberg R (1993) Optical properties of the ocelli of *Calliphora erythrocephala* and their role in the dorsal light response. J Comp Physiol A 173:143–149

Schwarz S, Albert L, Wystrach A, Cheng K (2011a) Ocelli contribute to the encoding of celestial compass information in the Australian desert ant *Melophorus bagoti*. J Exp Biol 214:901–906

Schwarz S, Wystrach A, Cheng K (2011b) A new navigational mechanism mediated by ant ocelli. Biol Lett 7:856–858

Sprint MM, Eaton JL (1987) Flight behavior of normal and anocellate cabbage loopers (Lepidoptera: Noctuidae). Ann Entomol Soc Am 80:468–471

Stange G (1981) The ocellar component of flight equilibrium control in dragonflies. J Comp Physiol A 141:335–347

Stange G, Howard J (1979) An ocellar dorsal light response in a dragonfly. J Exp Biol 83:351–355

Stange G, Stowe S, Chahl J, Massaro A (2002) Anisotropic imaging in the dragonfly median ocellus: a matched filter for horizon detection. J Comp Physiol A 188:455–467

Stavenga DG, Bernard GD, Chappell RL, Wilson M (1979) Insect pupil mechanisms. J Comp Physiol A 129:199–205

Taylor CP (1981a) Contribution of compound eyes and ocelli to steering of locusts in flight: I. Behavioural analysis. J Exp Biol 93:1–18

Taylor CP (1981b) Contribution of compound eyes and ocelli to steering of locusts in flight: II. Timing changes in flight motor units. J Exp Biol 93:19–31

Taylor GJ et al (2016) The dual function of oprchid bee ocelli as revealed by x-ray microtomography. Curr Biol 26:1319–1324

Toh Y, Sagara H, Iwasaki M (1983) Ocellar system of the insect: comparison of dorsal ocellus and lateral ocellus. Vis Res 23:313–323

van Kleef JP, Massy T, Maharbiz MM (2013) An ocellar-based flight control system for flying insects. IEEE EMBC Short papers 2664

Warrant EJ, Kelber A, Wallen R, Wcislo WT (2006) Ocellar optics in nocturnal and diurnal bees and wasps. Arthropod Struct Dev 35:293–305

Wellington WG (1953) Motor responses evoked by the dorsal ocelli of *Sarcophaga aldrichi* Parker and the orientation of the fly to plane polarized light. Nature 172:1177–1179

Wellington WG (1974) Bumblebee ocelli and navigation at dusk. Science 183:550–551

Wilby D, Aarts T, Tichit P, Bodey A, Rau C, Taylor G, Baird E (2019) Using micro-CT techniques to explore the role of sex and hair in the functional morphology of bumblebee (*Bombus terrestris*) ocelli. Vis Res 158:100–108

Wilson M (1975) Autonomous pigment movement in the radial pupil of locust ocelli. Nature 258:603–604

Wilson M (1978) The functional organisation of locust ocelli. J Comp Physiol A 124:297–316

Wunderer H, Jan De Kramer J (1989) Dorsal ocelli and light-induced diurnal activity patterns in the arctiid moth *Creatonotos transiens*. J Insect Physiol 35:87–95

Zeil J, Ribi W, Narendra A (2014) Polarisation vision in ants, bees and wasps. In: Horváth G (ed) Polarized light and polarization vision in animal sciences, Springer series in vision research, vol 2. Springer, Heidelberg

[illegible]
[illegible] and [illegible]
[illegible] (1969) [illegible]
[illegible]
[illegible]
[illegible]
[illegible] and their role in [illegible]
[illegible]
[illegible]
[illegible]
[illegible]
[illegible] National Academy of Sciences [illegible]
[illegible]
[illegible]
[illegible]
[illegible] Physiol A [illegible]
[illegible]
[illegible]
[illegible]
[illegible]
[illegible]
[illegible] neurons [illegible]
[illegible]
[illegible]
[illegible]
[illegible] and [illegible]
[illegible]
[illegible]
[illegible]
[illegible]
[illegible]
[illegible]
[illegible]
[illegible]

Chapter 9
The Cornucopia of Copepod Eyes: The Evolution of Extreme Visual System Novelty

Mireille Steck, Kristina C. Theam, and Megan L. Porter

Abstract Copepods are a miniscule but ecologically significant group of organisms that thrive in a multitude of aquatic habitats. Despite being toe-high to a grasshopper, multiple visually mediated behaviors exist that would suggest that vision is an important sensory mode to this subclass of crustaceans. Unlike many other crustacean lineages, adult copepods have tripartite naupliar eyes that in the most typical form have three fused ocellar cups, each made up of three parts: a retinal sphere, a tapetal layer, and a surrounding pigment cup. The form and function of naupliar eyes have not been well cataloged across the copepods, but the species studied thus far display an inordinate amount of diversification. Across species, modifications to the ocellar components of the typical naupliar eye structure can range from complete loss to extreme enlargement, separation of the cups into three independent eyes, and the addition of an astonishingly diverse array of focusing structures, including multiple crystalline or cuticular lenses. Modifications to the typical copepod naupliar eye structure have been histologically identified in four of the ten currently described copepod orders; additional eye diversity is likely to exist among the remaining unstudied groups as well. In this review, we assemble all of the currently available data on copepod naupliar eye function and structure to highlight the extreme diversity of visual modifications in the group, underscore how much is still unknown, reinvigorate research of copepod eyes, and provide a comprehensive evolutionary framework for future studies. Although the eyes of many species are not yet fully characterized, the visual modifications described here indicate that despite their minute size, copepods are often highly visual creatures with eyes that are an evolutionary playground of diversity.

Keywords Copepoda · Naupliar eye · Lens diversity · Gicklhorn's organ

M. Steck · K. C. Theam · M. L. Porter (✉)
School of Life Sciences, University of Hawai'i at Mānoa, Honolulu, HI, USA
e-mail: mlporter@hawaii.edu

E. Buschbeck, M. Bok (eds.), *Distributed Vision*, Springer Series in Vision Research, https://doi.org/10.1007/978-3-031-23216-9_9

9.1 Introduction

The crustacean subclass Copepoda is highly diverse taxonomically, morphologically, and ecologically. The most current estimates of species diversity contain >14,000 valid species distributed among ten orders and 241 families (Boxshall and Halsey 2004; WoRMS Editorial Board 2021) (Fig. 9.1a). These mostly miniscule crustaceans are ecologically significant on a global scale. Copepods are the most abundant multicellular animal group on the planet (Bron et al. 2011) and form the basis of most aquatic food webs by serving as the essential link between photosynthetic phytoplankton and larger, heterotrophic organisms. Because of this ecological importance, copepods are important indicators of aquatic ecosystem health (Hooff and Peterson 2006). Copepods are also ecologically diverse and can be found in just about every habitat on the planet, from marine systems, including pelagic, coral reefs, coastal waters, and tide pools, to freshwater systems, such as glacial meltwaters, hot springs, and hypersaline lakes (Huys 1988; Moeschler and Rouch 1988; Boxshall 1992; Suárez-Morales 2011); as either commensals of freshwater invertebrates or parasites of fish, mollusks, tunicates, and other aquatic animals (Wilson 1911; Boxshall and Defaye 2008; Hendler and Dojiri 2009); and even in semi-terrestrial habitats, like damp moss and leaf litter in humid forests (Boxshall and Defaye 2008). The diversity of copepod ecologies drives morphological diversity within the group, resulting in drastically different body plans, modified feeding appendages, and specialized sensory structures (Fig. 9.2).

Copepod ecological and morphological diversity is also reflected in the diversity of visual systems described from the group. Within the Crustacea, most of the major evolutionary lineages have visual systems based on compound eyes, composed of

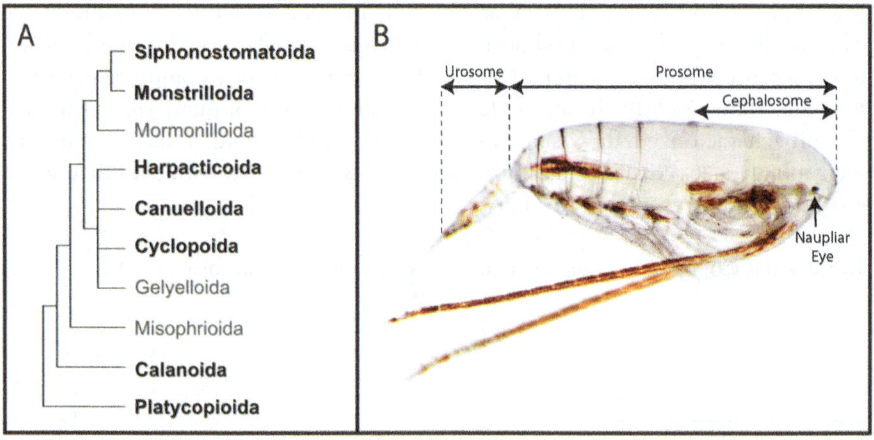

Fig. 9.1 (**a**) Phylogenetic relationships among the ten accepted copepod orders (Adapted from Ho 1994). The presence (bold) or absence (light gray) of eyes is indicated for each order. (**b**) General anatomy of a copepod highlighting the position of a typical naupliar eye, illustrated on an image of the calanoid *Calanus finmarchicus*. (Image by M. Bok)

Fig. 9.2 A selection of images illustrating the morphological diversity of species found within the copepods. (Image collage by D. Fenolio)

repeated ommatidial units consisting of optical components (e.g., facets, crystalline cones) that sit above and focus light onto an underlying set of light-sensitive receptor cells that make up the retina (Cronin and Porter 2008). Although there is diversity in the optical mechanisms used to focus light onto the retina among crustacean compound eyes, the general arrangement of cells devoted to optics and light detection (e.g., four crystalline cone cells, eight light-sensitive receptor cells) is conserved (Cronin 1986; Cronin and Porter 2008). Adult copepods, in comparison, have relatively simpler **naupliar eyes**, generally composed of three ocellar units often without any extra-retinal optical structures for focusing light. While typical naupliar eyes are commonly found in larval crustaceans and as the primary eyes in the adults of a few crustacean lineages (e.g., barnacles, ostracods), copepod visual systems offer a cornucopia of forms, where different visual components have been modified, added, and reorganized seemingly at random to create an endless diversity of eye structures among species (Boxshall et al. 2014). The ancestral lineage of copepods almost certainly had some form of naupliar eye (Boxshall et al. 1984), which, in the most basic form, is composed of three fused ocelli each containing a few receptor cells in front of a layer of reflective tapetum encased in screening

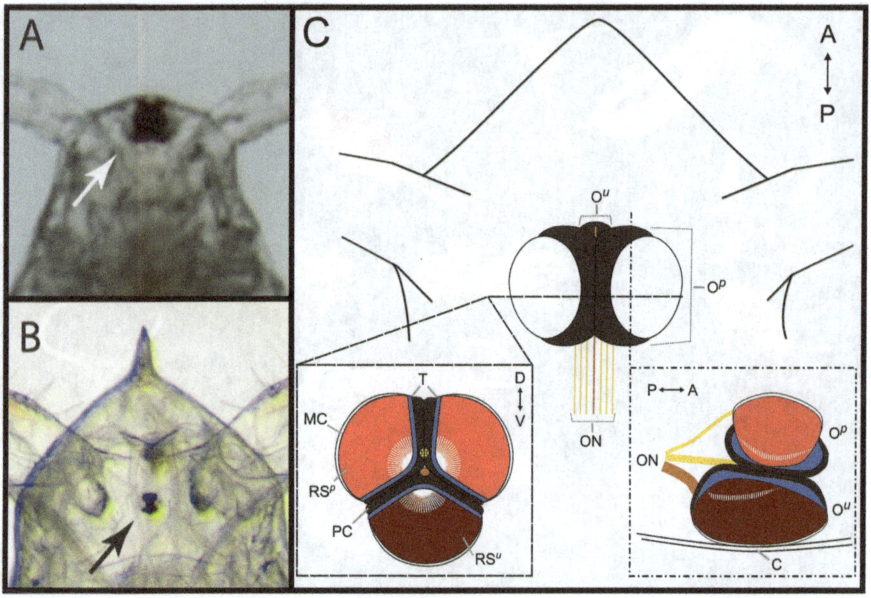

Fig. 9.3 Images of the cephalosomes of (**a**) *Acartia (Acanthacartia) tumida* and (**b**) *Neocalanus cristatus*, demonstrating variation in the size and placement of the naupliar eye (indicated by arrows) in each species (Images by R. Hopcroft). (**c**) Diagram of the anatomical arrangement of a typical copepod naupliar eye based on histology from *Eucalanus elongatus* (Vaissiere 1961). The larger schematic is a dorsal view of the cephalosome, while the boxes represent transverse (left) and parasagittal (right) sections through the naupliar eye. Each schematic is oriented along either the anterior (A) – posterior (P) or dorsal (D) – ventral (V) axis, with the rhabdomeres indicated by white stripes. *C* cuticle, *MC* membrane cell (white), O^p paired ocelli, O^u unpaired ocellus, *ON* optic nerve (paired ocelli nerves, light yellow; unpaired ocellus nerve, brown), *PC* pigment cup (black), RS^p paired retinal photoreceptor sphere (light red), RS^u unpaired retinal photoreceptor sphere (dark red), *T* tapetum (light blue). This schematic will be used as the basis for naupliar diagrams throughout the chapter

pigments (Fig. 9.3). When considering all extant species, however, the diversity of eyes is overwhelming; the size and position of the naupliar eye in the cephalosome (head) is highly variable (Figs. 9.1b, 9.3a, b); numerous species have highly modified naupliar eyes, where a subset of the three ocelli is enlarged and/or has specialized structures; many species have completely separated the three ocelli, which comprise a typical naupliar eye, into three distinct eyes (Fig. 9.4); the number of receptor cells, the retinal volume and shape, and rhabdomere arrangement are highly variable across species; and, perhaps most strikingly, there is extreme variability in the structures used to focus light onto the retina, with species using thickened cuticle, cells modified into pseudofacets, or even having up to three lenses in a single eye (Vaissiere 1961; Elofsson 1966, 2006; Land 1984).

Although studies of copepod vision are scattered throughout the literature, the disparate information has never been assembled together in one place. In this review, we bring together information on copepod visual ecology, physiology, and anatomy

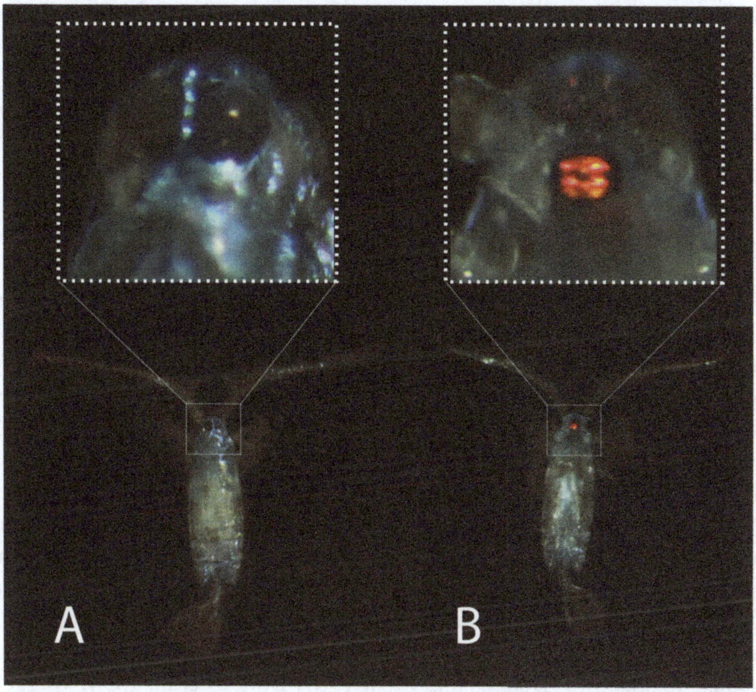

Fig. 9.4 Images of a *Pontella* sp. illustrating the arrangement of its three highly modified eyes. (**a**) Dorsal view with an enlarged photo of the lenses of the paired eyes. (**b**) Ventral view with an enlarged photo of the unpaired eye, with the bright red region showing light reflecting from the tapetum. (Images by M. Steck)

for the most complete overview of eye evolution within the group. By highlighting what is known about copepod visually guided behaviors, copepod visual system function, and the dizzying array of copepod eye morphological diversity within an evolutionary framework, we aim to unify, reframe, and reinvigorate studies of visual evolution in this minute but significant order of crustaceans.

9.2 Visually Mediated Behaviors

Copepods exhibit a multitude of visually mediated behaviors common to many animals, including the maintenance of orientation in the water column (Land 1988), finding a host (Novales Flamarique et al. 2000), mate recognition (Land 1988), prey capture (Gophen and Harris 1981), and predator avoidance (Buskey and Hartline 2003; Buskey et al. 2012). Some of the simplest visually mediated behaviors – orientation in space – have been described from species with some of the more complex copepod visual systems. In *Labidocera* species, light input to the dorsal eyes is coupled to tail and body movements to maintain a stable orientation relative to a light source; in *Pontella* species, illumination also affected swimming rates, which helped direct phototaxis behavior (Land 1988).

Perhaps the most important and well-known copepod behavior, however, is diel vertical migration (DVM), where some pelagic species spend most of the day deep in the water column and migrate to shallower waters at night to feed (Haney 1988; Cohen and Forward 2002, 2005a, b). Because of the sheer number of copepods in open ocean systems, this mass movement has significant ecosystem-level effects across the water column (McLaren 1974; Pearre 1979; Fancett and Kimmerer 1985; Bollens and Frost 1989; Longhurst and Harrison 1989). Although there is still debate over the proximate controls driving DVM, the circadian cycle of the behavior suggests that light, and therefore light detection, plays a role, with several competing hypotheses suggesting that visual systems are directly involved in controlling DVM (Forward 1988; van Gool and Ringelberg 1997; Frank and Widder 2002; Cohen and Forward 2009).

Diel vertical migration up and down through the water column is linked to both finding food in shallower, more productive waters at night and avoiding pelagic predators that are visual hunters by retreating to deeper, darker water during the day (Bollens et al. 1994). Both of these behaviors (finding prey and avoiding predators) are also directly mediated by vision in many copepod species. Some species of calanoid copepods exhibit strong predator avoidance behaviors, such as extreme jumps, in response to shadows (Bollens and Frost 1989; Bollens et al. 1994; Waggett and Buskey 2006). Other copepod species exhibit unique swarming behaviors to minimize predation risk, with some swarms across coral reefs reaching densities of 500,000–1,500,000 individuals per cubic meter (Hamner and Carleton 1979). Visual detection of conspecifics, habitat, and potential threats during swarming appears to be essential to preventing the swarm from being pushed into vulnerable positions (Buskey et al. 1996).

Copepod predation of other copepods is also fairly common in species with highly specialized eyes. The modified eyes of *Ditrichocorycaeus anglicus*, a diminutive cyclopoid, are thought to aid in the capture of their, often larger, copepod prey species (Landry et al. 1985). Many three-eyed pontellid copepods feed on smaller calanoid copepod nauplius stages (Landry 1978). Even species with a typical naupliar eye are found to prey on the nauplii of other copepod species (Lonsdale et al. 1979; Yen 1988). In all cases investigated, carnivorous copepods were size selective in their prey choices, suggesting a visually mediated behavior (although see Caparroy et al. 2000 for a strictly hydrodynamic explanation of prey selectivity). This idea is supported by the modification of predatory search patterns in response to changes in prey concentrations (Cowles and Strickier 1983; Buskey 1984).

Although many species with typical naupliar eyes tend to recognize mates based on chemical and mechanical signals (Bagøien and Kiørboe 2005; Kiørboe and Bagøien 2005; Kiørboe et al. 2005; Kiørboe 2007), the use of vision in mate recognition has been suggested for several groups of species, mainly based on the presence of highly sexually dimorphic eyes (e.g., Pontellidae – Land 1988; Buskey 1998). In bioluminescent *Pleuromamma* species, there are differences in the intensity of light produced in males versus females, also suggesting a potential role in visual signaling, and therefore vision, in mate recognition (Ohtsuka and Huys 2001).

Perhaps unsurprisingly, copepod species with highly modified eyes often have more visually based, distinct mate recognition behaviors. For example, many species in the three-eyed pontellid family have sexually dimorphic eyes, with males having much larger dorsally facing eyes than females. This is coupled with the observation that females are often much larger in size and brightly colored ventrally, likely to increase mate recognition for the males (Ohtsuka and Huys 2001). An extreme example of potentially visually guided mate recognition in copepods can be found in the family Sapphirinidae. Sapphirinds have unique, double-lensed, conical eyes (Fig. 9.5), and in the genera *Sapphirina* and *Copilia*, females have much larger eyes than their male counterparts (Wolken and Florida 1969; Downing 1972; Takahashi et al. 2015; Kimura et al. 2020). Additionally, male *Sapphirina* species have an intensely **iridescent** cuticle (Fig. 9.2), while females in the genus are mostly transparent (Chae and Nishida 1994, 2004). The sexual dimorphism seen in the family Sapphirinidae is further intensified by the fact that male sapphirinids are

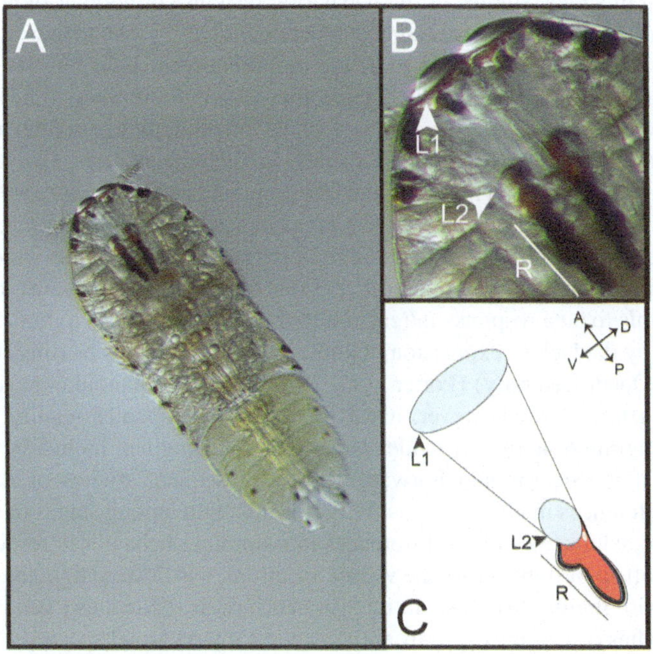

Fig. 9.5 Photo of *Sapphirina metallina* by R. Hopcroft. (**a**) Whole body and (**b**) enlarged view of eyes showing the position of the two lenses (L1 and L2) and retina (R). (**c**) Schematic of a longitudinal section of a single sapphirinid eye, with the lenses (L1, L2) and retina (R) indicated. The colors of the retina indicate the photoreceptor cells (red), phaosome (light orange), tapetum (blue), screening pigments (black), and crystalline cone (white)

usually much larger than females on average (Hirst and Kiørboe 2014; Takahashi et al. 2015). The current hypothesis posits that the larger female eyes coupled with the larger, iridescent male cuticles contribute to finding mates in the open ocean (Heron 1973; Chae and Nishida 1994).

9.3 Visual Function

Despite a relatively large body of literature describing the anatomy of copepod eye diversity, few studies have directly tested functional aspects of vision, such as visual acuity or spectral sensitivity. Studies of copepod visual physiology are likely hampered due to the difficulty in taking physiological measurements from such minute eyes and from capturing a diversity of species from the broad habitats that copepods occupy. Despite these limitations, physiological measurements from a naupliar eye have been done in at least one species. In the mesopelagic bioluminescent copepod *Gaussia princeps*, measurements of temporal and spectral sensitivity suggest that the naupliar eye has slow photoreceptor dynamics and is tuned for detecting bioluminescence, with a single visual pigment peak at 496 nm (Cohen and Frank 2019). Supporting this idea, studies of the behavioral response of another bioluminescent copepod, *Metridia longa*, to simulated copepod bioluminescence indicated that not only is the visual system tuned to detect bioluminescent light but that it may be recognized as a warning signal by conspecifics (Buskey and Swift 1985).

As a proxy for direct measurements of visual function, more research has focused on behavioral responses to light and visually mediated behaviors. Early studies on wavelength-specific behavior in the siphonostomatoid copepod *Lepeophtheirus salmonis* and the calanoid *Calanus finmarchicus* indicated that migration responses to visible light were minimal, unless ultraviolet (UV) wavelengths were included (Aarseth and Schram 1999). Rather than a function of vision, however, the authors proposed this to be a response in reaction to the radiation of UV light, a hypothesis supported by the lack of expression of known UV-sensitive opsins (this chapter; see Sect. 9.6, Opsin Diversity) (Porter et al. 2017). For the calanoid copepod *Acartia (Acanthacartia) tonsa*, however, it was observed that females readily performed vertical migration across a wide range of wavelengths, including UV light (380–700 nm) (Stearns and Forward 1984). Subsequent studies of *L. salmonis* showed differences in absolute sensitivity to white light among stages (e.g., nauplii, copepodids, adults) as well as differences in swimming behavior in response to the onset and offset of light across the visible spectrum, which were hypothesized to be involved with finding fish hosts in this parasitic species (Novales Flamarique et al. 2000). Studies of *Calanus* spp. from the Arctic showed broad spectral sensitivities (blue through green), as well as sensitivities to light intensities allowing DVM during the polar night using the night sky, the moon, and even the aurora borealis as cues (Båtnes et al. 2013). Interestingly, the peak behavioral responses to different wavelengths of light for DVM varied significantly among species (*Centropages typicus* – 500 nm, *Anomalocera ornata* – 520 nm, *Calanopia americana* – 480 and 520 nm) (Cohen and Forward 2002). Clearly, the interplay between light, visual sensitivities, and DVM behavior is intricate and needs continued study in the ecologically and visually diverse copepods.

Another aspect of copepod vision elucidated by behavioral studies is the detection of polarized light. Based on behavioral studies, copepod species within four orders exhibit some degree of polarotaxis, orienting themselves toward linearly polarized light (e.g., *Pontella karachiensis* (Manor et al. 2009); *Acanthocyclops vernalis*, *Tisbe furcata*, *Caligus rapax* (Umminger 1968); *Calanus* spp. (Lerner and Browman 2016)), while studies of an additional parasitic species (*L. salmonis*) showed a lack of orientation to polarized light (Novales Flamarique et al. 2000). Correspondingly, anatomical studies of the eyes in *Pontella karachiensis* and *Acanthocyclops vernalis* also found orthogonally oriented microvilli (Umminger 1968; Manor et al. 2009). Given the variation in photoreceptor cell numbers and rhabdomere arrangements across species, as well as among ocelli in the naupliar eye within a single species, much more research into the anatomy and physiology of polarization vision in copepods is needed to understand the variation in function, how different species are detecting and processing the information, and what behaviors are linked to this aspect of vision.

9.4 Copepod Eye Morphology – Overview

The majority of copepod species where eye morphology has been described have a tripartite naupliar eye (Table 9.1). Compared to the more common crustacean compound eye type, the naupliar eye is relatively simple. In the most common copepod form, typified by the calanoid species *Eucalanus elongatus* (Esterly 1908; Vaissiere 1961; Elofsson 1966), the naupliar eye is composed of three fused ocelli or "cups," with two cups generally oriented more dorsally and one positioned more ventrally (Fig. 9.3). Although simple, this basic structure has served as the foundation for significant evolutionary tinkering, leading to a diverse array of visual structures that vary in the arrangement and number of all components: the presence, positioning, and number of lenses; the shape of the retina; the type of reflecting structures; the position of the naupliar eye within the head; and even the number of "eyes" are variable. In this review, we endeavor for the first time to assemble all of the descriptions of eye anatomy from individual species to construct a unified evolutionary framework for copepod visual systems.

Part of the difficulty of compiling a thorough review of copepod visual systems, however, is the lack of consistent terminology in the literature for describing visual structures in the group. For example, naupliar eyes themselves are also referred to as "frontal" (Elofsson 1966) or "tripartite" (Esterly 1908; Elofsson 2006) eyes, and the two ocellar cups on the top of the eye are often referred to, based on position, as the "dorsal," "lateral," or "dorsolateral" eyes. We propose a scheme to unify this confusing history of terminology to lay the groundwork for future studies of copepod visual systems. To avoid confusion from positional-based terminology, we will refer to the three retinal cups as two sets of ocelli – the two cups that generally mirror each other in morphology we define as the **paired ocelli**; the remaining single cup we define as the **unpaired ocellus**. In species where the three ocelli have been divided into separate visual units, we use the term **eye** in place of ocellus to indicate the separation, which is often associated with evolutionary specializations for increased visual capabilities. As a foundation for understanding the evolution of eye diversity within the copepods, described later in the chapter, we use this framework to first describe the basic components of copepod naupliar eyes (Fig. 9.3) – retinal cells, pigment cells, lenses, and reflectors – including references to synonymous historical terminology where possible.

9.4.1 Retinular Cells

Within each paired and unpaired ocellus, the light-sensitive retina is most commonly situated atop a reflective **tapetum** and surrounded by a screening **pigment cup**. In the copepod naupliar eye, the retinal cells tend to be extremely clear and were once called the "crystal cells" (cellules cristallines) by Vaissiere (1961) or "retinular cells" by Elofsson (1966). Generally, the paired ocelli of copepod naupliar eyes each have nine retinal cells, while the unpaired ocellus has ten, though this number varies among taxa (Table 9.1). The retinal cells tend to form a spherically shaped retina in both paired and unpaired ocelli, most commonly formed by either a layering of the cells or a "crown" of cells arranged in a ring.

Each retinal cell contains a **rhabdomere**. As in many arthropods, the rhabdomeres are composed of clusters of microvilli oriented perpendicular to incident light. In each of the three ocelli, the rhabdomeres from multiple retinal cells tend to be clustered in the interior of the retina, forming a triangular plate just above the tapetum (Esterly 1908; Fasten 1916; Vaissiere 1961). When retinal cells within a retinal sphere are layered, the rhabdomeres in the distal layer tend to form lines running parallel to the proximal rhabdomeres, the pigment cup, and the tapetum (Elofsson 2006). However, the eyes of several species have been described where the position of the rhabdomeres within each cell and relative to other cells, in both the paired and unpaired ocelli, varies greatly from this typical arrangement (Vaissiere 1961; Elofsson 1966, 2006). We describe the retina of some of the species with the most exceptional retinular rearrangements below (see sections on Calanoida, Cyclopoida, Siphonostomatoida, and Harpacticoida).

Table 9.1 Information on eye anatomy from histological studies of diverse copepod species

Order/family		Species	Eye type	Ecology	# of receptor cells		# Lenses/cup		Tapetum		References
					Un-paired	Paired	Un-paired	Paired	Un-paired	Paired	
Order Platycopioida											
1			No eyes	Epibenthic	0	0	0	0	0	0	Fosshagen and Iliffe (1985)
Order Calanoida											
2	Aetideidae	*Aetideopsis armata*	Enlarged (paired)	Pelagic	10 L	9	0	2	---	---	Elofsson (1966)
3	Augaptilidae	*Augaptilus* sp.	Typical naupliar	Marine	10 L	---	---	---	---	---	Elofsson (1966)
4	Calanidae	*Calanus finmarchicus*	Typical naupliar	Bathypelagic	10 L	9	0	0	y	y	Elofsson (1966)
5	Calanidae	*Calanus hyperboreus*	Typical naupliar	Epi- to bathypelagic	10 L	9	0	0	y	y	Elofsson (1966)
6	Calanidae	*Neocalanus gracilis*	Typical naupliar	Pelagic	10 L	9	0	0	y	y	Elofsson (1966)
7	Candaciidae	*Candacia ethiopica*	Enlarged	Pelagic	6	9	0	0	---	---	Vaissiere (1961)
8	Centropagidae	*Centropages typicus*	Enlarged (unpaired)	Epipelagic	5 L	10	1	0	---	---	Vaissiere (1961)
9	Eucalanidae	*Eucalanus elongatus*	Typical naupliar	Pelagic	10	9	0	0	y	y	Vaissiere (1961)
10	Euchaetidae	*Paraeuchaeta norvegica*	Typical naupliar	Pelagic, polar	10 L	9	0	0	y	y	Elofsson (1966)
11	Phaennidae	*Cephalophanes tectus*	Reflector	Mesopelagic	0	4	---	---	---	n	Nishida (2002)

(continued)

Table 9.1 (continued)

	Order/family	Species	Eye type	Ecology	# of receptor cells		# Lenses/cup		Tapetum		References
					Un-paired	Paired	Un-paired	Paired	Un-paired	Paired	
12	Pontellidae	*Pontella karachiensis*	Y eye	Epipelagic	6	---	3 (M) 2 (F)	1	---	---	Manor (2009)
13	Pontellidae	*Pontella mediterranea*	Y eye	Epipelagic	6	7 L	---	---	---	---	Elofsson (1966)
14	Pontellidae	*Pontellopsis regalis*	Y eye	Epipelagic	6	---	---	0	---	---	Elofsson (2006)
Order Misophrioida											
15	Pseudocyclopiidae	*Stygocyclopia australis*	No eye	Anchialine	0	0	0	0	0	0	Jaume et al. (2001)
16	Speleophriidae	*Mexicophria cenoticola*	No eye	Hyperbenthic or anchialine	0	0	0	0	0	0	Boxshall et al. (2014)
17	Speleophriidae	*Speleophria bunderae*	No eye	Anchialine	0	0	0	0	0	0	Jaume et al. (2001)
18	Speleophriidae	*Speleophria germanyanezi*	No eye	Anchialine	0	0	0	0	0	0	Suarez-Morales (2017)
19	Speleophriidae	*Speleophria nullarborensis*	No eye	Hyperbenthic or anchialine	0	0	0	0	0	0	Karanovic Eberhard (2009)
Order Canuelloida											
20	Longipediidae	*Longipedia gonzalezi*	No eye[a]	Mesophotic reef	0	0	0	0	0	0	Schizas et al. (2015)
Order Gelyelloida											
21	Gelyellidae	*Gelyella droguei*	No eye	Karstic systems	0	0	0	0	0	0	Ito (1985)
Order Cyclopoida											
22	Corycaeidae	*Agetus flaccus*	Telescopic	Epimesopelagic	9	7 L	0	2	---	---	Elofsson (2006)

23	Corycaeidae	*Agetus limbatus*	Telescopic	Epimesopelagic	9	7 L	0	2	---	---	Elofsson (2006)
24	Corycaeidae	*Corycaeus speciosus*	Telescopic	Epimesopelagic, sometimes bathypelagic	9	7 L	0	2	y	---	Elofsson (1966, 2006)
25	Corycaeidae	*Farranula carinata*	Telescopic	Pelagic	9	7 L	0	2	---	---	Elofsson (2006)
26	Cyclopidae	*Macrocyclops albidus*	Enlarged (paired)	Freshwater	5	14 L	0	0	y	y	Fahrenbach (1964)
27	Notodelphyidae	*Doropygus seclusus*	Typical naupliar	Parasitic (marine ascidian)	8	8	---	---	---	---	Elofsson (2006)
28	Sapphirinidae	*Copilia mirabilis*	Telescopic	Epimesopelagic	9	7 L	0	2	---	---	Elofsson (2006)
29	Sapphirinidae	*Copilia quadrata*	Telescopic	Epipelagic	---	5	---	2	---	---	Woken and Florida (1969)
30	Sapphirinidae	*Sapphirina ovatolanceolata*	Telescopic	Epipelagic	7 L	7 L	---	---	---	---	Elofsson (1966)
31	Sapphirinidae	*Sapphirina* spp.	Telescopic	Pelagic	9	7 L	0	1 or 2	y	Double	Elofsson (1966, 2006)
32	Thaumatopsyllidae	*Caribeopsyllus amphiodiae*	Y eye	Parasitic (marine brittle star)	---	---	1	1	y	y	Hendler and Dojiri (2009)
Order Harpacticoida											
33	Miracidae	*Miracia efferata*	Telescopic	Epipelagic	3	3	0	2	---	---	Elofsson (1966, 2006)
34	Miraciidae	*Macrosetella gracilis*	Typical naupliar	Pelagic, tropical	---	5–7	---	---	---	---	Elofsson (1966)
35	Harpacticidae	*Tigriopus californicus*	Typical naupliar	Tide pools	7	9	---	---	y	y	Martin (2000)

(continued)

Table 9.1 (continued)

	Order/family	Species	Eye type	Ecology	# of receptor cells		# Lenses/cup		Tapetum		References
					Un-paired	Paired	Un-paired	Paired	Un-paired	Paired	
36	Harpacticidae	*Tigriopus fulvus*	Typical naupliar	Coastal benthic	6	6	---	---	---	---	Vaissiere (1961)
37	Harpacticidae	*Tisbe furcata*	Typical naupliar	Coastal benthic	6	6	---	---	---	---	Vaissiere (1961)
Order Mormonilloida											
38			No eye	Mesobathypelagic	0	0	0	0	0	0	Ivanenko and Defaye (2006)
Order Monstrilloida											
39	Monstrillidae	*Cymbasoma danae*	Typical naupliar	Marine	9	9	---	---	y	y	Elofsson (1966)
Order Siphonostomatoida											
40	Caligidae	*Caligus savala*	Telescopic	Marine	10 L	9	0	1	---	---	Elofsson (1966)
41	Caligidae	*Lepeophtheirus nordmannii*	Telescopic	Marine parasite (sunfish)	7	7	0	2	---	---	Vaissiere (1961)
42	Lernaeopodidae	*Salmincola edwardsii* (LARVAL)	Telescopic	Parasitic (freshwater fish)	5	9	0	0	---	---	Fasten (1916)
43	Pennellidae	*Pennella filosa*	Enlarged (paired)	Parasitic	7	7	0	1	---	---	Vaissiere (1961)

Information for each species includes taxonomy (order and family), eye type as defined in this review, ecology, and details on the structure of each eye, including receptor cell number and presence of layering (L), number of lenses present per ocellar cup, sexual dimorphism when present (male – M, female – F), and presence of tapetal structures (present – y, absent – n, or number of layers). Missing data indicated as '---'

[a]The description of the genus indicates that there are no eyes present

9.4.2 Pigment Cells

The screening pigments in copepod naupliar eyes form cup-like structures with a distinct x-shape when viewed dorsally (Fig. 9.3). These cups are located proximal to the retina in each ocellus, optically separating all three ocelli from each other. Each pigment cup consists of usually two (but sometimes one) cells filled with carotenoid pigments ranging in color from oranges to reds (Esterly 1908; Vaissiere 1961; Elofsson 1966, 2006). Some species have black or blue screening pigments, indicating that other pigments, such as melanins, may be used (Vaissiere 1961; Elofsson 1966). The general shape of the pigment cup and the number of cells that make up the structure vary among species. The cyclopoid species *Macrocyclops albidus*, for example, has modified pigment cups, where the cup surrounding the unpaired ocellus is made from two cells, while those surrounding the paired ocelli consist of a single cell (Fahrenbach 1964). In the cyclopoid suborder Ergasilida, the pigment cups for the paired ocelli are formed from a pair of layered cells that are conical in shape (Elofsson 1971). In the harpacticoid species *Doropygus seclusus*, glial cells were found to act as accessory pigment cells, covering the eyes marginally (Dudley 1972). The variation in the color, size, cell number, and shape of the naupliar eye-screening pigment cups in just these few species indicates that there is likely a much larger diversity found across the Copepoda.

9.4.3 Lenses and Other Light-Refracting Structures

Although most copepod naupliar eyes lack distinct optical structures, several families contain visual systems with a spectacular variety of **lenses**, ranging from large circular crystalline spheres to telescopic lenticular doublets or triplets (Vaissiere 1961; Elofsson 1966; Land 1984). Lenses in copepod visual systems are often found in species where the naupliar eye has been separated into three separate visual structures. Within the calanoids, species in the family Pontellidae commonly have three separate ocellar eyes (two dorsal, one ventral), and each can have one or more lenses (Fig. 9.4). There is a large diversity among species in both the number and shape of lenses. Some species in the genera *Labidocera* and *Epilabidocera* have large, spherical lenses atop each dorsolaterally facing paired eye but no lens on the unpaired eye. In addition to the pair of lensed eyes, species in the genus *Pontella* have added one, two, or even three lenses to the ventral unpaired eye (Land 1984; Elofsson 2006). In the genus *Anomalocera*, the paired eyes have two lenses each, one anterior and one posterior, and one lens on the unpaired eye (Vaissiere 1961). Studies on lens proteins in the pontellid *Anomalocera ornata* suggest that copepods use novel proteins as crystallins, which require further study (Cohen et al. 2005, 2007). Within the cyclopoid suborder Ergasilida and the harpacticoid family Miraciidae, large, double, front-facing lenses are commonplace (Claus 1863; Vaissiere 1961; Huys and Böttger-Schnack 1994). These telescopic eyes have an anterior lens at the cuticular level, as well as a second lens directly in front of the retina (Claus 1863; Vaissiere 1961; Huys and Böttger-Schnack 1994). In contrast to large lens structures, some calanoid species have made use of thin, flattened cells over the top of the paired ocelli to create a focusing structure (e.g., *Macrocyclops albidus albidus, Candacia ethiopica, Megacyclops gigas, Centropages typicus* – Vaissiere 1961; Elofsson 1966). The parasitic order Siphonostomatoida has a unique cuticular lens atop layers of cuticular thickening, distal to a crystalline type lens, which is situated directly on top of the paired ocelli (e.g., *Lepeophtheirus nordmannii, Caligus* sp. – Vaissiere 1961).

Numerous copepod species also make use of intracellular light-refracting material formed from condensed smooth endoplasmic reticulum in structures known as **phaosomes** (or dictyosomes) and **whorls**. Phaosomes are stacks of membranes, generally rod-like or lenticular in shape, which may refract light and are typically located near the nucleus or in front of the rhabdomeres (Vaissiere 1961; Elofsson 1966). These phaosome structures can condense or disperse, possibly based on exposure to light (Vaissiere 1961; Elofsson 1966, 1971). Whorls are similar, in that they are stacks of membranes that refract light; however, whorls are often found within the microvilli of the rhabdomere and are often divided and highly irregular in shape (Elofsson 1970). Although these structures are typically associated with rhabdomeres and are thought to interact with light, their function relative to vision is unknown.

9.4.4 Light-Reflecting Structures

There are several types of reflecting structures found in copepod visual systems. The most common reflecting structure found is a **tapetum** – a layer of reflective cells found directly behind the retina. Tapeta are found in a diversity of animals, not just copepods, and aid in light capture by reflecting light back onto the retina after it has already passed through the retinal cells. In copepods, the tapetum is found in all typical naupliar eyes and is often composed of two cells that form a cup shape, positioned between the retinal cells and the pigment-cup cells in each ocellus. Within the tapetal cells are tightly packed, refractive, square-shaped guanine platelets. The platelets generally are oriented flat sides toward the inner surface of the optic cup (Fahrenbach 1964). Tapeta are also found in separated ocellar eyes, where light reflection can be dramatic (Fig. 9.4).

In the deep-sea calanoid *Cephalophanes refulgens*, the tapetal and pigment cells are replaced by a large reflective mirror in the paired eyes (Nishida et al. 2002). Anatomically, the reflectors are composed of stacks of thin plates of unknown material that appear "softer" than typical aquatic animal guanine crystals (Nishida et al. 2002). These parabolic mirrors are optimized to direct light back to the retinal cells from all frontal angles, likely to aid in foraging in low light conditions (Boxshall 1992; Nishida et al. 2002).

9.4.5 *Nonvisual Copepod Light-Sensing Structures: Gicklhorn's Organ*

For a complete review of copepod photoreception, we must also include a discussion of nonvisual light-sensing structures, which in copepods have a long and convoluted history in the scientific literature. In addition to naupliar eyes, several other distinct neural structures have been posited to be light sensitive among crustaceans (Elofsson 1966, 1970; Arendt and Wittbrodt 2001). We attempt to untangle the complex history of terminology related to nonvisual photoreception in copepods, particularly for the often anatomically obvious, yet still poorly understood, Gicklhorn's organ.

A sense organ by many names, Gicklhorn's organ has been a topic of confusion for several decades. First described by German researcher Josef Gicklhorn in 1930, this set of paired globular organs was found in the cephalic region of the copepod *Cyclops strenuus*, at the anterior base of the antennae (Gicklhorn 1930). Though quite distant from the brain, these paired globules were well innervated from the central nervous system (CNS), with each consisting of two large, binucleated cells. Interest in the form and function of this structure did not resurface until several decades later when a series of papers by Patricia Dudley of Columbia University and Rolf Elofsson at the University of Lund compared the ultrastructure of this organ to the X-organ described in *Artemia* (Elofsson 1970, 1971; Dudley 1972). Dudley took a developmental approach, investigating this organ through the life stages of the cyclopoid *Doropygus seclusus*, while Elofsson went for a taxonomic approach, investigating the presence of this organ across several calanoid and cyclopoid species, coining the term "Gicklhorn's organ" in the process (Elofsson 1970, 1971).

Though the structural components of Gicklhorn's organ are conserved, the general size, shape, and structure of this organ vary between species. In some calanoids, this organ was so noticeable that it was used to differentiate between three species of *Calanus* based on its size and relative translucent coloration (Frost 1974). Additionally, in the genera *Eucalanus* and *Chiridius*, this organ terminates at the cuticle in a triangular shape, while in *Calanus finmarchicus*, the shape is that of a ventrally recurved horn (Elofsson 1970). Based on the presence of microvilli, and of potentially light-interacting phaosome and whorl structures also observed in the naupliar eye, Elofsson (1966, 1970) predicted that the Gicklhorn's organ was light sensitive. However, even the structure of microvilli varies among species. In *Calanus finmarchicus*, the Gicklhorn's organ microvilli form a distinct rhabdom on adjoining cell borders, similar in shape to those of the crustacean compound eye rhabdom; in other species the microvilli were not arranged in a way that would respond to incident light, suggesting variability in function. Additionally, in both *Euchaeta* sp. and *C. finmarchicus*, the organ is innervated by a neurosecretory axon, suggesting that the Gicklhorn's organ may have a multisensory role (Elofsson 1966, 1970). Following investigations of the cyclopoid *Sapphirina*, Elofsson (1971) proposed three possible functions for the Gicklhorn's organ based on cellular morphology:

(1) an internal chemosensor; (2) an external sensor, although he did not specify for what cue; or (3) still unknown, leaving the mystery of the Gicklhorn's organ function wide open.

Based on neuroanatomical studies, Elofsson (1970) hypothesized that the Gicklhorn's organ was not homologous to any other type of crustacean eye based on a lack of central innervation in the brain. In 2020, this organ resurfaced in the scientific literature, this time rebranded as an ancestral compound eye (i.e., Frase and Richter 2020). An important piece of evidence for arthropod compound eye homology is an innervation to the CNS, which includes a protocerebral bridge and either a central body or two accessory lateral lobes (Richter et al. 2010). In calanoid copepods, although studies of the neuroanatomical structure had previously been ambiguous or controversial, support for a central body was reported in the harpacticoid genus *Tigriopus* (Lacalli 2009; Andrew et al. 2012). Studies of the calanoid *Calanus finmarchicus* using antibody neural tracing to elucidate neural connections also suggested that the paired receptors of the Gicklhorn's organ may be homologous to the arthropod compound eye (Frase and Richter 2020). Tantalizing as these new studies may be, without any conclusive behavioral or physiological evidence, the Gicklhorn's organ function and its evolutionary origins remain a mystery. Unraveling these mysteries, as well as conclusively demonstrating that this structure is light sensitive, will be critical to understanding the overall photosensitivity and light-mediated behaviors in copepods.

9.5 Evolutionary Diversity of Copepod Eyes

Since some of the first comparative descriptions of copepod eyes published in 1863 (Claus 1863), anatomical studies of eyes with varying degrees of detail have been published for at least 43 species representing 31 genera, 25 families, and nine of ten orders across the group (Table 9.1). Although the ancestral copepod lineage certainly had some form of naupliar eye (Boxshall et al. 1984), the vast array of eye designs that have evolved from this simple eye holds more diversity in form than that of the more complex compound eye types found across the rest of the Crustacea.

Within the copepods, naupliar eyes vary in size, shape, relative position within the cephalsome, and the number of receptor cells in each ocellus. There are also numerous examples among copepod species where the common naupliar eye design has been significantly modified into enlarged paired ocelli, an enlarged unpaired ocellus, or, in the most extreme visual systems, three separate eyes. In species with three independent eyes, variations in retinal shape, the number and shape of lenses (or other light-focusing structures), reflector presence or absence, and the composition of pigment cups are common. In some species, these eyes are also sexually dimorphic. However, the plethora of copepod visual system morphologies has never been considered within the context of eye evolution across the entire group. Below we discuss the eye types for each of the ten currently recognized copepod orders in more detail, followed by an analysis of copepod-expressed opsin gene diversity. The orders are discussed in a phylogenetic sequence based on Fig. 9.1a, and the eyes have been categorized into eight general morphological types (Table 9.1): **(1) typical naupliar eye** – the ancestral eye type, as seen in *E. elongatus* (Fig. 9.3); **(2) enlarged (naupliar, paired, or unpaired) eyes** – relative to the typical naupliar eye, either the whole eye or a subset of the ocelli (paired or unpaired) are bigger in relation to the volume of the cephalosome, and in some species, the enlarged structures are associated with light-refractive structures, such as lenses or pseudofacets; **(3) reflector eyes** – eyes with modified mirror-like parabolic reflectors; **(4) Y eye** – the typical naupliar eye has separated into three distinct, often lensed, eyes but retains neural connections that form a Y shape; **(5) telescopic eyes** – paired ocelli with double lenses oriented in the same light path with one distal to the other, where the paired ocelli may be separated into two eyes, which are often, but not always, separated from an unpaired, nonlensed ocellus or eye; **(6) no eye** – complete loss of the naupliar eye. Together, the diversity of eye morphology and opsin gene expression serves as a foundation for understanding the evolution of copepod eye morphology and function, elucidating evolutionary patterns leading to eye diversity and highlighting the vast gaps that still exist in our knowledge of copepod vision.

9.5.1 *Platycopioida (No Eyes)*

The order Platycopioida is the most basal lineage of the copepods and is composed of a single family, four genera, and 11 species. Most species in the group were collected from anchialine caves or hyberbenthic habitats and do not appear to have eyes (Fosshagen and Iliffe 1985). The one exception is the species *Platycopia tumida*, which is sexually dimorphic, with females having small and males having large "eyespots" (Wilson 1946).

9.5.2 *Calanoida*

The order Calanoida comprises a taxonomically diverse (~2700 species distributed among 43 families) group that encompasses a range of copepod ecological niches (freshwater, brackish, and marine species from pelagic, benthic, and benthopelagic habitats (Huys and Boxshall 1991)). Based on previously published anatomical descriptions from species representing ten families, most calanoids possess a typical naupliar eye that sits anterior to the brain rather than directly upon it (Figs. 9.1b, 9.3 and 9.6), although variations in the size and placement of the eye within the cephalosome have been observed within the order. Members of the family Calanidae tend to have smaller eyes than those of Euchaetidae, Eucalanidae, and Peltidiidae (subfamily Clytemnestrinae) (Elofsson 1966). The general retinal cell arrangement for the calanoid eye is a crown formation in the paired ocelli, oriented dorsolaterally, and a two-layered ventrally facing sphere, with plates of rhabdomeres lining the dorsal most region of the unpaired ocellus (Vaissiere 1961; Elofsson 1966). Despite the predominance of naupliar eyes described from calanoid species, extreme eye modifications and morphologies have been documented in species from six families. Below, we highlight some of the most extreme eye types in this order by family.

9.5.2.1 Phaennidae (Reflector Type)

The deep-sea genus *Cephalophanes* has a modification to their eyes that is unique among the copepods – large, bilateral, frontally facing, semi-parabolic, mirror-like reflectors encircle each paired ocellus (Figs. 9.6a and 9.7); the unpaired ocellus is significantly reduced or missing altogether (Nishida et al. 2002). The reflectors, situated directly behind the retinal sphere, are parabolic when viewed dorsally but spherical in lateral view and, at the thickest (posterior) end of the reflector, made up of stacks of ~150 thin plates of putative cuticle material. The *Cephalophanes* reflector is highly efficient only when the light source is incident and originates anterior to the copepod. When these conditions are met, ~85% of light at wavelengths from 350 to 700 nm is reflected toward the rhabdomeres of a pair of retinal cells (Nishida et al. 2002). It was also presumed, based on muscle anatomy, that *Cephalophanes* can control reflector direction, making these eyes one of the most effective broadband, directional animal light detectors described. While these eyes are unlikely to have any image-resolving power with only two receptor cells in each reflector focal area, the ability to move the eyes suggests the potential for increasing the field of view as suggested in other aquatic arthropods with scanning eye movements (e.g., copepods – Downing 1972; Land 1988; diving beetles – Buschbeck et al. 2007). Regardless of whether the *Cephalophanes* can control eye movements, it is clear that their eyes are optimized for collecting light in deep-sea habitats. From the forward-facing orientation of these eyes, it is unlikely that these were developed for mate recognition or predator avoidance; thus, it was suggested that the primary use of these large eyes is foraging in low light conditions (Steuer 1928; Boxshall 1992). Investigations of the gut contents of two *Cephalophanes* species indicated that these detritivores feed primarily on the "rain of carcasses (Leichenregen)," as predicted by Steuer (1928), potentially aided by bioluminescent bacteria commonly found on these carcasses (Nishida et al. 2002). An additional oddity of these unique copepod eyes is the presence of an additional pair of receptor cells located posterior to each of the reflectors; this strange placement of the photoreceptors defies current understandings of animal eye anatomy, and even the authors originally describing the structure claim to have "no plausible explanation" for their function (Nishida et al. 2002).

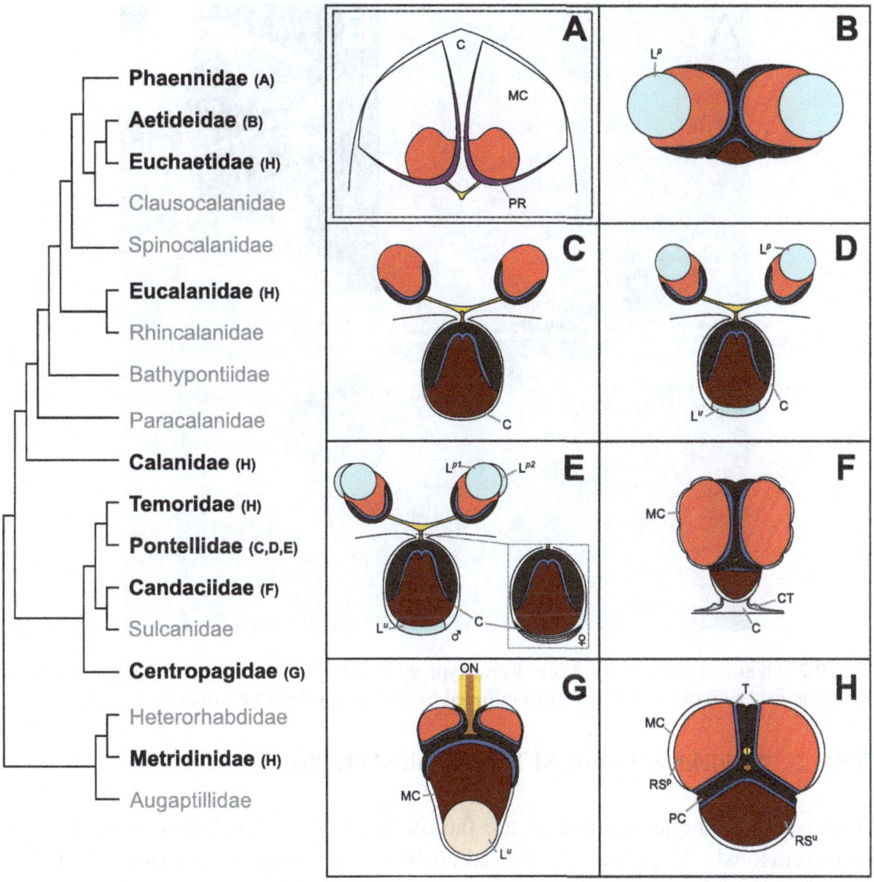

Fig. 9.6 Reduced family-level calanoid cladogram based on Blanco-Bercial et al. (2011), with families containing species with modified eye morphologies in bold. Families without histological information on eye anatomy are in light gray. Dorsal section (**a**) and frontal sections (**b–h**): (**a**) parabolic reflector eye of *Cephalophanes refulgens*, modified from Nishida et al. (2002); (**b**) enlarged paired ocelli of *Aetideopsis armata*, modified from Elofsson (1966); (**c**) frontal section of *Pontellopsis* spp. Y-type eyes without lenses; (**d**) Y-type eye with single lens per eye found in *Labidocera wollastoni*; (**e**) frontal section of *Anomalocera patersoni* Y-type eye with double lenses on each paired eye; (**f**) enlarged eye of *Candacia oethicopa* with pseudofacets on the paired ocelli and a lens-like cuticular thickening on the unpaired ocellus; (**g**) enlarged unpaired eye of *Centropages typicus*; (**h**) typical naupliar eye type based on *Eucalanus elongatus*. (**c–h** Modified from Vaissiere 1961). *C* cuticle, *MC* membrane cell, *PR* parabolic reflector (purple), L^p paired ocellus lens (^{p1} and ^{p2} are in different planes in the case of *Anomalocera*), L^u unpaired ocellus lens, *CT* cuticular thickening that secretes crystallin-like substances (light blue), *ON* optic nerve (light yellow = paired neurons, dark yellow = unpaired neurons), *T* tapetum (dark blue), *PC* pigment cup (dark gray), RS^p paired retinal sphere containing four to ten retinal cells (light red), RS^u unpaired retinal sphere containing five to ten retinal cells (dark red)

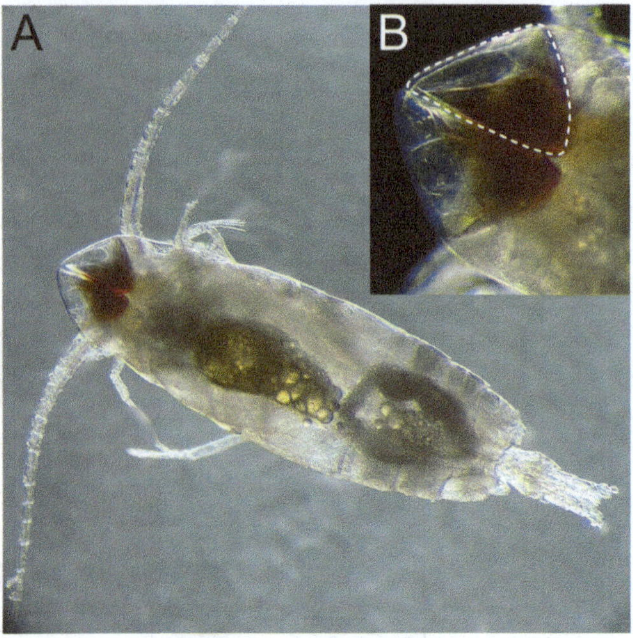

Fig. 9.7 Photos of *Cephalophanes* reflector-type eyes. (**a**) Whole animal. (**b**) Enlarged photo of reflector eyes with one of the reflectors outlined by a white dashed line. (Images by R. Hopcroft)

9.5.2.2 Aetideidae (Enlarged Type – Paired Ocelli)

The eyes in a single member of the family Aetideidae, *Aetideopsis armata*, have been previously described. The paired ocelli are oriented laterally rather than dorsally in this species and are significantly enlarged compared to the unpaired ocellus. In each paired ocellus, a lens composed of two large cells is encased by the retinal cells, which form a bowl shape (Fig. 9.6b) (Elofsson 1966). This modification may help *A. armata* forage for plant and animal matter at the 200 m depths in Arctic fjords where it lives (Schøyen and Kaartvedt 2004).

9.5.2.3 Pontellidae (Y-Eye Type)

The family Pontellidae has separated the three ocelli of the naupliar eye into three distinct, often lensed, eyes: two dorsal and one ventral (Figs. 9.4 and 9.6c–e). The paired eyes are formed by deep photoreceptor cups near the dorsal surface of the head. The unpaired eye protrudes ventrally under the rostrum and is shaped like the finger of a glove. When lensed, the photoreceptors of the unpaired eye form a posterior cap to the lens, and the whole eye is directed at a 30° angle anteroventrally (Vaissiere 1961). In many pontellid species, visual system sexual dimorphism is very apparent, with males having much larger paired eyes than females and occasionally larger or more intricate unpaired eyes.

Perhaps most notable about the modifications in pontellid eyes across species is the extreme variation in the number, shape, and optical mechanisms for focusing the light of the lenses. In some species (i.e., *Pontellopsis* spp.), none of the eyes have any lens structures (Fig. 9.6c) (Vaissiere 1961). In the genus *Labidocera*, the paired eyes each have a single, highly refractive spherical lens (Fig. 9.6d), and the paired eyes are joined medially with the ability to scan a visual field of about 40° in an anterior-posterior direction (Land 1984, 1988). In *Anomalocera*, the paired eyes have two lenses each (one anterior, the other posterior), while the unpaired eye has a single lens. The females of *Anomalocera patersonii* do not have a lens in the unpaired eye; instead they have multiple chitinous thickenings rather than the same refractive materials of the male lens (Fig. 9.6e) (Vaissiere 1961). This sexual dimorphism in lens materials can also be observed in the species *Labidocera wollastoni* (Vaissiere 1961).

In the most extreme example, male *Pontella spinipes* paired eyes do not have a lens, but the unpaired eye has an assembly of three lenses, resembling a high-power microscope setup (Land 1984). The front-most lens has a parabolic profile, likely to reduce spherical aberration. This eye type has been interpreted as a spot-and-surround detector and may provide a means for mate recognition (Boxshall 1992).

9.5.2.4 Candaciidae (Enlarged Type)

In a brief description by Vaissiere (1961), the pelagic calanoid *Candacia ethiopica* was noted to have a modified ommatidia-like grid over the paired ocelli and a ventral unpaired ocellus with a cuticular thickening that secretes a lens-like substance (Fig. 9.6f). These structural modifications are another unusual visual modification that is unique among copepods. Because this eye type has not been the subject of intensive study, little is known about its visual function. The genus *Candacia* is thought to selectively feed on pelagic salps and larvaceans and has modified feeding appendages for this purpose (Ohtsuka and Onbé 1989; Ohtsuka and Nishida 2017). It may be that the modifications in this eye type aid in the prey capture of transparent organisms.

9.5.2.5 Centropagidae (Enlarged Type – Unpaired Ocellus)

The brackish-to-marine epipelagic copepods in the genus *Centropages* have a uniquely large eye (Fig. 9.6g) that sits within a large clear, bag-like zone in the center of the cephalosome and can rotate left-right and "dorsalward" (Krishnaswamy 1948; Boxshall and Halsey 2004). Histological data on *Centropages typicus* suggested that the paired ocelli were significantly reduced in this group, consisting only of retinal cells encased in a pigment cup (Vaissiere 1961). The unpaired ocellus is significantly enlarged and has a large lens made of similar material to phaosomes pointing anteroventrally (Vaissiere 1961).

9.5.2.6 Acartiidae (Enlarged Type – Whole Eye)

The members of this group have a distinctly enlarged and frontally placed eye compared to the typical naupliar eye (Fig. 9.3a). Slight movements of this eye, tilting up-down and left-right, have been mentioned briefly (Claus 1863). The drawings of this eye type indicate large lens or window-like structures in the paired (one or two "windows") and unpaired ocelli (one window). No histological work has yet confirmed the structure of this eye type, though modifications might be expected as this group hunts bioluminescent prey in the polar seas (Esaias and Curl 1972).

9.5.3 Misophrioida (No Eyes)

The Misophrioida copepods, so far, consist only of 37 species in three families; all are from low-light hyperbenthic or no-light anchialine habitats. In cave animals, eye reduction or complete loss is often associated with adaptation to living in the dark (Porter and Crandall 2003). It is unsurprising then that the descriptions for species in the group have either noted the absence of eyes or not included the eye as a descriptive feature at all (Jaume et al. 2001; Karanovic and Eberhard 2009; Suárez-Morales 2011; Boxshall et al. 2014). Based on the ecology of the order and the few instances where the absence of eyes was noted in descriptions, it is assumed that most of the species in this order have no eyes.

9.5.4 Gelyelloida (No Eyes)

The order Gelyelloida is composed of only two species (*Gelyella droguei* and *G. monardi*) in one family found in subterranean waters of high-elevation karstic systems in Switzerland. Previously thought to be cave-adapted harpacticoids, there was no mention of eyes in the original species' descriptions (Moeschler and Rouch 1988). In a redescription of the order, Huys (1988) noted the absence of eyes in the species *G. droguei*, confirming the lack of eyes in this group.

9.5.5 Cyclopoida

The evolution of eye diversity is quite striking in the cyclopoids, which consist of ~4,000 species across ~100 families. Most descriptions of eyes within the group have focused on the extreme structures within the large suborder Ergasilida, where species across at least three closely related families (Khodami et al. 2019) have completely separated the naupliar ocelli into three distinct eyes with elongated and lensed paired eyes, with some lineages also gaining a second lens (Fig. 9.8d, e). The

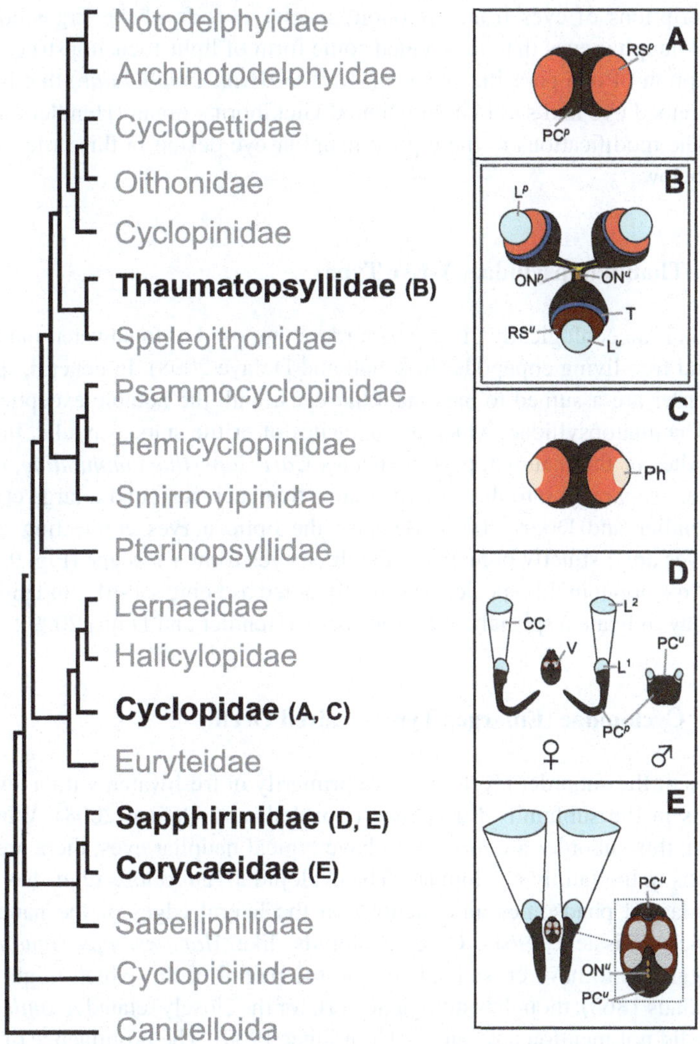

Fig. 9.8 Cyclopoida cladogram based on Khodami et al. (2019), with families with documented eye morphology in bold. Described cyclopoid eye anatomies represent either dorsal views (**a**, **c**–**f**) or frontal sections (**b**). (**a**) Typical naupliar eye of *Eucalanus elongatus*; (**b**) Y-type eye of *Caribeopsyllus amphiodiae* from Hendler and Dojiri (2009); (**c**) enlarged paired ocelli with enlarged phasomes from *Megacyclops gigas*, modified from Vaissiere (1961); (**d**) telescopic eyes of *Copilia* spp. modified from Eloffson (1966); (**e**) telescopic eyes of *Sapphirina* spp. modified from Eloffson (1966). Structures are indicated either by color, label, or both: *CC* crystalline cell; *light blue* paired (L^p) or unpaired (L^u) lenses, *ON^p/yellow* paired optic nerve, *ON^u/dark orange* unpaired optic nerve, *Ph/light orange* phasomes, *black* paired (PC^p) or unpaired (PC^u) pigment cup, *RS^p/light red* paired photoreceptors, *RS^u/dark red* unpaired photoreceptors, *T/dark blue* tapetum, *V* vacuole

few descriptions of eyes from cyclopoid species outside of the Ergasilida are of modified naupliar eyes that have added some form of light-focusing structure, with the exception of one parasitic species (*Caribeopsyllus amphiodiae*) that has a distinct Y-shaped eye and sac-like, pigmented Gicklhorn's organ (Hendler and Dojiri 2009). The modifications to the typical naupliar eye design in this order are highlighted below.

9.5.5.1 Thaumatopsyllidae (Y-Eye Type)

Oithonida is an ecologically diverse suborder composed of freshwater, marine, parasitic, and free-living copepods (Boxshall and Defaye 2008). In general, species in the suborder are assumed to have naupliar eyes, with the notable exception of the family Thaumatopsyllidae, which are parasites of brittle stars as adults. In ontogenetic studies of the thaumatopsyllid species *Caribeopsyllus amphiodiae*, the three ocelli were separated into three distinct cups that each surround a large crystalline lens (Hendler and Dojiri 2009). Because the optic nerves connecting the three lensed eyes are distinctly pigmented, the three eyes form a Y shape (Fig. 9.8b). The Gicklhorn's organ in this species was briefly noted as being sac-like and pigmented, which may indicate a specialization for vision (Hendler and Dojiri 2009).

9.5.5.2 Cyclopidae (Enlarged Type – Paired Ocelli)

Members of the suborder Cyclopida live primarily in freshwater, with a few brackish forms in the subfamily Euryteinae (Boxshall and Defaye 2008). While most species in this suborder are thought to have typical naupliar eyes, there are several exceptions in the family Cyclopidae. The cyclopid *Megacyclops giga*s has concentrated enlarged phaosomes and vacuoles at the lateral edges of the naupliar eye (Fig. 9.8c) (Vaissiere 1961). Other cyclopids, like *Macrocyclops signatus*, have been suggested to have crystal spheres on the paired ocelli, which might suggest lenses (Claus 1863), though histological work for the closely related *Acanthocyclops vernalis* did not mention any lenses (Umminger 1968). The prominence of this eye type has not been deeply investigated in this family, although the large naupliar eye is noticeable enough in most cyclopid species to suggest that this may be a more common eye type than previously described.

9.5.5.3 Ergasilida (*Sapphirina*, *Corycaeus*, and *Copilia*) (Telescopic Type)

Ergasilids have a unique life cycle among copepods; naupliar and adult stages inhabit the pelagic, and only fertilized females seek out and infest hosts as parasites (Boxshall and Defaye 2008). While most species in this group are typically found in freshwater and on fish hosts, the unique and highly modified eyes discussed here are exclusive to marine species that parasitize tunicates in the genera *Sapphirina*,

Corycaeus, and *Copilia* (Heron 1973; Lopes et al. 2007). Species in this group have highly modified telescopic paired eyes, with two sets of lenses that are separated by a clear zone (Figs. 9.5 and 9.8d, e). The genera *Corycaeus* and *Sapphirina* have tubular retinas that have been elongated longitudinally along the body. The unpaired eye is located between the paired eyes and has window-like openings facing dorsally in the form of vacuoles concentrating at the ventral-most parts of the cells. Species in the genus *Copilia* also have sexually dimorphic eyes, where *Copilia* females have double-lensed paired eyes distal to elongated boomerang-shaped retinas that curve proximally, while males only have an enlarged naupliar eye with the paired ocelli having small crystalline lenses that are focused anteriorly (Fig. 9.8d) (Vaissiere 1961). The eyes in these three genera are thought to aid in mate recognition due to the iridescence of the males of these species and the sexual dimorphism observed in *Copilia* (Heron 1973; Chae and Nishida 1994; Takahashi et al. 2015).

9.5.6 *Canuelloida*

Canuelloids are a free-living, benthic, intertidal group (George et al. 2018) with two families and ~80 species; some species live mutualistically within the shells of gastropods and hermit crabs (Ho 1988; Boxshall and Hayes 2019). Relative to vision, species descriptions from the order range from no mention of eyes to "great" naupliar eyes in the adults of the genus *Canuella* (Vincx et al. 1979). A few developmental studies of canuelloid species indicate that naupliar stages have obvious eyes, but adults may lose them, or the eyes are otherwise not described (Gurney 1930; Nicholls 1945; Schizas et al. 2015). Anatomical studies for this group are still needed to elucidate the evolutionary importance of vision in the Canuelloida.

9.5.7 *Harpacticoida*

This order has over 4500 species in 52 families. Unlike the eye diversity found in the calanoids and cyclopoids, harpacticoid species generally have simpler naupliar eye structures. The most detailed descriptions of harpacticoid visual behaviors come from the tide-pool-dwelling genus *Tigriopus*, which displays phototactic behaviors, including aggregation in shady areas during mid-day (Smith and Baker 1979; Martin et al. 2000; Andrew et al. 2012). In the species *Tigriopus californicus*, the adults retain a red-orange naupliar eye, in which the spheroid unpaired ocellus and paired ocelli are similar in size (Andrew et al. 2012). The setup of the photoreceptors in this species is quite different from the groups previously described. The photoreceptors are composed of microvilli packed centrally in the ocelli, on adjoining cell borders – rather than organized in plates or stacks, like species in other copepod orders. This arrangement is analogous to the rhabdomere orientation of crustacean compound eyes. There is some evidence for lensed or windowed structures in both the paired and unpaired ocelli in the naupliar eye of species in this order (Claus 1863), but detailed histology to confirm these early morphological drawings has not yet been done.

9.5.7.1 Miraciidae (Telescopic Type)

One notable exception of extreme eye modification in the harpacticoids is in the family Miraciidae. This family has very large, forward-facing naupliar eyes (Fig. 9.9). In the paired ocelli of species in this group, two lenses of unknown substance are linearly organized with a clear zone between them, similar in arrangement to the telescopic eyes of the Ergasilida. However, unlike ergasilid eyes, in the Miraciidae, the naupliar eye has not separated into three distinct structures. Despite the modifications to the paired ocelli, all three ocelli are still oriented in a typical naupliar eye shape. The genera of this family can be identified quickly by the distance between the exterior lenses of the paired ocelli (Huys and Böttger-Schnack 1994).

9.5.8 Mormonilloida (No Eyes)

There are four species in one family in the order Mormonilloida. There were no mentions of eyes, or indications of eye spots, in any of the descriptions for these species (Huys et al. 1992; Ivanenko and Defaye 2006). Several of the species in this order are meso- to bathy-pelagic and, based on the morphology of the feeding structures, likely feed on small particles (Boxshall 1985; Huys et al. 1992). Based on habitat alone, these species may not have well-developed eyes or any eyes at all.

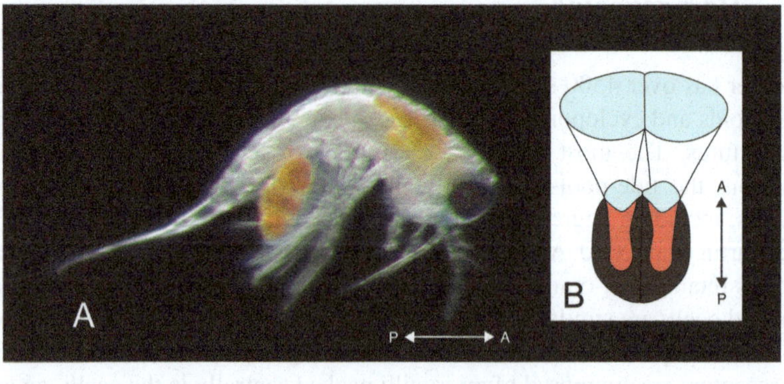

Fig. 9.9 The modified naupliar eyes of *Distioculus minor.* (**a**) Image of the entire copepod by R. Hopcroft. (**b**) Diagrammatic representation of the dorsal view of the naupliar eye of *D. minor* modified from Huys and Böttger-Schnack (1994). *A* anterior, *P* posterior, *light blue* lens, *light red* paired photoreceptors, *black* pigment cell

Future studies should specifically confirm the presence or absence of eyes in this group.

9.5.9 *Monstrilloida*

The monstrilloids (173 species in a single family) are semi-parasitic copepods of marine benthic invertebrates that are abundant in reef-related habitats (Suárez-Morales 2011). Their eyes have yet to be fully characterized anatomically or investigated histologically, though species descriptions of this group indicate that the naupliar eye is typically present. Based on more recent species descriptions, the naupliar eye has separated into three distinct eyes in numerous species (Grygier and Ohtsuka 1995; Suárez-Morales 2011; Suárez-Morales and McKinnon 2014, 2016; Ohtsuka and Nishida 2017). The degree of separation of the paired eyes is often included as a species-specific characteristic, which would likely indicate that these copepods have more complex eyes than the typical naupliar form. Detailed anatomical studies across the group are needed to elucidate the degree of evolutionary diversification in the visual system.

9.5.10 *Siphonostomatoida*

There are >2000 species of siphonostomatoid copepods across 41 families. Most of the siphonostomatoids are parasitic; as such, visual systems in this order have not been extensively studied as the adults in many species tend to have either significantly reduced or completely lost eyes. Despite this trend, the presence of large cuticular lenses has been suggested in the chalimus and adult specimens of the few species studied thus far, rather than typical naupliar eyes. The larvae of *Salmincola edwardsii* (Lernaeopodidae) were documented to have tripartite naupliar eyes, but histology documented a cuticular layering above the paired ocelli that resembles ommatidia (Fasten 1916). Reportedly, the adults of this species have three reduced, separated, and unpigmented eyes (Wilson 1946), though the histology could not be confirmed in this review. The only groups in the order with detailed eye histology are members of the families Caligidae and Pennellidae (Vaissiere 1961), which detailed cuticular lenses in Caligidae but not in Pennellidae. Without histological studies from other species, it is difficult to say conclusively whether species in this group have only modified and reduced eyes or whether there are typical naupliar eyes remaining in the adult stages.

9.5.10.1 Caligidae (Telescopic Type)

Although there were few detailed anatomical studies from siphonostomatoid species, the eyes of ectoparasitic species in the family Caligidae (i.e., *Lepeophtheirus nordmannii* and *Caligus diaphanus*) have complex naupliar eyes with unusual focusing structures. The large X-shaped pigment cups (~260 μm wide) of the paired ocelli are quite apparent when viewing the animal dorsally. Two pigment types make up the black-brown color of the fused cups: melanins and carotenoids (Vaissiere 1961). The chitin above the naupliar eye is differentiated into two symmetrical cuticular lens-like thickenings, with layers of refractive thickenings underneath (Fig. 9.10 CL, CT). The paired ocelli are separated from this cuticular structure by a thin membrane cell, and each has an associated crystalline lens (Fig. 9.10b, L^p).

9.5.10.2 Pennellidae (Enlarged – Paired Ocelli)

The chalimus and adults of *Pennella filosa* and other *Pennella* species were observed to have eyes very similar to those of Caligidae, except that the cuticular layers and cuticular lens were not present (Vaissiere 1961).

9.6 Copepod Opsin Diversity

The diversity of eye morphologies among the ten orders of Copepoda is overwhelming. From no eyes to three separated eyes, and everything in between, the evolution of copepod visual systems appears to be a tangle of intricacies waiting to be woven into a coherent web. The variety of modifications and cellular structures involved in the described diversity is unprecedented among crustaceans. As such, it would not be surprising that the diversity of the proteins responsible for light detection, opsins, would also be diverse among this group.

At the foundation of all visual systems is the ability to detect light, which is mediated by visual pigments formed by a light-sensitive chromophore bound to an opsin protein. Most of the variations in the wavelengths of light detected by a particular species are due to variation in amino acid composition among different copies of the opsin protein. Therefore, studies of the diversity of opsin genes expressed in a species can provide insights into visual function, particularly for a group like copepods, where physiological studies of vision are scarce. As a functional complement to the evolution of anatomical diversity in copepod eyes, we used Trinity software to assemble all previously published copepod RNAseq data from the NCBI SRA database that met our quality criteria (paired end Illumina data with over 250 k read depth, up to three SRA data sets per species, whole body or head tissue data sets only, available prior to December 2019), in addition to using all available, already assembled NCBI data. From these transcriptomes, we identified all expressed opsin genes for each of 29 species representing 19 families across four

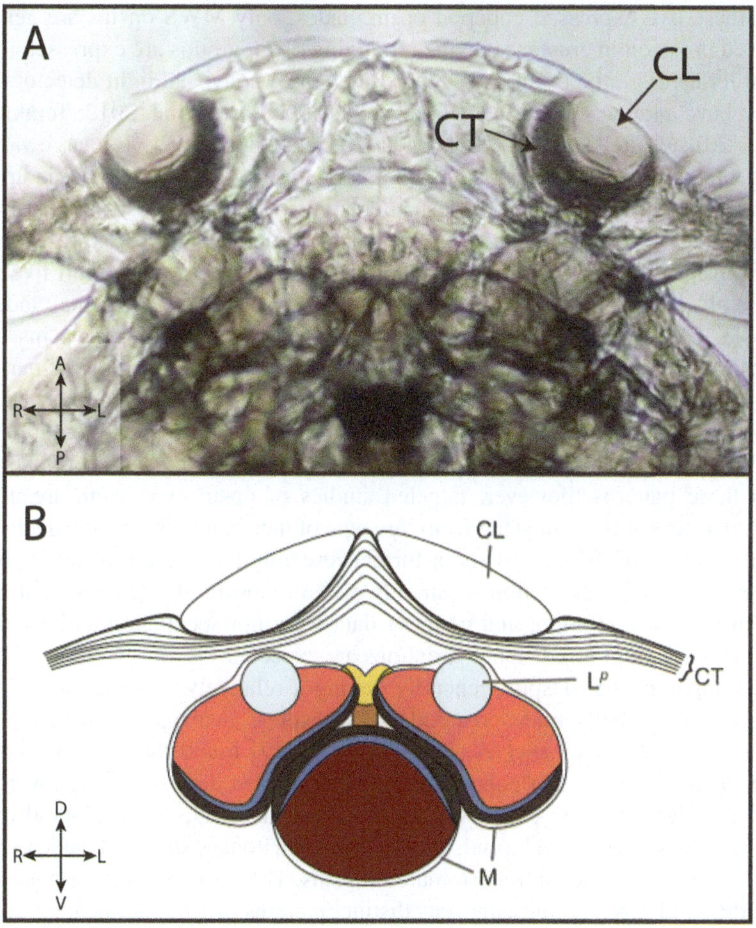

Fig. 9.10 (**a**) Ventral view of *Caligus olsoni* eyes (Image by P. Bryant). (**b**) Diagrammatic representation of a frontal section of *Lepeophtheirus nordmannii* cuticular lensed eye, modified from Vaissiere (1961). Structures are indicated as *CL* cuticular lens, *CT* layered cuticular thickening, L^p crystalline lens of paired ocelli, *M* membrane cells, *light red* paired retinal sphere, *dark red* unpaired retinal sphere, *dark blue* tapetum, *dark gray* pigment cups, *light yellow* paired optic neurons, *dark yellow* unpaired optic neurons

orders of copepods (Table 9.2). Although these data represent expression patterns among different sexes, times of day, and seasons, among other variables, in aggregate, they present broad patterns of opsin diversity within copepods, including the diversity of opsins that may be involved in vision (Fig. 9.11).

Based on these data, copepods express opsins from at least five major clades, representing four of the nine bilaterian opsin lineages (Ramirez et al. 2016): (1) middle-wavelength-sensitive (MWS) and (2) rhodopsin 7 (Rh7) genes from the canonical R-opsin group, (3) peropsins and (4) neuropsins from the tetraopsin cluster, and (5) pteropsins from the canonical C-opsin group (Fig. 9.11 and Table 9.2).

From these five expressed copepod opsin clades, only MWS opsins are generally involved in vision in crustaceans; the remaining opsin groups are expressed in nonretinal tissue (e.g., the brain) and either function as nonvisual light detectors or do not yet have a known function (Velarde et al. 2005; Porter et al. 2012; Terakita and Nagata 2014; Battelle et al. 2016; Ni et al. 2017). Although there is a large variation in the total number of opsins expressed among species, ranging from one up to 18 transcripts, within these five clades, interesting patterns of opsin expression are emerging at the level of orders. All four of the orders examined expressed visual MWS opsins. However, only the cyclopoids expressed opsins from all five major clades; all of the other orders were missing opsins from at least one clade. For example, only harpacticoid and cyclopoid species expressed neuropsins, while peropsins were expressed in all of the orders except the harpacticoids. Of particular note, we did not recover any Rh7 or neuropsins in the Siphonostomatoida. This pattern is intriguing as the siphonostomatoids are generally parasites thought to have reduced or lost eyes, making the loss of some types of opsin genes seem likely. For all of these patterns, however, targeted studies of opsin expression are needed. Genes that appear to be missing from this type of data cannot be conclusively interpreted as a loss of the gene without further research due to the transient nature of gene expression. Transcriptomes can only give a snapshot of expression patterns at a given time or condition, and as these data were not specifically collected with visual expression in mind, some variations are expected.

It is apparent that despite generally having a relatively "simple" naupliar eye with a few retinal cells, many copepod species express an abundance of MWS opsin genes. In a study of 12 species, Porter et al. (2017) found three major clades of MWS opsins expressed in copepods. In our expanded data set, we recovered the same three large clades: clade A – calanoid, harpacticoid, and cyclopoid opsins; clade B – harpacticoid, cyclopoid, and siphonostomatoid opsins; and clade C – calanoid, harpacticoid, and siphonostomatoid opsins. The stability of these clades with the addition of more species suggests distinct patterns of gene duplication and loss at the level of orders (Fig. 9.11). The number of expressed MWS opsin genes also varies across species, from one MWS opsin in *Metridia lucens* to 13 in *Acartia tonsa* in the calanoids alone. As many copepod species have ten or fewer receptor cells in each ocellus or eye (Table 9.1), the number of MWS opsins expressed suggests either coexpression within single photoreceptor cells, differential expression patterns among ocelli/eyes, or that some of these typically visual opsins are used in nonvisual contexts. As further support for the use of MWS opsins in nonvisual contexts in some copepods, we recovered a diversity of opsins from siphonostomatoid and cyclopoid species without eyes, including several copies of the visual MWS opsin gene (Table 9.2).

In addition to the number of opsins expressed in copepod eyes, it is also unusual that copepods express opsins from a single visual opsin spectral clade (MWS). Based on measured photoreceptor sensitivities from a diverse array of species, most crustaceans minimally express opsins from two different visual clades: (1) MWS and/or long-wavelength-sensitive (LWS) opsins and (2) short-wavelength-sensitive opsins (SWS – encompassing violet or ultraviolet sensitivity) (Marshall et al. 1999;

Table 9.2 Copepod opsin diversity

	Order/Family	Species	Eye type	R-ops		C-ops	Tetra		Total	Bioproject (SRA Runs)
				MWS	*Rh7*	*Pter*	*Per*	*Neur*		
Order Calanoida										
1	Acartiidae	*Acartia (Acanthacartia) fossae*	Enlarged	13		2	1		16	PRJNA275311 (SRR1805707)
2	Acartiidae	*Acartia (Acanthacartia) tonsa*	Enlarged	12	1	3	1		17	PRJEB20069, PRJNA407266
3	Calanidae	*Calanus finmarchicus*	Typical	5	1	10	1		17	PRJNA236983, PRJNA236528, PRJNA231164
4	Calanidae	*Neocalanus flemingeri*	Typical	4		4	1		9	PRJNA324453, PRJNA496596
5	Clausocalanidae	*Pseudocalanus acuspes*	Typical	1					1	PRJNA296544 (SRR2478563, SRR2478562)
6	Diaptomidae	*Gigantodiaptomus amblyodon*	Enlarged	2					2	PRJNA254268
7	Metridinidae	*Metridia lucens*	Typical	1					1	PRJNA449123 (SRR6956663)
8	Metridinidae	*Pleuromamma robusta*	Typical	1					1	PRJNA416202 (SRR6232731, SRR6232732)
9	Metridinidae	*Pleuromamma xiphias*	Typical	1		2	1		4	PRJNA352670
10	Pontellidae	*Labidocera madurae*	Y eye	9	1	6	1		17	PRJNA324849
11	Pseudodiaptomidae	*Pseudodiaptomus annandalei*	Enlarged	4			2		6	PRJNA558682
12	Rhincalanidae	*Rhincalanus gigas*	Typical	2			1		3	PRJNA666170
13	Temoridae	*Eurytemora affinis*	Enlarged	10	1	4	3		18	PRJNA242763, PRJNA278152
14	Temoridae	*Temora longicornis*	Typical	2	1	3	1		7	PRJNA473764
Order Harpacticoida										
15	Harpacticidae	*Tigriopus californicus*	Typical	6	2	4		1	13	PRJNA504307, PRJNA263967, PRJNA263967
16	Harpacticidae	*Tigriopus japonicus*	Typical	6	1	5		1	13	PRJNA274317
17	Harpacticidae	*Tigriopus kingsejongensis*	Typical	6	2	5		1	14	PRJNA283925
18	Laophontidae	*Platychelipus littoralis*	Typical	5	2	3			10	PRJNA575120

(continued)

Table 9.2 (continued)

Order/Family		Species	Eye type	R-ops		C-ops	Tetra			
				MWS	*Rh7*	*Pter*	*Per*	*Neur*	Total	Bioproject (SRA Runs)
19	Tisbidae	*Tisbe furcata furcata*	Typical	2					2	PRJNA254316
20	Tisbidae	*Tisbe holothuriae*	Enlarged	8					8	PRJEB23629
Order Cyclopoida										
21	Cyclopettidae	*Paracyclopina nana*	Typical	2	2	2	1	1	8	PRJNA268783
22	Cyclopidae	*Apocyclops royi*	Typical	2	1	3	1	1	8	PRJEB28764
23	Cyclopidae	*Eucyclops serrulatus*	Typical	5					5	PRJNA231234
24	Lernaeidae	*Lernaea cyprinacea*	Unknown	3	2	2	1	1	9	PRJNA232511 (SRR1107498)
25	Mytilicolidae	*Mytilicola intestinalis*	Unknown				1		1	PRJNA430138 (SRR6513728, SRR6513724, SRR6513718)
26	Oithonidae	*Oithona nana*	Typical	8	1	5	1	1	16	PRJEB18938 (ERR1794860)
Order Siphonostomatoida										
27	Caligidae	*Caligus rogercresseyi*	Telescopic	3		1			4	PRJNA234316
28	Caligidae	*Lepeophtheirus salmonis*	Telescopic	3		2			5	PRJNA73431
29	Lernaeopodidae	*Tracheliastes polycolpus*	Unknown	1			2		3	PRJNA476682

For each species surveyed, data include the eye type, the accession number for the RNA-seq data set used, and the number and classification for all identified opsins

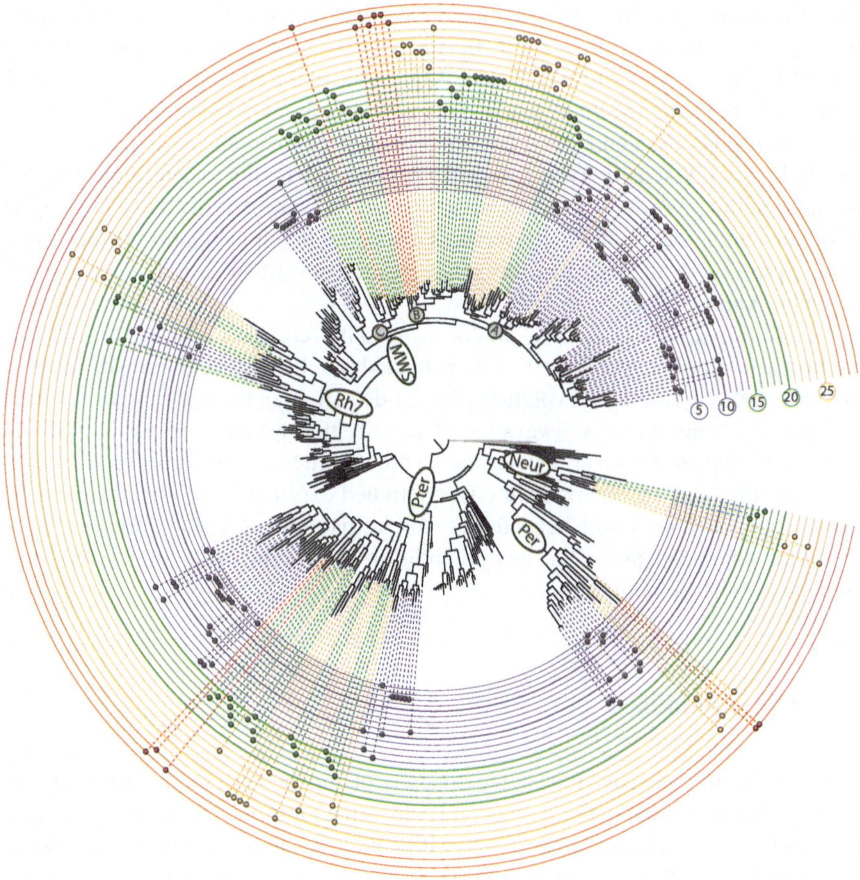

Fig. 9.11 Maximum likelihood molecular phylogeny of copepod opsins with reference opsins from other crustacean species. Each ring represents one species of copepod, as numbered in Table 9.2. The five major copepod opsin clades include middle-wavelength-sensitive (MWS) opsins, rhabdomeric opsin 7 (Rh7), pteropsins (Pter), neuropsins (Neur), and peropsins (Per). The three major clades of copepod MWS opsins identified by Porter et al. (2017) are labeled as A, B, and C. The copepod order for each species is indicated as *purple* Calanoida, *yellow* Cyclopoida, *green* Harpacticoida, *red* Siphonostomatoida

Cronin and Porter 2008). This was confirmed by Henze and Oakley (2015), who suggested that ancestral pancrustaceans had four potentially visual opsins – one LWS, two MWS, and one SWS – and also demonstrated that most crustaceans express at least one SWS opsin, making the lack of SWS opsin expression in copepods noteworthy. Additional studies are needed to determine what variation, if any, there is in copepod MWS opsin spectral characteristics.

Interestingly, the expression of multiple MWS opsins opens the possibility that copepods could be using multiple spectrally different opsins as a depth gauge. In the planktonic larva of the annelid *Platynereis*, the ratio of light input to ciliary photoreceptors expressing UV-sensitive c-type opsins and rhabdomeric photoreceptors expressing blue-sensitive r-type opsins may be used as a depth gauge driving vertical migration (Verasztó et al. 2018). While none of the copepod MWS sequences identified here contained the lysine residue typically thought to confer UV sensitivity in arthropods (Salcedo et al. 2003), further strengthening the hypothesis that copepods lack UV sensitivity, it will be particularly intriguing to determine whether the large numbers of MWS opsins in some species represent diversity in spectral absorbance that could potentially serve as a spectral depth gauge or whether every copy expressed in the eye has the same spectral absorbance, making copepods monochromats.

As additional data become available from other copepod orders, tracking the taxonomic composition of these major opsin clades will continue to provide insights into and enigmas about the evolution of light detection and visual system function in copepods. While there is always the possibility that we have missed identifying an expressed opsin in a given species due to the specifics of the extraction, sequencing, assembly, and annotation methods, continued exploration of opsin expression across copepod orders will help identify broad patterns of visual gene evolution within the copepods, as well as any potential links to some of the highly modified eye types described above.

9.7 Conclusions

By taking a comprehensive look at the diversity of copepod eye morphology relative to the evolution of each major lineage, it is clear that multiple, independent lineages have evolved trinocular visual systems, with a wide range of convergent morphologies, especially with regard to the appearance of lenses. The diversity among copepod visual systems offers a striking comparative evolutionary system to rival the better-studied crustacean compound eyes. Although there is a fairly large base of literature on copepod eye anatomy, the information is widely dispersed and deeply buried and has never been brought together for a broader overview of eye evolution within the entire group. Furthermore, studies of copepod eyes have been hampered by a confusing maze of terminology. Although much work has been done, there are many groups that have not yet had a thorough anatomical description of the eye, and the current understanding of diversity suggests that there may still be much to discover. There is correspondingly even less known about the development and physiological function of copepod eyes and how they are evolutionarily linked to the diversity of naupliar eyes in other crustacean lineages. By bringing all of this information together in one place, and unifying terminology, we hope to inspire continued work on these fascinating systems and provide a coherent framework for future studies of copepod visual system function and evolution.

Key Terms

(a) **Copepod naupliar eye**: a tripartite eye composed of two sets of ocelli – two ocelli that mirror each other in morphology, which are often viewed as a paired set, and an unpaired ocellus.
(b) **Crystallin**: refractive proteins that may aggregate within the cell to form dimers or heterodimers to aid in focusing light.
(c) **Crystallin lens**: transparent structures made up of crystallins with refractive properties used to focus light onto the retina.
(d) **Cuticular lens**: optically clear, refractive focusing structures, which are made up of cuticular or chitinous material.
(e) **Gicklhorn's organ**: a sac-like organ containing microvilli in the cephalosome situated at the anterior cuticle, near the antennae; may be photosensitive, chemosensory, or secretory, depending on the species.
(f) **Iridescence**: change in the color of a surface depending on the viewing angle.
(g) **Paired eyes**: paired ocelli that have become separated in space, which are generally more developed than in a typical naupliar eye and are often situated dorsolaterally.
(h) **Paired ocelli**: paired photoreceptive units joined together with an unpaired ocellus.
(i) **Phaosomes**: folds or layers of smooth endoplasmic reticulum, often situated around the cell nucleus or in front of the rhabdomere; sometimes also referred to as dictyosomes.
(j) **Pigment cup**: carotenoid- or melanin-filled cells that surround each ocellus in the naupliar eye to block light.
(k) **Rhabdomere**: photosensitive layers of microvilli within an ocellus.
(l) **Tapetum**: a reflective layer composed of guanine platelets positioned behind the retina to redirect light back to the rhabdomere.
(m) **Unpaired eye**: unpaired ocellus that has become separated in space from the paired ocelli, which is often situated anteroventrally and more developed than in a typical naupliar eye.
(n) **Unpaired ocelli**: singular photoreceptive unit attached, often ventrally, to the paired ocelli.
(o) **Whorls**: folds or layers of mitochondria or smooth endoplasmic often situated within the rhabdomere, found mainly in the Gicklhorn's organ.

Acknowledgments We would like to thank Mike Bok, Peter Bryant, Danté Fenolio, and Russel Hopcroft for sharing their exceptional images of copepods for this chapter; Petra Lenz, Lauren Block, and Leocadio Blanco-Bercial for their expertise in copepod taxonomy and help with species identification; and Petra Lenz for introducing us to copepods in the first place. This research was funded by grants from the National Science Foundation (OIA-1738567 and DEB-1556105) to MLP and is based upon work supported by the National Science Foundation under Grant Nos. DBI-1062432 2011, ABI-1458641 2015, and ABI-1759906 2018 to Indiana University. Any opinions, findings, and conclusions or recommendations expressed in this material are those of the authors and do not necessarily reflect the views of the National Science Foundation, the National Center for Genome Analysis Support, or Indiana University. This is publication 165 from the School of Life Sciences, University of Hawai'i at Mānoa.

References

Aarseth K, Schram T (1999) Wavelength-specific behaviour in *Lepeophtheirus salmonis* and Calanus finmarchicus to ultraviolet and visible light in laboratory experiments (Crustacea:Copepoda). Mar Ecol Prog Ser 186:211–217. https://doi.org/10.3354/meps186211

Andrew DR, Brown SM, Strausfeld NJ (2012) The minute brain of the copepod *Tigriopus californicus* supports a complex ancestral ground pattern of the tetraconate cerebral nervous systems. J Comp Neurol 520:3446–3470. https://doi.org/10.1002/cne.23099

Arendt D, Wittbrodt J (2001) Reconstructing the eyes of Urbilateria. Philos Trans R Soc Lond 356:1545–1563. https://doi.org/10.1098/rstb.2001.0971

Bagøien E, Kiørboe T (2005) Blind dating – mate finding in planktonic copepods. I. Tracking the pheromone trail of *Centropages typicus*. Mar Ecol Prog Ser 300:105–115

Båtnes A, Miljeteig C, Berge J et al (2013) Quantifying the light sensitivity of *Calanus* spp. during the polar night: potential for orchestrated migrations conducted by ambient light from the sun, moon, or aurora borealis? Polar Biol 38:51–65. https://doi.org/10.1007/s00300-013-1415-4

Battelle B-A, Ryan JF, Kempler KE et al (2016) Opsin repertoire and expression patterns in horseshoe crabs: evidence from the genome of *Limulus polyphemus* (Arthropoda: Chelicerata). Genome Biol Evol 8:1571–1589. https://doi.org/10.1093/gbe/evw100

Blanco-Bercial L, Bradford-Grieve J, Bucklin A, (2011). Molecular phylogeny of the calanoida (Crustacea: Copepoda). Mol Phylogenet Evol, 59(1), pp.103–113. S105579031100025X https://doi.org/10.1016/j.ympev.2011.01.008

Bollens SM, Frost BW (1989) Predator-induced diel vertical migration in a planktonic copepod. J Plankton Res 11:1047–1065

Bollens SM, Frost BW, Cordell J (1994) Chemical, mechanical and visual cues in the vertical migration behavior of the marine planktonic copepod *Acartia hudsonica*. J Plankton Res 16:555–564. https://doi.org/10.1093/plankt/16.5.555

Boxshall GA (1985) The comparative anatomy of two copepods, a predatory calanoid and a particle-feeding mormonilloid. Philos Trans R Soc Lond Ser B 311:303–377. https://doi.org/10.1098/rstb.1985.0155

Boxshall GA (1992) Copepoda. Microscopic anatomy of invertebrates. Crustaceana 9:347–384

Boxshall GA, Defaye D (2008) Global diversity of copepods (Crustacea: Copepoda) in freshwater. Hydrobiologia 595:195–207. https://doi.org/10.1007/s10750-007-9014-4

Boxshall GA, Halsey SH (2004) An introduction to copepod diversity. Ray Society, Andover

Boxshall G, Hayes P (2019) Biodiversity and Taxonomy of the Parasitic Crustacea. In: Smit NJ, Bruce NL, Hadfield KA (eds) Parasitic Crustacea. Springer, pp 73–134

Boxshall GA, Ferrari FD, Tiemann H (1984) The ancestral copepod: towards a consensus of opinion at the first international conference on Copepoda. Crustaceana 7:68–84

Boxshall GA, Zylinski S, Jaume D et al (2014) A new genus of speleophriid copepod (Copepoda: Misophrioida) from a cenote in the Yucatan, Mexico with a phylogenetic analysis at the species level. Zootaxa 3821:321–336. https://doi.org/10.11646/zootaxa.3821.3.2

Bron JE, Frisch D, Goetze E et al (2011) Observing copepods through a genomic lens. Front Zool 8:22. https://doi.org/10.1186/1742-9994-8-22

Buschbeck EK, Sbita SJ, Morgan RC (2007) Scanning behavior by larvae of the predacious diving beetle, *Thermonectus marmoratus* (Coleoptera: Dytiscidae) enlarges visual field prior to prey capture. J Comp Physiol Neuroethol Sens Neural Behav Physiol 193:973–982. https://doi.org/10.1007/s00359-007-0250-x

Buskey EJ (1984) Swimming pattern as an indicator of the roles of copepod sensory systems in the recognition of food. Mar Biol 79:165–175. https://doi.org/10.1007/BF00951825

Buskey EJ (1998) Components of mating behavior in planktonic copepods. J Mar Syst 15:13–21. https://doi.org/10.1016/S0924-7963(97)00045-6

Buskey EJ, Hartline DK (2003) High-speed video analysis of the escape responses of the copepod *Acartia tonsa* to shadows. Biol Bull 204:28–37. https://doi.org/10.2307/1543493

Buskey EJ, Swift E (1985) Behavioral responses of oceanic zooplankton to simulated bioluminescence. Biol Bull 168:263–275. https://doi.org/10.2307/1541239

Buskey EJ, Peterson JO, Ambler JW (1996) The swarming behavior of the copepod *Dioithona oculata*: in situ and laboratory studies. Limnol Oceanogr 41:513–521. https://doi.org/10.4319/lo.1996.41.3.0513

Buskey EJ, Lenz PH, Hartline DK (2012) Sensory perception, neurobiology, and behavioral adaptations for predator avoidance in planktonic copepods. Adapt Behav 20:57–66. https://doi.org/10.1177/1059712311426801

Caparroy P, Thygesen U, Visser A (2000) Modelling the attack success of planktonic predators: patterns and mechanisms of prey size selectivity. J Plankton Res 22:1871–1871. https://doi.org/10.1093/plankt/22.10.1871

Chae J, Nishida S (1994) Integumental ultrastructure and color patterns in the iridescent copepods of the family Sapphirinidae (Copepoda: Poecilostomatoida). Mar Biol 119:205–210. https://doi.org/10.1007/BF00349558

Chae J, Nishida S (2004) Swimming behaviour and photoresponses of the iridescent copepods, *Sapphirina gastrica* and *Sapphirina opalina* (Copepoda: Poecilostomatoida). J Mar Biol Assoc U K 84:727–731. https://doi.org/10.1017/S0025315404009816h

Claus C (1863) Die frei lebenden Copepoden: mit besonderer Berücksichtigung der Fauna Deutschlands, der Nordsee und des Mittelmeeres

Cohen JH, Forward RB (2002) Spectral sensitivity of vertically migrating marine copepods. Biol Bull 203:307–214

Cohen JH, Forward RB (2005a) Photobehavior as an inducible defense in the marine copepod *Calanopia americana*. Limnol Oceanogr 50:1269–1277. https://doi.org/10.4319/lo.2005.50.4.1269

Cohen JH, Forward RB (2005b) Diel vertical migration of the marine copepod *Calanopia americana*. II. Proximate role of exogenous light cues and endogenous rhythms. Mar Biol 147:399–410. https://doi.org/10.1007/s00227-005-1570-4

Cohen JH, Forward RB (2009) Zooplankton diel vertical migration – a review of proximate control. Oceanogr Mar Biol Annu Rev 47:77–109

Cohen JH, Frank TM (2019) Eyes and vision in a mesopelagic copepod. In: Front physiol conference abstract: international conference on invertebrate vision

Cohen JH, Piatigorsky J, Ding L et al (2005) Vertebrate-like by-crystallins in the ocular lenses of a copepod. J Comp Physiol A 191:291–298. https://doi.org/10.1007/s00359-004-0594-4

Cohen JH, Piatigorsky J, Ding L et al (2007) ERRATUM: vertebrate-like βγ-crystallins in the ocular lenses of a copepod. J Comp Physiol A 193:573–574. https://doi.org/10.1007/s00359-007-0221-2

Cowles TJ, Strickier J (1983) Characterization of feeding activity patterns in the planktonic copepod *Centropages typicus* Kroyer under various food conditions. Limnol Oceanogr 28:106–115. https://doi.org/10.4319/lo.1983.28.1.0106

Cronin TW (1986) Optical design and evolutionary adaptation in crustacean compound eyes. J Crustac Biol 6:1–23. https://doi.org/10.1163/193724086X00686

Cronin TW, Porter ML (2008) Exceptional variation on a common theme: the evolution of crustacean compound eyes. Evol Educ Outreach 1:463–475. https://doi.org/10.1007/s12052-008-0085-0

Downing AC (1972) Optical scanning in the lateral eyes of the copepod *Copilia*. Perception 1:247–261. https://doi.org/10.1068/p010247

Dudley PL (1972) The fine structure of a cephalic sensory receptor in the copepod *Doropygus seclusus* Illg (Crustacea: Copepoda: Notodelphyidae). J Morphol 138:407–431. https://doi.org/10.1002/jmor.1051380403

Elofsson R (1966) The nauplius eye and frontal organs of the non-Malacostraca (Crustacea). Sarsia 25:1–128. https://doi.org/10.1080/00364827.1966.10409568

Elofsson R (1970) A presumed new photoreceptor in copepod crustaceans. Z Für Zellforsch 109(3):316–326

Elofsson R (1971) The ultrastructure of a chemoreceptor organ in the head of copepod crustaceans. Acta Zool 52:299–315. https://doi.org/10.1111/j.1463-6395.1971.tb00565.x

Elofsson R (2006) The frontal eyes of crustaceans. Arthropod Struct Dev 35:275–291. https://doi.org/10.1016/j.asd.2006.08.004

Esaias WE, Curl HC (1972) Effects of dinoflagellate bioluminescence on copepod ingestion rates. Limnol Oceanogr 17:901–906. https://doi.org/10.4319/lo.1972.17.6.0901

Esterly CO (1908) Third report on the Copepoda of the San Diego region. The University Press

Fahrenbach WH (1964) The fine structure of a nauplius eye. Z Für Zellforsch 62:182–197

Fancett MS, Kimmerer WJ (1985) Vertical migration of the demersal copepod *Pseudodiaptomus* as a means of predator avoidance. J Exp Mar Biol Ecol 88:31–43. https://doi.org/10.1016/0022-0981(85)90199-6

Fasten N (1916) The eye of the parasitic copepod, *Salmincola edwardsii* Olsson (Lernaeopoda, Edwardsii Olsson). Biol Bull 31:407–418. https://doi.org/10.2307/1536319

Forward RB Jr (1988) Diel vertical migration: zooplankton photobiology and behavior. Oceanogr Mar Biol Annu Rev 26:361–393

Fosshagen A, Iliffe TM (1985) Two new genera of Calanoida and a new order of Copepoda, Platycopioida, from marine caves on Bermuda. Sarsia 70:345–358. https://doi.org/10.1080/00364827.1985.10419688

Frank T, Widder E (2002) Effects of a decrease in downwelling irradiance on the daytime vertical distribution patterns of zooplankton and micronekton. Mar Biol 140:1181–1193. https://doi.org/10.1007/s00227-002-0788-7

Frase T, Richter S (2020) The brain and the corresponding sense organs in calanoid copepods – evidence of vestiges of compound eyes. Arthropod Struct Dev 54:100902. https://doi.org/10.1016/j.asd.2019.100902

Frost BW (1974) *Calanus marshallae*, a new species of calanoid copepod closely allied to the sibling species *C. finmarchicus* and *C. glacialis*. Mar Biol 26:77–99. https://doi.org/10.1007/BF00389089

George KH, Pointner K, Packmor J (2018) The benthic Copepoda (Crustacea) of Anaximenes Seamount (eastern Mediterranean Sea) – community structure and species distribution. Prog Oceanogr 165:299–316. https://doi.org/10.1016/j.pocean.2018.06.006

Gicklhorn J (1930) Zur kenntnis der frontalorgane von *Cyclops strenuus* Fischer. Zool Anz 90:209–216

Gophen M, Harris RP (1981) Visual predation by a marine cyclopoid copepod, *Corycaeus anglicus*. J Mar Biol Ass 61:391–399. https://doi.org/10.1017/S0025315400047020

Grygier MJ, Ohtsuka S (1995) SEM observation of the nauplius of *Monstrilla hamatapex*, new species, from Japan and an example of upgraded descriptive standards for monstrilloid copepods. J Crustac Biol 15:703–719. https://doi.org/10.1163/193724095X00118

Gurney R (1930) The larval stages of the copepod *Longipedia*. J Mar Biol Ass 16:461–474. https://doi.org/10.1017/S0025315400072866

Hamner WM, Carleton JH (1979) Copepod swarms: attributes and role in coral reef ecosystems: copepod swarms. Limnol Oceanogr 24:1–14. https://doi.org/10.4319/lo.1979.24.1.0001

Haney JF (1988) Diel patterns of zooplankton behavior. Bull Mar Sci 43:583–603

Hendler G, Dojiri M (2009) The contrariwise life of a parasitic, pedomorphic copepod with a non-feeding adult: ontogenesis, ecology, and evolution. Invertebr Biol 128:65–82. https://doi.org/10.1111/j.1744-7410.2008.00164.x

Henze MJ, Oakley TH (2015) The dynamic evolutionary history of pancrustacean eyes and opsins. Integr Comp Biol 55:830–842. https://doi.org/10.1093/icb/icv100

Heron AC (1973) A specialized predator-prey relationship between the copepod *Sapphirina angusta* and the pelagic tunicate *Thalia democratica*. J Mar Biol Assoc U K 53:429–435. https://doi.org/10.1017/S0025315400022372

Hirst AG, Kiørboe T (2014) Macroevolutionary patterns of sexual size dimorphism in copepods. Proc Biol Sci 281:1791. https://doi.org/10.1098/rspb.2014.0739

Ho J (1988) Cladistics of Sunaristes, a genus of harpacticoid copepods associated with hermit crabs. Hydrobiologia 167(1):555–560. https://doi.org/10.1007/BF00026352
Ho J (1994) Copepod phylogeny: a reconsideration of Huys & Boxshall's parsimony versus homology'. Hydrobiologia 292/293:31–39. https://doi.org/10.1007/BF00229920
Hooff RC, Peterson WT (2006) Copepod biodiversity as an indicator of changes in ocean and climate conditions of the northern California current ecosystem. Limnol Oceanogr 51:2607–2620. https://doi.org/10.4319/lo.2006.51.6.2607
Huys R (1988) Gelyelloida, a new order of stygobiont copepods from European karstic systems. Hydrobiologia 167–168(1):485–495. https://doi.org/10.1007/BF00026343
Huys R, Böttger-Schnack R (1994) Taxonomy, biology and phylogeny of Miraciidae, Harpacticoida, Copepoda. Sarsia 79:207–283
Huys R, Boxshall GA (1991) Copepod evolution. The Ray Society, London
Huys R, Boxshall G, Boettger-Schnack R (1992) On the discovery of the male of *Mormonilla* Giesbrecht 1892 (Copepoda, Mormonilloida). Bull Nat Hist Mus Zool 58:157–170
Ito T (1985) A new subspecies of Longipedia andamanica Wells from the Pacific coast of Japan, with reference to the morphology of L. coronata Claus (Copepoda: Harpacticoida). Publications of the Seto marine biological Laboratory, 30(4–6), pp.307–324.
Ivanenko VN, Defaye D (2006) Planktonic deep-water copepods of the family Mormonillidae Giesbrecht, 1893 from the East Pacific Rise (13°N), the Northeastern Atlantic, and near the North Pole (Copepoda, Mormonilloida). Crustaceana 79:707–726. https://doi.org/10.1163/156854006778026861
Jaume D, Boxshall GA, Humphreys WF (2001) New stygobiont copepods (Calanoida; Misophrioida) from Bundera Sinkhole, an anchialine cenote in North-Western Australia. Zool J Linnean Soc 133:1–24. https://doi.org/10.1111/j.1096-3642.2001.tb00620.x
Karanovic T, Eberhard SM (2009) Second representative of the order Misophrioida (Crustacea, Copepoda) from Australia challenges the hypothesis of the Tethyan origin of some anchialine faunas. Zootaxa 2059:51–68. https://doi.org/10.5281/ZENODO.186777
Khodami S, Mercado-Salas NF, Tang D, Martinez Arbizu P (2019) Molecular evidence for the retention of the Thaumatopsyllidae in the order Cyclopoida (Copepoda) and establishment of four suborders and two families within the Cyclopoida. Mol Phylogenet Evol 138:43–52. https://doi.org/10.1016/j.ympev.2019.05.019
Kimura T, Takasaki M, Hatai R et al (2020) Guanine crystals regulated by chitin-based honeycomb frameworks for tunable structural colors of sapphirinid copepod, *Sapphirina nigromaculata*. Sci Rep 10:1–7. https://doi.org/10.1038/s41598-020-59090-4
Kiørboe T (2007) Mate finding, mating, and population dynamics in a planktonic copepod *Oithona davisae*: there are too few males. Limnol Oceanogr 52:1511–1522
Kiørboe T, Bagøien E (2005) Motility patterns and mate encounter rates in planktonic copepods. Limnol Oceanogr 50:1999–2007. https://doi.org/10.4319/lo.2005.50.6.1999
Kiørboe T, Bagøien E, Thygesen UH (2005) Blind dating – mate finding in planktonic copepods. II. The pheromone cloud of *Pseudocalanus elongatus*. Mar Ecol Prog Ser 300:117–128
Krishnaswamy S (1948) A preliminary note on the eye of *Centropages furcatus* Dana. Curr Sci 17:190–191
Lacalli TC (2009) Serial EM analysis of a copepod larval nervous system: Naupliar eye, optic circuitry, and prospects for full CNS reconstruction. Arthropod Struct Dev 38:361–375. https://doi.org/10.1016/j.asd.2009.04.002
Land MF (1984) Crustacea. In: Photoreception and vision in invertebrates. Plenum Press, New York
Land MF (1988) The functions of eye and body movements in *Labidocera* and other copepods. J Exp Biol 140:381–391
Landry MR (1978) Predatory feeding behavior of a marine copepod, *Labidocera trispinosa* 1: marine copepod predation. Limnol Oceanogr 23:1103–1113. https://doi.org/10.4319/lo.1978.23.6.1103

Landry MR, Lehner-Fournier JM, Fagerness VL (1985) Predatory feeding behavior of the marine cyclopoid copepod *Corycaeus anglicus*. Mar Biol 85:163–169. https://doi.org/10.1007/BF00397435

Lerner A, Browman HI (2016) The copepod *Calanus* spp. (Calanidae) is repelled by polarized light. Sci Rep 6:1–7. https://doi.org/10.1038/srep35891

Longhurst AR, Harrison GW (1989) The biological pump: profiles of plankton production and consumption in the upper ocean. Prog Oceanogr 22:47–123. https://doi.org/10.1016/0079-6611(89)90010-4

Lonsdale DJ, Heinle DR, Siegfried C (1979) Carnivorous feeding behavior of the adult calanoid copepod *Acartia tonsa* Dana. J Exp Mar Biol Ecol 36:235–248. https://doi.org/10.1016/0022-0981(79)90119-9

Lopes RM, Dam HG, Aquino NA et al (2007) Massive egg production by a salp symbiont, the poecilostomatoid copepod *Sapphirina angusta* Dana, 1849. J Exp Mar Biol Ecol 348:145–153. https://doi.org/10.1016/j.jembe.2007.04.005

Manor S, Polak O, Saidel WM et al (2009) Light intensity mediated polarotaxis in *Pontella karachiensis* (Pontellidae, Copepoda). Vis Res 49:2371–2378. https://doi.org/10.1016/j.visres.2009.07.007

Marshall J, Kent J, Cronin T (1999) Visual adaptations in crustaceans: spectral sensitivity in diverse habitats. In: Archer SN, Djamgoz MBA, Loew ER et al (eds) Adaptive mechanisms in the ecology of vision. Springer, Dordrecht, pp 285–327

Martin GG, Speekmann C, Beidler S (2000) Photobehavior of the harpacticoid copepod *Tigriopus californicus* and the fine structure of its nauplius eye. Invertebr Biol 119:110–124. https://doi.org/10.1111/j.1744-7410.2000.tb00179.x

McLaren IA (1974) Demographic strategy of vertical migration by a marine copepod. Am Nat 108:91–102

Moeschler P, Rouch R (1988) Découverte d'un nouveau représentant de la famille des Gelyellidae (Copepoda, Harpacticoida) dans les eaux souterraines de Suisse. Crustaceana 55:1–16

Ni JD, Baik LS, Holmes TC, Montell C (2017) A rhodopsin in the brain functions in circadian photoentrainment in *Drosophila*. Nature 545:340–344

Nicholls AG (1945) A new Calanoid copepod from Australia. Ann Mag Nat Hist 12:501–514. https://doi.org/10.1080/00222934508654753

Nishida S, Ohtsuka S, Parker A (2002) Functional morphology and food habits of deep-sea copepods of the genus Cephalophanes (Calanoida: Phaennidae): perception of bioluminescence as a strategy for food detection. Mar Ecol Prog Ser 227:157–171. https://doi.org/10.3354/meps227157

Novales Flamarique I, Browman HI, Bélanger M, Boxaspen K (2000) Ontogenetic changes in visual sensitivity of the parasitic salmon louse *Lepeophtheirus salmonis*. J Exp Biol 203:1649–1657

Ohtsuka S, Huys R (2001) Sexual dimorphism in calanoid copepods: morphology and function. Hydrobiologia 453:441–466. https://doi.org/10.1023/A:1013162605809

Ohtsuka S, Nishida S (2017) Copepod biodiversity in Japan: recent advances in Japanese copepodology. In: Motokawa M, Kajihara H (eds) Species diversity of animals in Japan. Springer Japan, Tokyo, pp 565–602

Ohtsuka S, Onbé T (1989) Evidence of selective feeding on larvaceans by the pelagic copepod *Candacia bipinnata* (Calanoida: Candaciidae). J Plankton Res 11:869–872. https://doi.org/10.1093/plankt/11.4.869

Pearre SJ (1979) Problems of detection and interpretation of vertical migration. J Plankton Res 1:29–44. https://doi.org/10.1093/plankt/1.1.29

Porter ML, Crandall KA (2003) Lost along the way: the significance of evolution in reverse. Trends Ecol Evol 18:541–547. https://doi.org/10.1016/S0169-5347(03)00244-1

Porter ML, Blasic JR, Michael John B et al (2012) Shedding new light on opsin evolution. Proc R Soc B Biol Sci 279:3–14. https://doi.org/10.1098/rspb.2011.1819

Porter ML, Steck M, Roncalli V, Lenz PH (2017) Molecular characterization of copepod photoreception. Biol Bull 233:96–110. https://doi.org/10.1086/694564

Ramirez MD, Pairett AN, Pankey MS et al (2016) The last common ancestor of most bilaterian animals possessed at least nine opsins. Genome Biol Evol 8:3640–3652. https://doi.org/10.1093/gbe/evw248
Richter S, Loesel R, Purschke G et al (2010) Invertebrate neurophylogeny: suggested terms and definitions for a neuroanatomical glossary. Front Zool 7:1–49. https://doi.org/10.1186/1742-9994-7-29
Salcedo E, Zheng L, Phistry M et al (2003) Molecular basis for ultraviolet vision in invertebrates. J Neurosci 23:10873–10878
Schizas NV, Dahms H-U, Kangtia P et al (2015) A new species of *Longipedia* Claus, 1863 (Copepoda: Harpacticoida: Longipediidae) from Caribbean mesophotic reefs with remarks on the phylogenetic affinities of Polyarthra. Mar Biol Res 11:789–803. https://doi.org/10.1080/17451000.2015.1013556
Schøyen M, Kaartvedt S (2004) Vertical distribution and feeding of the copepod *Chiridius armatus*. Mar Biol 145:159–165
Smith RC, Baker KS (1979) Penetration of UV-B and biologically effective dose-rates in natural waters. Photochem Photobiol 29:311–323. https://doi.org/10.1111/j.1751-1097.1979.tb07054.x
Stearns DE, Forward RB (1984) Photosensitivity of the calanoid copepod *Acartia tonsa*. Mar Biol 82:85–89. https://doi.org/10.1007/BF00392766
Steuer A (1928) On the geographical distribution and affinity of the appendiculate trematodes parasitizing marine plankton copepods. J Parasitol 15:115–120. https://doi.org/10.2307/3271344
Suárez-Morales E (2011) Diversity of the Monstrilloida (Crustacea: Copepoda). PLoS One 6:e22915. https://doi.org/10.1371/journal.pone.0022915
Suárez-Morales E, McKinnon AD (2014) The Australian Monstrilloida (Crustacea: Copepoda) I. *Monstrillopsis* Sars, *Maemonstrilla* Grygier & Ohtsuka, and *Australomonstrillopsis* gen. nov. Zootaxa 3779:301–340. https://doi.org/10.11646/zootaxa.3779.3.1
Suárez-Morales E, McKinnon D (2016) The Australian Monstrilloida (crustacea: Copepoda) II. *Cymbasoma* Thompson, 1888. Zootaxa 4102:1–129. https://doi.org/10.11646/zootaxa.4102.1.1
Suárez-Morales E, Cervantes-Martínez A, Gutiérrez-Aguirre MA, Iliffe T (2017) A new Speleophria (Copepoda, Misophrioida) from an anchialine cave of the Yucatán Peninsula with comments on the biogeography of the genus. bms 93(3):863–878. https://doi.org/10.5343/bms.2017.1012
Takahashi K, Ichikawa T, Tadokoro K (2015) Diel colour changes in male *Sapphirina nigromaculata* (Cyclopoida, Copepoda). J Plankton Res 37:1181–1189. https://doi.org/10.1093/plankt/fbv088
Terakita A, Nagata T (2014) Functional properties of opsins and their contribution to light-sensing physiology. Zool Sci 31:653–659
Umminger BL (1968) Polarotaxis in copepods. II. The ultrastructural basis and ecological significance of polarized light sensitivity in copepods. Biol Bull 135:252–261. https://doi.org/10.2307/1539632
Vaissiere R (1961) Morphologie et histologie comparée des yeux des crustaceens copepodes. Arch Zool Exp Gen 100:1–125
van Gool E, Ringelberg J (1997) The effect of accelerations in light increase on the phototactic downward swimming of *Daphnia* and the relevance to diel vertical migration. J Plankton Res 19:2041–2050
Velarde RA, Sauer CD, Walden KK et al (2005) Pteropsin: a vertebrate-like non-visual opsin expressed in the honey bee brain. Insect Biochem Mol Biol 35:1367–1377
Verasztó C, Gühmann M, Jia H et al (2018) Ciliary and rhabdomeric photoreceptor-cell circuits form a spectral depth gauge in marine zooplankton. elife 7:e36440. https://doi.org/10.7554/eLife.36440
Vincx M, Heip C, Scott A (1979) Larval development and biology of *Canuella perplexa* T. Cah Biol Mar 20:281–299

Waggett RJ, Buskey EJ (2006) Calanoid copepod escape behavior in response to a visual predator. Mar Biol 150:599–607. https://doi.org/10.1007/s00227-006-0384-3
Wilson CB (1911) North American parasitic copepods – part 9. The lernaeopodidae. Proc U S Natl Mus 39:189–241
Wilson MS (1946) The species of *Platycopia* Sars (Copepoda, Calanoida). Smithson Misc Collect 106:1–20
Wolken JJ, Florida RG (1969) The eye structure and optical system of the crustacean copepod, *Copilia*. J Cell Biol 7
WoRMS Editorial Board (2021) World register of marine species. https://doi.org/10.14284/170
Yen J (1988) Directionality and swimming speeds in predator-prey and male-female interactions of *Euchaeta rimana*, a subtropical marine copepod. Bull Mar Sci 43:395–403

Chapter 10
Distributed Vision in Spiders

Alex M. Winsor, Nathan I. Morehouse, and Elizabeth M. Jakob

Abstract We examine the distributed visual system of spiders, an ancient and diverse lineage of predators. Across families, prey-capture strategies include active pursuit, sit-and-wait predation, and the use of prey-capture webs. Spiders also have rich communicatory repertoires, using visual, vibratory, and chemical signals to communicate with potential mates, rivals, and social partners. Some species even demonstrate impressive problem-solving capabilities. Accompanying this behavioral diversity is impressive morphological variation, especially with respect to their visual systems. This variation includes the size of the eyes and their arrangement, eye anatomy and optical properties, photoreceptor structure, and underlying brain neuromorphology. Spiders have up to four pairs of "camera-type" eyes, any of which can exhibit specializations to overcome specific visual challenges. In this chapter, we will first examine vision in a well-studied family: the elegant, compact, and tightly integrated distributed visual system of jumping spiders (family Salticidae). From this example, we then expand our scope to a review of other spider families' vision while making the case for the importance of additional phylogenetically informed work.

Keywords Spider vision · Jumping spiders · Salticids · Principal eyes · Eye movements · Retinal specializations · Visual acuity · Visual sensitivity

---

A. M. Winsor (✉)
Graduate Program in Organismic and Evolutionary Biology, University of Massachusetts Amherst, Amherst, MA, USA
e-mail: amwinsor@umass.edu

N. I. Morehouse
Department of Biological Sciences, University of Cincinnati, Cincinnati, OH, USA
e-mail: morehonn@ucmail.uc.edu

E. M. Jakob
Biology Department, University of Massachusetts Amherst, Amherst, MA, USA
e-mail: ejakob@umass.edu

E. Buschbeck, M. Bok (eds.), *Distributed Vision*, Springer Series in Vision Research, https://doi.org/10.1007/978-3-031-23216-9_10

10.1 Why Spiders?

Spiders provide a stellar opportunity for studying the evolution of a distributed visual system. Most members of this group, the Araneae, have eight eyes, but beyond that, there is extraordinary diversity in everything from their visual ecology and eye arrangement to their eye function and the underlying neural circuitry. To illustrate this diversity, first consider the jumping spiders (family Salticidae). Charismatic, alert, and reactive, quickly pivoting to direct their prominent anterior-facing eyes at an approaching human, jumping spiders have been a natural object of study for behavioral scientists for many years. Take, for example, this description of the courtship display of a jumping spider by two pioneers in the study of salticid behavior, George and Elizabeth Peckham, in the late 1800s:

> On the first day of June of the present year we were so fortunate as to discover on a hot, stony hillside, large numbers of males and females of a new species of *Habrocestum* having a modification of the third leg... As it was their mating season, we had now a welcome opportunity of seeing what use the active little male, which is further beautified by having his first legs of a delicate light green color, with a fringe of white hairs along the outer side, makes of this adornment in paying his addresses to the female. When they are put into a mating-box together, the male notices the female at a distance of from six to eight inches, and rapidly approaches her. When within three or four inches, he begins to move from side to side, with his handsome first legs pointed downward and somewhat outward, his palpi [small appendages near the mouth used in sperm transfer] extended parallel with them, and his third legs raised above the first and second in such a way as to show the apophyses on the patellae. Frequently, in these preliminary movements, he bends the ends of the first legs inward—the bend being at the tibia—so as to put them into the form of a diamond, meanwhile moving the palpi rapidly up and down. As he approaches the female, she all the time eying him most intently, he raises the first pair of legs, swaying them backward and forward, still keeping the third pair well up, seeming as eager to display them as the first pair. In this way he approaches to within about two inches, when she rushes at him and he retreats. The whole performance is repeated (Peckham and Peckham 1890).

How can one fail to be charmed? It is no wonder that jumping spider vision has attracted a growing number of researchers, especially as novel techniques emerge. But what is also remarkable is how different other spider families can be. For many years, one of us (EMJ) had studied pholcids—typical cobweb spiders that one might find in the basement. Pholcids seem driven by vibration, easily fooled into thinking they have captured a fly by a tuning fork touching their web, but they barely react to changes in light. Other families illustrate yet more permutations: wolf spiders court with vibratory and visual signals, crab spiders wait on flowers to grab their prey, and net-casting and bolas spiders snatch their prey from the air. Thus, across spiders, we find an extraordinary diversity of visual capabilities and visually driven behaviors; the potential for comparative studies is enormous and has barely been tapped.

In this chapter, we will begin with a deep dive into salticids because of a particularly rich literature on jumping spider behavior and visual ecology, not to mention our own research interests. Next, we expand our view across the Araneae, in which we contextualize key concepts about spider vision. Throughout, we have tried to include enough detailed explanations to satisfy arachnologists who are new to vision research as well as visual ecologists who are new to spiders.

10.2 Jumping Spiders: A High-Performing, Compact Distributed Visual System

We begin by focusing on the Salticidae, whose visual systems have been studied far more than those of other spider families and thus can serve as a point of comparison later in the chapter. With the advent of exciting new techniques, recent years have seen a spike in the number of laboratories around the world devoted to studying jumping spider vision. Excellent and detailed primers on jumping spider vision are available (Harland and Jackson 2004; Harland et al. 2012; Land and Nilsson 2012; Morehouse et al. 2017; Morehouse 2020; Hill 2022). We especially wish to acknowledge the contributions of Michael Land, who became interested in salticids as a graduate student at Berkeley in the 1960s. His work forms the foundation for much of what we know about jumping spider vision (reviewed in Jackson and Harland 2009) and directly inspired much of our own work (e.g., Jakob et al. 2018; Zurek et al. 2015). In this section, we briefly describe first the interesting range of visually based behavior demonstrated by salticids, and then how that behavior is enabled by their distributed visual system. We begin our discussion with a historical context outlining how the field has approached the study of jumping spider vision and conclude with an overview of physiological techniques, only recently applied to spiders, that can be incorporated into tests of hypotheses about visual function.

10.2.1 Vision-Based Behavior of Jumping Spiders

10.2.1.1 Methods for Studying Vision-Based Behavior in Jumping Spiders

In order to understand spider vision, it is essential to first grasp its function. To this end, behaviorists have developed increasingly elegant methods for interrogating visually guided behaviors and associated cues. For example, to identify which visual features spiders attend to when classifying an object as conspecific, prey, or predator, researchers create flat or three-dimensional stimuli and observe the spiders' responses (e.g., Crane 1949; Drees 1952; Forster 1985; Harland and Jackson 2000; Rößler et al. 2021). Jumping spiders will display to their own reflection, and thus one can quantify the distance at which they can visually identify their reflection as a spider (Harland et al. 1999). To determine how spiders orient toward and track moving stimuli, spiders, like many insects (e.g., Taylor et al. 2015), can be tethered so that they would walk on a trackball that reconstructs their fictive path. Spiders attempting to turn toward a visual stimulus will rotate the trackball (e.g., Zurek et al. 2010; De Agrò et al. 2021), thereby allowing researchers to probe visual functions, like motion perception, spatial acuity, and contrast thresholds. A particularly valuable discovery came about by accident. One night, while watching a video of courtship behavior in the lab, David Clark noticed that a female spider was also watching the display. When Clark scaled down the video to life size, the female approached it and gave a receptive signal (D. Clark, personal communication). Clark and Uetz (1990) went on to show that spiders did not appear to distinguish between living crickets and a live video feed of those crickets. Since then, there has been widespread use of video playback and animation techniques in jumping spider studies, including in our own labs, to study courtship, predation, and other behaviors (a few examples of many include Clark and Morjan 2001; Harland and Jackson 2002; Bednarski et al. 2012; McGinley and Taylor 2016). Particularly ingenious is a virtual reality setup in which tethered spiders on a trackball navigate through a digital environment; in a virtual world they approach beacons indicating the location of their nest sites (Peckmezian and Taylor 2015a) just as they do in the field (Hoefler and Jakob 2006). Finally, the development of a spider-specific eye tracker (Canavesi et al. 2011, based on a design by Land 1969b) allows the precise measurement of the gaze direction of spiders as they view video stimuli, possible even while simultaneously recording their brain activity (Menda et al. 2014).

10.2.1.2 Behavioral Contexts in Which Jumping Spiders Use Vision

Jumping spider predatory behavior is highly visual and especially amenable to study, allowing researchers to probe the visual cues that these animals use to detect, identify, and respond to potential prey. Jumping spiders do not build prey-capture webs but are cat-like hunters, stalking and pouncing upon insects and smaller spiders (reviewed in Forster 1982; Jackson and Pollard 1996). They attack even unrealistic "prey," such as a tuft of wool, as long as it is moving (Heil 1936; Drees 1952). When a spider detects a moving stimulus toward the side or rear, it turns, in either a large turn or a series of small turns, so that its body axis faces toward the stimulus. The spider often orients immediately to the stimulus (termed "fixation" by Land 1971). If the spider does not fixate, it may rotate again if the stimulus moves. After the spider fixates, it then seems to evaluate the stimulus and will either turn and run, court it, or attack it (reviewed in Land 1971). Both local motion (leg, head, and antennal movement) and global motion (movement of the entire body) by prey elicit attack in *Phidippus* jumping spiders (Bednarski et al. 2012). In addition, jumping spiders attend to shapes that hold particular relevance. For example, mosquito-eating spiders will attack abstract representations of prey, provided the abstract representations contain lines at the proper angles (Dolev and Nelson 2014). Spiders can use visual cues alone to distinguish prey from nonprey and among different types of prey (Edwards and Jackson 1993; Harland et al. 1999; Harland and Jackson 2000), assess the direction of movement and direct a predatory strike at the head of the prey (Bartos and Minias 2016), and avoid dangerous insects by sight (Nelson and Jackson 2006).

Vision is also used extensively in intraspecific communication. Courtship displays, like those observed by the Peckhams, have been described for dozens of species. Visual courtship elements may include waving of different pairs of outstretched legs, both together and in alternation; "knee pops" in which a bent leg is raised so that the patella is displayed to the female; sidling from side to side; palp waving;

and lifting and wagging of the abdomen. In many species, visual signals are accompanied by substrate-borne vibrations (e.g., Elias et al. 2012). Females assess courtship displays to identify males as conspecifics—especially important to males, which risk being attacked and eaten during courtship interactions—as well as evaluate the traits of prospective conspecific suitors. In several salticid groups, displays have rapidly diversified, as exemplified by isolated populations of *Habronattus pugilis* in the "sky islands" of southwestern mountaintops (e.g., Maddison and McMahon 2000; Masta and Maddison 2002; Hebets and Maddison 2005) and the speciose, tiny, colorful peacock spiders (Girard and Endler 2014; Girard et al. 2015; Girard et al. 2018). (A YouTube search for peacock spider displays yields results more gratifying than any verbal description could provide.) Jumping spiders also signal to conspecifics of the same sex. For example, contest dynamics between males are often largely mediated by vision (e.g., Taylor et al. 2001; Elias et al. 2008; Tedore and Johnsen 2015).

In spite of having sesame-seed-sized brains, jumping spiders are quite capable of cognitive tasks that often rely heavily on vision, such as vision-based learning and problem-solving (reviews in Cross and Jackson 2006; Jackson and Cross 2011; Jakob et al. 2011; Jakob and Long 2016; Aguilar-Arguello and Nelson 2021). To take just a few examples, spiders learn to avoid visual cues associated with aversive stimuli, such as shock, vibration, or heat (Drees 1952; Nakamura and Yamashita 2000; Bednarski et al. 2012; Long et al. 2015; Peckmezian and Taylor 2015b, 2017); avoid or locate food associated with particular colors or contextual stimuli (Skow and Jakob 2005; Jakob et al. 2007; Taylor et al. 2016; Vickers and Taylor 2018; Winsor et al. 2020); and solve a confinement problem (Jackson et al. 2001; Cross and Jackson 2015). Many species have been presented with variations of a detour problem, where a spider can see a goal but must follow an indirect path to reach it. Some salticids that are particularly good at solving this problem are from the genus *Portia*. These unusual salticids prefer to prey on other spiders, including stealthily stalking web-building spiders. To reach its prey, *Portia* may take elaborate detours, during which it may lose sight of its prey. Before embarking on its approach, *Portia* can select among complete and incomplete routes to the prey, examining a prospective detour path by visually tracing out routes from the target. If the potential route dead-ends, *Portia* will then look back at the target and begin again until it identifies a complete route (Tarsitano and Jackson 1994, 1997; Tarsitano 2006; Cross and Jackson 2019). Jumping spiders also use visual cues to select microhabitats to use as hunting and resting sites (de Omena and Romero 2010; Tedore and Johnsen 2016). Many salticids build silken retreats, to which they return at night or during inclement weather, and can learn the characteristics of prominent nearby features (beacons) to help them return to these retreats (Hoefler and Jakob 2006).

Jumping spiders do not attend to all incoming visual stimuli equally but selectively prioritize certain information in a process called visual attention (Dukas 2002). Visual attention is often categorized into two types. "Bottom-up" processes are driven by certain features of the stimulus itself. Similar to how we attend to movement in an otherwise static scene, spiders are more likely to attend to moving dots of particular sizes and speeds (Zurek et al. 2010; De Agrò et al. 2021). As in humans, spiders' attention to a visual stimulus wanes over time. Spiders habituate to repeated visual stimuli; evidence suggests that this visual decrement is a result of central nervous system (CNS) modulation rather than simple receptor fatigue (Humphrey et al. 2018; Humphrey et al. 2019; Melrose et al. 2019; Nelson et al. 2019). Also driving visual attention are "top-down," or goal-directed, processes. For example, in humans, experimental participants might be given instructions to look for a blue *x* among a field of letters of different colors and would focus their search accordingly. In spiders, individuals that are primed by a stimulus in another modality, such as the odor of conspecifics, are then more likely to detect an obscured visual stimulus of the same type (Cross and Jackson 2009, 2010; Carvell et al. 2017). Current work in one of our labs includes how cross-modal priming with sound or odor influences the gaze direction of the principal eyes. Many topics in visual attention that have been studied in humans, such as visual search, object recognition, and navigation, are now being addressed by studying the gaze direction of spiders (Winsor et al. 2021).

10.2.2 Modular Vision: Two Eye Types

The alert, responsive behavior of jumping spiders has naturally led to research on the visual system that underlies it. Jumping spiders have appealing faces that feature large forward-facing eyes, but those noticeable large eyes are only part of the story. Like other spiders, jumping spiders typically have four pairs of simple "camera-type" eyes, named for their relative positions on the cephalothorax, which provide a near-360° view of their surroundings (Fig. 10.1a). These eyes are of two types: the large principal eyes, also called the anterior median (AM) eyes, and three pairs of secondary eyes. While they are similar in external appearance, the principal and secondary eyes have distinct evolutionary histories, developmental pathways, internal structures, and neural connectivity to higher brain regions (reviewed in Morehouse et al. 2017).

10.2.2.1 Secondary Eyes of Jumping Spiders

Of the two eye types, the secondary eyes of jumping spiders have a simpler morphology than the principal eyes but are still impressively capable. The posterior lateral (PL) eyes and posterior median (PM) eyes are directed toward the side and rear of jumping spiders, while a pair of the forward-facing anterior lateral (AL) eyes share a field of view with the principal (AM) eyes (Fig. 10.1). The PM eyes are reduced or even absent in some species (Land 1985a).

The AL and PL eyes have large retinas and wide fields of view (Fig 10.1b) and serve as excellent motion detectors. When these eyes detect movement, a spider will turn to orient its forward-facing eyes toward the stimulus, even when other eyes are masked (Land 1971; Zurek and Nelson 2012a, b). While Land (1971) demonstrated

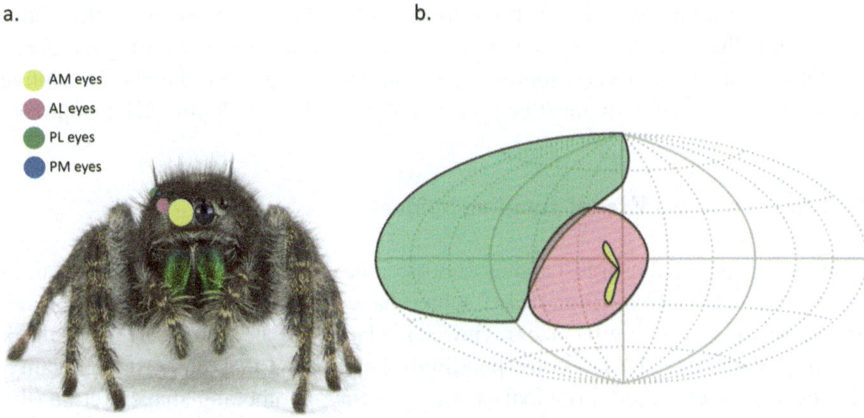

Fig. 10.1 Modular visual system of the jumping spider *Phidippus audax*. (**a**) The principal (AM) eyes are shown in yellow, while the secondary eyes are shown in reddish purple (AL eyes), bluish green (PL eyes), and blue (PM eyes). Reduced PM eyes are situated dorsally between the AL and PL eyes. (**b**) Orthographic projection mapping approximate visual fields of the AM eyes, AL eyes, and PL eyes of *Servaea incana*, a spider with similar eye size and arrangement to *P. audax*. The reduced PM eyes are not shown but view a small dorsal strip of overlap between the AL eyes and PL eyes. The point of origin is between the principal-eye retinas (i.e., figure corresponds to an anterior view of the spider's face). The principal-eye retina visual fields are boomerang shaped and overlap the AL-eye visual fields. The boomerang-shaped retinas can be moved. (Image recreated with permission from Morehouse 2020)

that salticids make orientation turns in response to stimulus movements of about 1°, which is close to the interreceptor angle of the PL eyes and was long considered to be the limit of motion detection, it has since been found that even smaller stimulus movements can be detected, a phenomenon known as motion hyperacuity (Zurek and Nelson 2012a). The AL eyes also appear to be responsible for detecting biological motion, the repetitive movement patterns characteristic of living organisms (De Agrò et al. 2021). Data suggest that, at least in some species, the AL and PL eyes have only a single peak in spectral sensitivity, making their vision monochromatic (Yamashita and Tateda 1976a; Terakita and Nagata 2014). The function of the PM eyes, much reduced or even missing in salticids, is unclear, and some authors have suggested that they are vestigial (Eakin and Brandenburger 1971). However, Terakita and Nagata (2014) point out that in *Hasarius adansoni*, the PM eyes express ultraviolet (UV) and blue-sensitive visual pigments and may therefore be specialized for detecting objects against the sky or changes in the brightness of the sky.

The photoreceptors in the secondary eyes have inverted rhabdomeres, which means that their photoreceptive segments lie below their cell bodies. Thus, light entering the inverted retinas of the secondary eyes must typically traverse the cell bodies before being absorbed in the rhabdomere. The result is lower light capture, compared to the principal eyes, due to scattering and inutile absorption by cell body constituents (Land 1985a), an effect partially ameliorated in some jumping spiders by the repositioning of the cell body to the side. The disadvantage of this latter approach is that it impacts the maximum acuity of these eyes by limiting the dense spatial packing of photoreceptors in the retinal mosaic (as both the photoreceptive segment and the cell body must be accommodated side by side for each receptor).

10.2.2.2 Principal Eyes of Jumping Spiders

In jumping spiders, the largest and most noticeable eyes are the principal eyes (anterior median (AM) eyes) (Fig. 10.1a), which have the highest spatial acuity of any animal with eyes of a similar size (reviewed in Harland et al. 2012). One might think that these large eyes have a correspondingly large retina that supports this acuity. However, jumping spider principal eyes have a tiny, boomerang-shaped retina (the visual field is shown in Fig. 10.1b). The extraordinary capabilities of these eyes result from a suite of interesting traits: eye tubes with a telescope-like structure, the ability to actively direct the eye tubes toward objects of interest independently of the spider's body movement, and a layered retina.

The basic structure of the principal eye is as follows. At the exterior end of each principal eye tube is a nonmoving converging corneal lens, part of the carapace, similar to that of the secondary eyes (Fig. 10.2). At the internal end of each eye tube, deep within the cephalothorax, is the boomerang-shaped retina. The eye's focal length, and thus its ability to resolve distant objects, is increased by a pit distal to the receptors. The pit's refractive index allows it to act as a diverging lens at the rear of the eye, magnifying the image received by the retinal cells and creating a Galilean telescope-like effect (Williams and McIntyre 1980; Blest and Price 1984).

The principal-eye retina has fewer than 1500 receptors (in contrast to the 200 million receptors in the human eye), and maximum spatial resolution is confined to a roughly 200-receptor region at the center of the boomerang. The field of view is correspondingly small (0.8–5° in the horizontal dimension, depending on species) (Blest and Price 1984). However, the disadvantages of a small retina size are partly overcome by six dedicated muscles that allow the eye tube to be rotated and moved horizontally and vertically inside the cephalothorax by as much as 50° (Fig. 10.2) (Land 1969a, b, 1971, 1972; Williams and McIntyre 1980; Blest et al. 1990). As the tube moves, it samples the larger image provided by the corneal lens, as if shining a flashlight at different parts of the image. These eyes thus provide excellent vision in a fraction of the space required by a spherical eye with similar capabilities (Land 1974; Harland and Jackson 2004), albeit sampling of the full visual field afforded by the corneal lens can only be accomplished through retinal movements, which take time.

Retinal movements vary depending on the stimulus that the spider is viewing. Land (1969b), using an ophthalmoscope, described four behaviors of the principal eyes: spontaneous movement across a scene; saccades, or rapid shifts to different parts of the scene; tracking a moving object; and scanning, or a back-and-forth horizontal motion accompanied by rotation as the spider inspects an object of interest (Fig. 10.3). Scanning seems to be unique to salticids and is certainly involved in

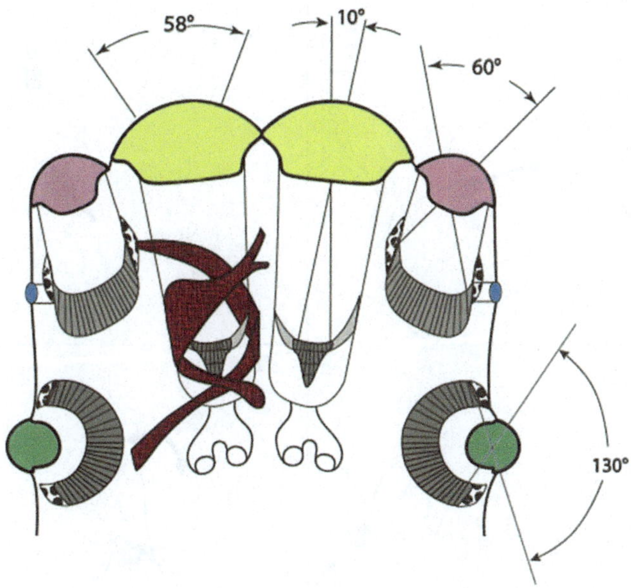

Fig. 10.2 A schematic horizontal section through the jumping spider head and eyes showing the internal structure of the visual system. The approximate fields of view for each eye pair are shown. The six muscles that control the principal eye tube are shown in dark red; the movements of the eye tube compensate for a small field of view. The moveable principal-eye retinas subtend about 10° of visual space at a given time within a maximum visual angle of about 58°. Retinas and their associated receptors are shown in dark gray. Cells containing pigment granules form a pseudo-iris outside the retina. Transparent vitreous cells (not shown) fill the space between the lens and retina. The optic nerve and first optic neuropil are shown for the principal eyes (see Sect. 10.3.6). (Image recreated with permission from Land 1969a)

object identification, as attested by the behavioral experiments described below. Land's findings have since been confirmed using a more advanced eye tracker developed over many years by an international team (Canavesi et al. 2011).

The retinal structure of the principal eyes is also complex. In contrast to the secondary eyes, the photoreceptors are everted (i.e., the photoreceptive segments are positioned toward incoming light, with the cell bodies below them), so light does not attenuate through the cell bodies and the receptors can be packed very closely (Blest 1985). Spatial acuity is greatest in the center of the boomerang, where the pit magnifies with minimal distortion closest to the optical axis and where receptors are more tightly organized. Acuity then falls off toward the boomerang tips (Blest and Price 1984). The central regions of the retina provide the highest known spatial acuity of any terrestrial invertebrate (Warrant and McIntyre 1993). For example, the principal eyes of the salticid *Portia* have spatial acuity greater than that of dragonflies, rivaling that of pigeons, and only a fifth that of humans (reviewed in Harland et al. 2012). Cells in the principal-eye retina are arranged in four tiers or layers. Layer I, furthest from the cornea, is specialized for resolving fine detail. It contains dense, tightly organized photoreceptors that function as light guides (Blest 1985;

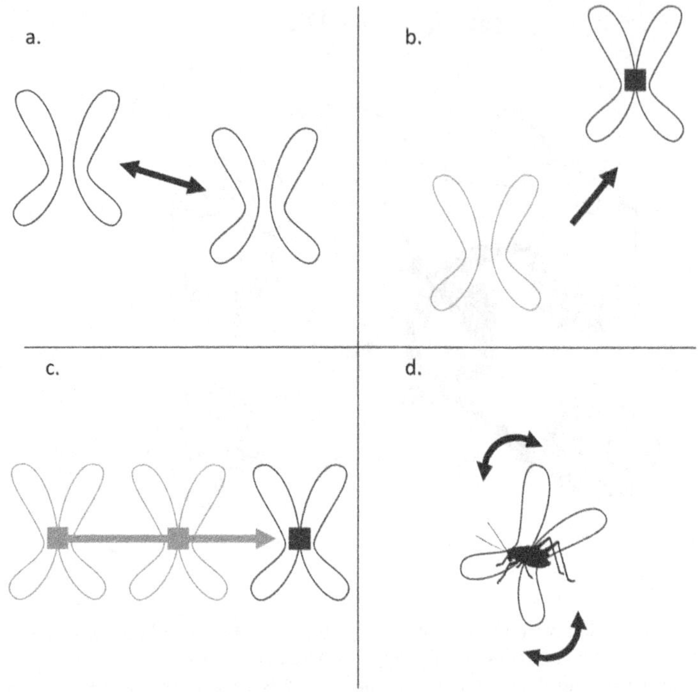

Fig. 10.3 The four categories of retinal movements: (**a**) exploratory movements, (**b**) saccades, (**c**) tracking, and (**d**) scanning. Double arrows represent back-and-forth movement, while single arrows represent movement in one direction. Opacity changes represent object displacement (lighter objects represent a starting position before displacement). Exploratory movements are spontaneous and can occur in any direction. The fields of view of the retinas converge when the spider is examining an object of interest. In the panel depicting scanning retinas, the cricket silhouette remains stationary, and the retinas exhibit torsional movements in either direction over it. (Image recreated with permission from Land 1969b)

Blest and Carter 1987; Blest et al. 1990; see Sect. 10.3.3.2). Layers II–IV have lower spatial acuity due to their larger, less densely packed photoreceptors, which allows greater passage of light to Layer I (Blest 1985). The tiered retina also helps solve an optical difficulty presented by the lens system. When light is transmitted through the pair of lenses, different wavelengths come into focus at slightly different distances behind the lens due to linear chromatic aberration. The best solution to this would be to position photoreceptors of different spectral sensitivities in the layers where the wavelengths they are maximally sensitive to are in best focus, and indeed, jumping spiders appear to do so, with short-wavelength-sensitive photoreceptors positioned in the distal two tiers (layers III and IV) and longer-wavelength-sensitive photoreceptors located in the proximal layers I and II (e.g., Nagata et al. 2012). In addition to this clever solution to chromatic aberration, the tiering of the principal eye retina provides another hidden benefit. Because the same region of space is sampled simultaneously by each tier, input from different tiers may be compared to extract color information without the loss in spatial acuity that typically accompanies color vision (reviewed in Harland et al. 2012; Morehouse 2020). Some jumping spiders have as many as four sensitivity peaks, ranging from UV to orange or red (Land 1969a; Yamashita and Tateda 1976a; Blest et al. 1981), sometimes augmented by intraretinal filters that shift the peak sensitivities of underlying photoreceptors (Zurek et al. 2015). In addition, evidence suggests that retinal tiering may allow spiders to estimate distance based on the relative degree of image defocusing on different layers (Nagata et al. 2012). Thus, these unusual eyes provide a moveable view of the world while supplying high acuity, color perception, and depth information.

10.2.2.3 Division of Labor in Jumping Spider Eyes

Beginning nearly a century ago, behavioral researchers masked different sets of eyes to deduce their functions and coordination. For example, spiders with secondary eyes masked failed to pivot toward a moving stimulus unless it was directly in front of their principal eyes (Homann 1928; Crane 1949; Land 1971). Those with only their principal eyes masked oriented to the stimulus but did not respond further. These results implied that the secondary eyes function as motion detectors and the moveable principal eyes are responsible for object identification. Forster (1979), using similar masking techniques, found that AL eyes are necessary for chasing prey, whereas principal-eye input is needed to initiate stalking behavior. Spiders made only short pounces onto prey when their secondary eyes were masked and required both principal and AL eyes to make long-range pounces. Spiders were less discriminatory when attacking faster-moving targets compared to slower-moving or stationary targets, suggesting that the principal eyes are primarily used to scan slower-moving or stationary targets. Later work showed that spiders back away from objects that appear to be looming closer; this behavior is driven by AL eyes, and the principal eyes are unnecessary (Spano et al. 2012). This result makes sense in light of the large field of view of the AL eye retina, which is necessary to detect the increasing subtended angle of the looming stimulus.

The AL eyes not only guide a spider's turning response to a stimulus but also guide the gaze direction of the moveable principal eyes. Using the updated eye tracker, we have documented that when the AL eyes are unmasked, the principal eyes effortlessly track moving stimuli; when the AL eyes are masked, the principal eyes are unable to locate suddenly appearing stimuli or track moving stimuli, although they can scan motionless images that appear directly in front of them (Jakob et al. 2018). Recent work confirms that jumping spiders can recognize stationary objects during an encounter (Rößler et al. 2021), a process likely mediated by the principal eyes.

The principal eyes do not automatically orient their gaze toward a stimulus detected only by the AL eyes (Bruce et al. 2021). If a spider is scanning a complex, biologically relevant image of a cricket with its principal eyes, it ignores a distractor oval appearing only in view of its AL eyes, but if it is scanning a less interesting oval, it does redirect its gaze toward the distractor. A spider examining a cricket image can, however, be distracted by a looming stimulus. This result is reminiscent of human visual behavior, when we are less likely to attend to a distractor appearing in our peripheral vision when we are examining closely a stimulus in our foveal vision (Savage et al. 2019).

10.2.3 Next Steps in the Study of Salticid Vision

We see at least two areas ripe for expanding research on salticid vision and visually guided behaviors. First, given the availability of new techniques, we expect that studies will increasingly incorporate both physiological and behavioral approaches rather than one or the other. This might include, for example, simultaneously recording neural and behavioral responses to visual stimuli to probe the neural underpinnings of visual cognition. Second, jumping spiders are an incredibly diverse family with over 600 genera and 6000 described species (World Spider Catalog 2022) and thus offer wonderful opportunities for comparative work. For example, both *Saitis barbipes* and *Habronattus pyrrithrix* are sexually dimorphic species, and males have red coloration. It would be tempting to conclude that the red color is a sexual signal, but Glenszczyk et al. (2022) found that *S. barbipes* lack long-wavelength-sensitive photoreceptors or spectral filters to perceive the color red. In contrast, the principal eyes of *H. pyrrithrix* have spectral filters that enable them to perceive longer wavelengths (Zurek et al. 2015). The retinal filters are confined to the center point of each retina, and using the eye tracker, we see that females direct them toward the center of the male display (D. Zurek, unpubl. data). This pair of studies illustrates the value of integrating physiology and behavior in a comparative context.

10.3 Distributed Visual Systems Across the Araneae

Beyond jumping spiders, there is enormous variety in the form that spider vision takes. This is perhaps unsurprising given that spiders are one of the world's most species-rich animal groups, with an estimated 80,000 extant species (Raven and Yeates 2007), of which only a little over half are described (nearly 50,000 species described to date; World Spider Catalog 2022). They are also an ancient lineage; the earliest spiders arose in the Devonian (Foelix 2011). Over the past 400 million years, these animals have evolved a remarkable array of lifestyles, behaviors, and ecological niches. Voracious predators as they are, they can be found in all of the world's major biomes and on every continent except Antarctica (Turnbull 1973). Although there are many reasons for the evolutionary success of spiders, their unique and remarkably elegant modular visual systems have certainly played a significant role.

10.3.1 Vision-Based Behavior Across Spiders

Visually guided behavior is widespread among spider families and is distributed across the phylogeny. We begin with a brief overview of the contexts in which different species use vision. Given that we surveyed many of the behaviors demonstrated by jumping spiders in the previous section, here we shift our emphasis to non-salticid species, though we periodically highlight salticids with unique traits and compare salticids with other groups.

Some spiders navigate using features of the environment such as visual landmarks and patterns of polarization. Similar to some jumping spiders, the Namib Desert spider *Leucorchestris arenicola* (family Sparassidae) uses visual beacons when navigating at night (Nørgaard et al. 2006), with nocturnal navigation being essential to avoid oppressively high daytime temperatures. The wolf spider *Lycosa tarentula* (family Lycosidae) requires visual input for path integration when homing (Ortega-Escobar 2002). The ground spider *Drassodes cupreus* (family Gnaphosidae) uses polarized light from the sky to navigate home after bouts of foraging (Dacke et al. 1999).

Spiders also visually assess their environment to increase the chance of capturing prey. For example, the nocturnal orb-web spider *Larinioides sclopetarius* (family Araneidae) builds its web near artificial lights where prey is more abundant. Spiders are not simply responding to the presence of prey; in the lab, naïve spiders sought out better-lit spots without the confound of prey (Heiling 1999). The orb-weaver *Nephila clavipes* (family Araneidae) spins webs of different spectral qualities depending on the properties of ambient light, such as brightness and wavelength composition, and the webs are thus harder for prey to see (Craig et al. 1996). The spider-eating specialist salticid *Portia labiata* exploits UV-reflecting silk stabilimenta in the webs of other spiders to locate them (Li and Lim 2005).

Other than for navigation and selecting foraging sites, many taxa use vision for prey capture. Arboreal green lynx spiders *Peucetia viridans* (family Oxyopidae) spend their daylight hours stalking prey that reside on the branches of plants, pouncing from the vantage point of a higher branch (Whitcomb and Eason 1965). Another cursorial hunter, *Tibellus macellus* (family: Philodromidae), uses its vision to capture a wide variety of small insect prey (Huseynov 2008). The crab spider *Misumena* (family Thomisidae) waits on flowers for arriving prey; it is so reliant on motion cues that it sometimes walks right over stationary prey (Morse 2007). The net-casting spiders (family Deinopidae) hold a small silken snare between their front legs and use enormous eyes, sensitive in dim light, to help them quickly scoop up prey (Robinson and Robinson 1971; Stafstrom and Hebets 2016).

Spiders have many predators, notably birds, wasps, and other spiders, including conspecifics, and have evolved many visually guided antipredator strategies (reviewed in Robledo-Ospina and Rao 2022). Crab spiders perch on a flower, ready to grab an unsuspecting pollinator, and some can select floral background colors that best complement their own (Heiling et al. 2005), a process presumably mediated by their visual system (Defrize et al. 2011). The ambulatory wolf spider

Schizocosa ocreata instead flees when a simulated bird shadow passes overhead (Lohrey et al. 2009).

Vision can also be used to assess mates during elaborate courtship displays and competitors during agonistic social encounters. For example, wolf spiders rely on vision during conspecific interactions (Rovner 1996). Extravagant visual displays have been thoroughly explored in wolf spiders of the genus *Schizocosa*: males display foreleg ornamentation to females during courtship, which improves mating success in some species (e.g., Hebets and Uetz 1999). As noted earlier, elaborate courtship displays involving color, pattern, and motion are widespread in jumping spiders as well, including the paradise spiders of North America (genus *Habronattus*, Elias et al. 2012) and the peacock spiders of Australia (genus *Maratus*, Girard et al. 2015). Fighting with a competitor can be costly, so mutual visual assessment of fighting ability can allow spiders to settle disputes unscathed. Many spiders assume a defensive posture by lifting their first pair of legs when visually presented with a conspecific competitor, which can be used for rank assessment (Riechert 1982). By eavesdropping on competing males, female *Thiania bhamoensis* jumping spiders show changes in preference between two potential mates (see Chan et al. 2008).

While many spiders rely on vision, it is worth noting that many of the visual cues described above are accompanied by signals and cues in other modalities. This multimodality is important to consider in the context of visual system evolution because it informs both neural integration of visual inputs and resulting behavioral responses. The most common is mechanoreception, the ability to detect vibrations. The strikingly diverse web-building spiders rely on vibratory cues from prey entangled in webs (e.g., Landolfa and Barth 1996). Other examples include *Cupiennius salei* (family Trechaleidae; this species is well represented in the literature but recently moved from Ctenidae; see Piacentini and Ramírez 2019), which, while it uses some visual cues, relies primarily on vibratory cues for localizing prey (reviewed in Barth 2002; Fenk et al. 2010), and fishing spiders of the genus *Dolomedes* (family Pisauridae), which detect vibrations borne on the water's surface (Bleckmann and Rovner 1984). Furthermore, many spiders rely on chemoreception mediated by receptors on their appendages (e.g., Tietjen and Rovner 1982; Persons and Uetz 1996; Foelix 2011). Other modalities, such as audition, are also important (Shamble et al. 2016; Stafstrom et al. 2020; Zhou et al. 2022). How different spider lineages have evolved to prioritize inputs from these various senses or integrate them with vision is an area ripe for deeper investigation, especially considering that many (but not all) groups that heavily rely on other senses exhibit reduced visual systems.

10.3.2 Origin and Evolution of Spider Eyes

Arachnids are among the few groups of arthropods that rely primarily on single-lens eyes (Land 1985a). Most other arthropods, including insects and crustaceans, use a pair of compound eyes as their primary visual organs (although the role of ocelli should not be understated; see Chap. 8 in this volume). Single-lens eyes can potentially support better resolution for eyes of their size (Land 1981; Land and Nilsson 2012; Nilsson 2021). Non-spider arachnids vary in eye types and number, but vision is reported to be relatively poor in many of these animals. For example, scorpions (Arachnida: Scorpiones) and whip spiders (Arachnida: Amblypygi), like many other arachnids, have two morphologically distinct eye types (Bedini 1967; Loria and Prendini 2014; Lehmann and Melzer 2018a; Sinakevitch et al. 2021), but it appears that other sensory modalities are more important (Miether and Dunlop 2016). Among the arachnids, spiders have undoubtedly evolved the greatest visual system diversity (Strausfeld 2012). In spiders, many lineages have poor vision (even lacking eyes altogether), while others have exceptional vision (e.g., the Salticidae and Deinopidae).

10.3.2.1 Origin and Development

As described in Sect. 10.2.2 for salticids, other spider families also have four pairs of eyes divided into two types (the principal and secondary eyes, Homann 1928), with specific eye pairs named for their anatomical position on the cephalothorax. Historically, the delineation between principal and secondary eyes was contingent on whether their retinula cells are everted (principal eyes) or inverted (secondary eyes). Unlike the principal eyes, the secondary eyes often possess a light-reflecting tapetum (see Sect. 10.3.3.4) and do not have muscles for movement (see Sect. 10.3.5.1). The two eye types also have different neural connectivity and patterns of development (Strausfeld and Barth 1993; Strausfeld et al. 1993). In spite of their name, the principal (AM) eyes are not always the primary visual organ for spiders; in some spider lineages, it is one of the secondary eye pairs (AL, PM, or PL) that is most prominent. These eye types also have different evolutionary histories (Morehouse et al. 2017). Here, we provide a brief overview.

The ancient Cambrian relatives of spiders likely had both single-lens and compound eyes (Paulus 1979). Fossil evidence supports this, including those of trilobites and horseshoe crabs, which possess single-lens "medial" and compound "lateral" eyes (Paulus 2000; Strausfeld et al. 2016; Lan et al. 2021). The medial eyes are evolutionarily related to the lateral eyes and might have derived from an ancestral visual organ before the diversification of arthropods (Zhou et al. 2016). Genetic evidence suggests that key mechanisms of eye development are mediated by ancient, deeply conserved gene regulatory networks (reviews in Friedrich 2006; Morehouse et al. 2017). The ancestral single-lens eye is thought to be homologous with the principal eyes of spiders, medial eyes of other arachnids, and ocelli of insects, while the ancestral compound eye is thought to be homologous with the secondary eyes of spiders, lateral eyes of other arachnids, and modern compound eyes of insects. The secondary eyes of spiders might have arisen from the subdivision and subsequent fusion of ommatidia from an ancestral compound eye or the enlargement of its individual ommatidia (e.g., Buschbeck 2014). How chelicerate eyes evolved has not been resolved, so it is unclear if this occurred once or multiple times in spiders and other arachnid lineages (Miether and Dunlop 2016), but it seems that gene duplication has played an important role in their visual system evolution (Gainett et al. 2020).

The principal and secondary eyes of spiders develop from separate areas of the ectoderm of the head: a median ectodermal groove and the lateral head ectoderm, respectively (Schomburg et al. 2015). During development, the principal eyes innervate the protocerebrum, and the secondary eyes innervate the lateral protocerebrum (Strausfeld and Barth 1993; Strausfeld et al. 1993). The principal eyes derive from their own progenitor cells, much like the ocelli of insects, and a bilateral pair of "eye-fields" fragment in a cluster to form the secondary eyes, much like the compound eyes of insects (Schomburg et al. 2015; Samadi et al. 2015). The development of eyes in *Drosophila melanogaster* is dictated by a core set of developmental genes—including sine oculis (*so*), eyes absent (*eya*), dachshund (*dac*), atonal (*ato*), and orthodenticle (*otd*)—and two Pax6 orthologs, which determine the eye field during early development—eyeless (*ey*) and twin of eyeless (*toy*)—all of which are found across arthropods (reviewed in Friedrich 2006; Morehouse et al. 2017). In insects, these retinal determination network genes regulate ocelli and compound eye development. In spiders, each eye type expresses a unique combination of these transcription factors (Schomburg et al. 2015; Samadi et al. 2015; Baudouin-Gonzalez et al. 2022).

Some conserved genes in spiders appear to serve similar functions as in insects or vertebrates, while important differences have been noted in others. For example, the *Drosophila* proneural gene *ato* is present in spiders (Samadi et al. 2015; Baudouin-Gonzalez et al. 2022). In both *Drosophila* and spiders, *ato* seems to initiate photoreceptor differentiation (Baudouin-Gonzalez et al. 2022). Recently, Baudouin-Gonzalez et al. (2022) found that similarly to vertebrates, Wnt signaling and potentially the gene hedgehog (*hh*) in spiders may restrict the expression of retinal determination genes around each eye primordium, providing a plausible mechanism underlying variation in eye number, placement, and size. In contrast, the expression of the ubiquitous eye development master control gene Pax6 does not seem to be expressed during eye development in the common house spider *Parasteatoda tepidariorum* (family: Theridiidae) (Schomburg et al. 2015; Baudouin-Gonzalez et al. 2022) or several other genera investigated so far (e.g., *Acanthoscurria*, *Pholcus*, *Marpissa*; L. Baudouin-Gonzalez and L. Sumner-Rooney, pers. comm.). However, Samadi et al. (2015) found late-stage expression of Pax6 in the principal eyes of *C. salei*. Recent work has shown that instead of Pax6, an ortholog of a different Pax gene called Pax2 is expressed in the spider secondary eye primordia (Janeschik et al. 2022). Thus, variation in spider eye arrangements seems to involve network components that are common to insects or vertebrates and others that are unique. Future work is needed to better understand the extent to which these ancient gene networks show conserved functions and how gene duplication and subsequent functional divergence impact spider eye development. Another interesting future direction is how the visual system function changes across later ontogenetic stages. For example, Goté et al. (2019) found that in jumping spiders, the smaller eyes of juveniles are likely less sensitive than those of adults but still benefit from high visual acuity.

10.3.2.2 Eye Arrangement and Visual Fields

When beginning to identify spiders, a novice first learns that families can be distinguished by the size and positions of their eyes (Fig. 10.4; Foelix 2011). For example, many visual hunters that stalk and pounce on prey, including jumping spiders and wolf spiders (family Lycosidae), exhibit forward-facing eyes with a prominent pair (AM and PM eyes, respectively). There are exceptions, however, such as the nursery web spiders (family Pisauridae), which are also active hunters but have equally sized eyes. The night-active net-casting spiders (family Deinopidae), which seize passing prey with a small web stretched between their legs, have an enormous pair of PM eyes and see exceptionally well in dim light. In fact, the PM eye of *Deinopis subrufa* is among the largest simple eyes of all arthropods, with a diameter exceeding 1 mm (Blest and Land 1977). The ambush-hunting crab spiders (family Thomisidae) have similarly sized eyes relative to one another—with slightly enlarged AL eyes—which are well distributed around the cephalothorax. The

Fig. 10.4 Typical spider eye patterns used for family identification. Starting in the upper left corner, from left to right, each row in turn: Salticidae (*Phidippus putnami*), Lycosidae (*Rabidosa rabida*), Pisauridae (*Pisaurina mira*), Deinopidae (*Deinopis aurita*), Thomisidae (*Synema parvulum*), and Pholcidae (*Pholcus phalangioides*). Principal (AM) eyes are shown in yellow, while the secondary eyes are shown in reddish purple (AL eyes), blue (PM eyes), and bluish green (PL eyes). (Images courtesy of Thomas Shahan (Salticidae, Lycosidae, Thomisidae), Jay Stafstrom (Deinopidae), and the USGS Bee Inventory and Monitoring Lab (Pholcidae))

web-building cellar spiders (family Pholcidae) have clusters of diminutive eyes. Exceptions to these family-level characteristics occur in genera and species with specialized lifestyles.

With the visual fields of the principal and secondary eyes combined, most spiders can see nearly 360°, but this is certainly not universal. The extent to which a spider can see the full hemisphere surrounding it is determined by the location of the eyes on the cephalothorax, which direction the eyes are facing, and their fields of view (FOV) or the solid angle of space outside of the animal that is imaged by the retina. The FOV of each eye is contingent on its size, the focal length of its lens, and the dimensions and position of the retina. The FOV can be calculated or measured using ophthalmoscopic techniques (e.g., Homann 1928; Land 1985b; Land and Barth 1992; Goté et al. 2019). The FOV size and shape are highly variable within and across species, even for a corresponding eye type. For instance, the principal eyes of crab spiders have a larger FOV than those of jumping spiders (Insausti et al. 2012).

Two transverse rows of similarly sized eyes are found in many spider families (Fig. 10.4; Homann 1971; Land 1985a). The bottom row is slightly recurved and comprised of the centrally located principal eyes flanked by the AL eyes, while the top row is slightly procurved and comprised of the PM and PL eyes. This pattern was hypothesized by Homann (1971) as the primitive state for spiders. Recent work across arachnids suggests that secondary eyes are usually in bilaterally symmetric triads in basal groups, while the principal eyes assume a central position (Miether and Dunlop 2016). In spiders, the secondary eye triads intermingle with the principal eyes to yield two basic ground patterns: either all eyes clustered together on a single tubercle (raised area), as usually seen in the Mygalomorphae, or eyes positioned in two rows, as often seen in the Araneomorphae (Fig. 10.5). Variation in these ground patterns is apparent across the phylogeny (Fig. 10.5). The presumed basal two-row eye pattern of araneomorphs is supported by recent molecular phylogenetic and unipartite directional network approaches. Within the speciose retrolateral tibial apophysis (RTA) clade (i.e., mostly ground-dwelling araneomorph spiders that are synapomorphic for a tibial projection on the pedipalps of males), the two-row pattern was supported as the ancestral state; the Ctenidae configuration (Fig. 10.5) independently evolved seven times, while the *Agelenopsis*, Oxyopidae, Lycosidae, Selenopidae, and Salticidae patterns (Fig. 10.5) each evolved once Hazzi and Hormiga 2022. Genes that determine the location and size of eyes may be conserved but differ in spatial or temporal expression patterns (Morehouse et al. 2017).

Fig. 10.5 (continued) Catalog 2022), shown are 40 that were selected for their reliance on vision, phylogenetic position, or within-family diversity. Note that *Cupiennius*, a genus well represented in the spider vision literature, was recently moved from Ctenidae to Trechaleidae. The visual abilities of the Deinopidae, Sparassidae, Oxyopidae, Pisauridae, Lycosidae, Thomisidae, Philodromidae, and Salticidae also have been relatively well studied (all of which are in the RTA clade, except the Deinopidae). Families that rely on vision to hunt often have enlarged eye pairs. Field guides and taxonomic keys (Ubick et al. 2005; Elliott 2006; Platnick 2020) were used to determine the most common eye pattern for each family

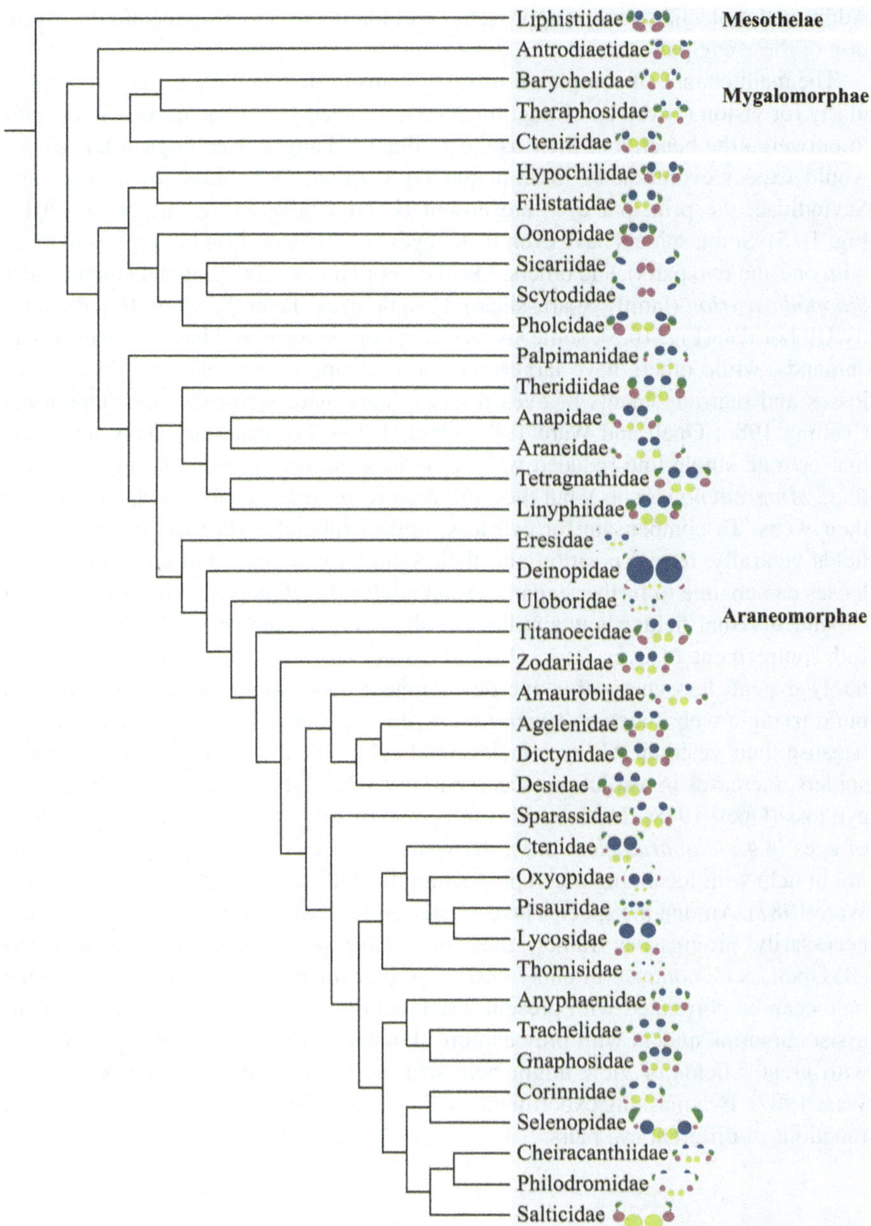

Fig. 10.5 Phylogeny with typical eye patterns for each family mapped. Tree topology inferred from Wheeler et al. (2017), with branch lengths not to scale. Principal (AM) eyes are shown in yellow, while the secondary eyes are shown in reddish purple (AL eyes), blue (PM eyes), and bluish green (PL eyes). Families are grouped in the suborder Mesothelae (which contains a single extant family) or Opisthothelae, the latter of which is subdivided into the infraorders Mygalomorphae (31 families) and Araneomorphae (99 families). Of the 131 currently valid families (World Spider (continued)

Additional molecular genetic approaches will likely provide insights into the evolution of these eye patterns.

The maintenance of complex sensory systems is energetically expensive, particularly for vision (Niven and Laughlin 2008); if the cost of maintaining eyes begins to outweigh the benefits of their use (e.g., due to changes in ecological niche), we would expect evolutionary loss. In the Dysderidae, Oonopidae, Sicariidae, and Scytodidae, the principal eyes are absent (Land 1985a; Morehouse et al. 2017; Fig. 10.5). Some spiders have even fewer eyes (e.g., Caponiidae has representatives with one and two pairs), and others, like the Laotian-cave-dwelling huntsman spider *Sinopoda scurion* (family Sparassidae), have no eyes (Jäger 2012). Within the family Uloboridae (Fig. 10.5), some species spin simple webs and have reduced visual demands, while others have larger and more complex webs. These spiders show losses and rearrangements of eyes in accordance with web reduction (Opell and Cushing 1986; Opell and Ware 1987; Opell 1988). For example, many uloborids that operate single-line reduced webs also show losses of both anterior eye pairs (e.g., *Miagrammopes* spp.), but they still require enough visual coverage to operate their webs. To compensate for eye loss, optical tubercles shift the PL-eye visual fields ventrally, retinal position and their symmetry change, and curvature of the lenses can change to further expand visual angles (Opell and Cushing 1986). Their expanded visual fields show similar overall coverage compared to species with a full complement of eyes (e.g., *Octonoba sinensis*), but spiders with fewer eyes likely expend less energy for eye development and maintenance. Uloborids that build triangle webs, such as *Hyptiotes cavatus*, appear to have six functional eyes because their vestigial AL eyes lack retinal cells (Opell and Ware 1987). In these spiders, increases in resolution (see Sect. 10.3.3.2) might also help compensate for eye loss (Opell 1988). In uloborids with complex orb webs and a full complement of eyes (e.g., *Uloborus glomosus*), the visual fields have overlapping patterns that might help with localizing prey approaching from different orientations (Opell and Ware 1987). Among the species investigated so far, visual system changes were not necessarily progressive; rather, they may have been independent adaptations (B. Opell, pers. comm.). In many cases, eye placement and their associated visual fields can be correlated with present-day function. For example, front-facing eyes assist cursorial spiders with prey capture (Forster 1979), and dorsally placed eyes with greater fields of view might help with aerial predator detection (Opell and Ware 1987). Eye masking experiments will be useful for ascertaining the behavioral functions of different eye pairs.

10.3.3 Structure and Optical Performance of Eyes

Spider eye anatomy was first described by pioneers such as Grenacher (1879), Bertkau (1886), Hentschel (1899), Widmann (1907,1908), and Scheuring (1913, 1914). Deeper investigations into the physiology and optics of spider eyes were later undertaken in the Lycosidae (Homann 1931; Bacetti and Bedini 1964),

Thomisidae (Homann 1934), and Pisauridae (Williams 1979). The eyes of many spider families were described by the renowned German arachnologist Heinrich Homann (e.g., Homann 1951, 1952, 1971; reviewed in Levi 1994). Starting in the late twentieth century, a collection of influential papers on spider eyes was published by David Blest and Michael Land (e.g., Blest and Land 1977). Detailed studies of the tiger wandering spider *Cupiennius salei* by Friedrich Barth, Axel Schmid, and colleagues (e.g., Land and Barth 1992; Schmid 1998) included not only the visual system but other sensory systems as well. Recently, there has been a resurgence of interest in comparative spider vision in a number of labs around the world.

The optical power of eyes determines their maximal potential performance, but realized performance depends on other factors, such as ambient light conditions, the ability of the retina to sample an image, and how the nervous system processes incoming information. Assessing the optics of animal eyes requires a combination of mathematical modeling and careful experimentation, the details of which are beyond the scope of this chapter. Here, we provide a cursory overview of spider eye optics, with an emphasis on resolution and sensitivity (for in-depth reviews, see Warrant and McIntyre 1993; Land 1985a; Land and Nilsson 2012; Cronin et al. 2014; Meece et al. 2021).

10.3.3.1 Corneal Lens Properties

The refractive properties of the corneal lens in combination with retinal placement determines how objects are focused on the retina. This varies both across species and between different eyes of the same individual. The first useful metric is focal length, which is defined as the distance from the nodal point of a lens to the point where light rays form a focused image (the focal point, Land and Nilsson 2012). Focal length is determined by the radius of curvature and the refractive power of the lens and can be measured using Homann's hanging drop technique (Homann 1928). The longest known focal length in spiders belongs to the principal eyes of the salticid *Portia fimbriata* and is 1.980 mm (without the pit lens, it would be 1.701 mm; Williams and McIntyre 1980). In contrast, the focal length of the PM eyes of the trechaleid *Cupiennius salei* is around 0.448 mm (Land and Barth 1992), and those of many web-building species are shorter still. While longer focal lengths are useful for greater magnification, they can increase the extent of chromatic aberrations, resulting in blurring when light rays of different wavelengths are not brought to a single focus. As discussed previously, the pit lens in the principal eyes of salticids provides a telephoto component (Williams and McIntyre 1980), but it also magnifies chromatic aberrations, which may be compensated for by retinal tiering (Land 1969a). Spherical aberration, another instance in which all light rays are not brought to a single focus, can occur in lenses with larger apertures and relatively shorter focal lengths. As another example of evolutionary corrective optics, spherical aberration is nearly eliminated in the PM eyes of *Deinopis* because of a precise gradient of refractive indices in their lens (from the center to the edges; Blest and Land 1977). Similar gradient-index optics are found in jumping spider principal eye lenses as well (Williams and McIntyre 1980), suggesting that this clever solution to spherical aberration may be widespread across spiders.

The minimum focusing distance is the nearest distance between an external object and the point at which light rays form a focused image on the retina. This can be calculated using focal length, lens diameter, and photoreceptor spacing (Land 1981), which also vary across species and eye pairs. For example, the principal eyes of *P. fimbriata* have a minimum focusing distance of about 20 cm (Williams and McIntyre 1980), while the principal eyes of *C. salei* have a minimum focusing distance of about 4 mm (Land and Barth 1992). Different eyes of the same animal usually have different focusing distances, which influences the behavioral utility of each eye pair. For example, in Lycosidae, anterior eyes are optimized for viewing close objects, while posterior eyes are focused further away. The principal (AM) eyes of *Lycosa leuckarti* have a minimum focusing distance of 4.5 mm, and the AL eyes have a minimum focusing distance of 2.7 mm (Clemente et al. 2010). Given how small these distances are, most close-range objects that the spider might encounter should be in focus. In contrast, the PL eyes of *L. leuckarti* have a minimum focusing distance of 24 mm, and the PM eyes have a focusing distance of 32 mm (Clemente et al. 2010). In these eye pairs, more distant objects will be in focus. Eyes that focus on close objects might facilitate prey capture and intraspecific communication, while eyes that only focus on objects several body lengths away might be better suited for long-distance detection and identification.

In addition to these focusing functions of lenses, the transmission properties of the cornea can influence wavelength sensitivity. For example, because UV light is not filtered out by the cornea in many visually hunting families, UV light perception is possible. However, spiders in dim environments, such as those from the families Atypidae and Ctenizidae (Fig. 10.5), often have corneas that block much of the incoming UV light, while species inhabiting open (i.e., not forested) areas often have UV-transmitting corneas (Hu et al. 2014). These differences may contribute to the use of UV light in a number of contexts, including communicatory behaviors. For example, the corneas of all investigated jumping spiders transmit at least some UV light above 290 nm (Hu et al. 2012), and some species attend to UV signals during sexual signaling (Li et al. 2008).

10.3.3.2 Resolution

Spatial acuity, or the ability to resolve fine details, varies across species and across eyes within individuals. For high-resolution vision (of static objects; see Sect. 10.2.2.1 for an explanation of motion hyperacuity), adjacent points in space must be resolved independently by different receptors (reviewed in Meece et al. 2021). Resolution depends on rhabdom density (see Sect. 10.3.4) and their associated interreceptor angles (denoted as $\Delta\Phi$), which can be calculated by dividing the space between the center of adjacent receptors by the focal length (Land 1985a). A smaller $\Delta\Phi$ often correlates with a smaller acceptance angle ($\Delta\rho$), which describes the maximum angle at which incident light can enter the receptor. Narrower interreceptor and acceptance angles confer better resolution vision at the expense of reduced light capture. These metrics can be used to quantify differences between species. Most spiders have an $\Delta\Phi$ of 1–5° across both eye types (Land 1985a). However, visually hunting lineages often exhibit higher resolution at least in a subset of their eyes. For example, huntsman spiders of the genus *Olios* (family Sparassidae) have an $\Delta\Phi$ of 1.8 in their AL eye (Land 1985a). The Salticidae have an unusually small $\Delta\Phi$ of 0.04–0.13° in the center of their principal eyes and a correspondingly narrow $\Delta\rho$ (0.15° for the typical salticid *Phidippus johnsoni*) (Land 1981). The secondary eyes of salticids have larger interreceptor and acceptance angles than the principal eyes (e.g., the AL eyes of *P. johnsoni* have an $\Delta\Phi$ of 0.5–1.5°; Land 1969a). While the principal eyes are often used to inspect objects, the spatial acuity of the secondary eyes can exceed that of the principal eyes in some cases. For example, the principal (AM) eyes of *Cupiennius salei* have an $\Delta\Phi$ of 2.9° and an $\Delta\rho$ of 5.4°, while the large PM eyes have an $\Delta\Phi$ of 1.0° and an $\Delta\rho$ of 2.0° (Land and Barth 1992; Grusch et al. 1997; Pirhofer-Walzl et al. 2007). Similarly, the principal (AM) eyes of ground crab spiders of the genus *Xysticus* have an $\Delta\Phi$ of 3.6°, while the slightly enlarged AL eyes have an $\Delta\Phi$ of 1.8–2.6° (Homann 1934; Land 1985a).

Some spider groups have evolved adaptations to increase the resolution of their everted principal eye photoreceptors. For example, the Salticidae have evolved narrower and more densely packed photoreceptors in their principal-eye retinas (although there is variation; see Blest et al. 1990). The physical isolation of rhabdomeres also improves resolution, as in the Oxyopidae, compared to the closely related Pisauridae (Fig. 10.5), which have contiguous rhabdomeres of adjacent receptors (Blest 1985). Resolution can be improved further with receptor pigment shielding, which absorbs stray light, neatly exemplified by the secondary eyes across much of the Salticidae (e.g., Cerveira et al. 2021). The light capture of photoreceptors can be enhanced when each rhabdomere is surrounded with material that has a lower refractive index (i.e., is less optically dense), which traps light by internal reflection (reviewed in Warrant and McIntyre 1993). This functions similarly to a fiber optic cable. A possible example is the salticid *Portia*, which lacks some organelles and other cellular components, such as microtubules in the cytoplasm of the receptors in the acute regions of their retinas, perhaps to increase the refractive index difference between each rhabdomere and its surroundings (Blest and Price 1984). While it appears that the morphology of rhabdoms is more conserved in comparison to dioptric structures, rapid modifications have occurred in some groups, such as the salticids, which is potentially related to their diversification. We recommend that readers consult Blest (1985) for a more comprehensive review of spider photoreceptor ultrastructure.

10.3.3.3 Sensitivity

Sensitivity, or the ability to capture light, also varies across species and across eyes within individuals. For low-light vision, nocturnal and crepuscular spiders must make the greatest use of relatively few photons available (reviewed in Meece et al.

2021), whereas diurnal species often show lower visual sensitivity, relying instead on light available during their active period. Sensitivity is influenced by the F-number, which divides the focal length by the aperture and describes the physical light-gathering ability of an eye. A lower F-number corresponds to a shorter focal length or wider aperture and optically confers higher sensitivity (Warrant and McIntyre 1993). For example, nocturnal net-casting spiders of the genus *Deinopis* have an F-number of 0.58 in their PM eyes (Blest and Land 1977), the nocturnal wolf spider *Arctosa variana* has an F-number of about 1 in their PM eyes (Land and Nilsson 2012), and the diurnal jumping spider *Portia fimbriata* has an F-number of 2.4 in their principal (AM) eyes (Warrant and McIntyre 1993). To measure sensitivity, the S-number can be used, which is a product of the relative aperture of the eye, the cross-sectional area of the receptor, and the proportion of light entering a receptor that is absorbed (Land 1985a). A higher S-number indicates a more sensitive eye; for example, the PM eyes of *Deinopis* have an extremely high S number of 101, while the principal eyes of *Phidippus* have a much smaller S-number of 0.04 (Land and Nilsson 2012). The proportion of light that is actually absorbed by a receptor depends on the dimensions of the photoreceptive segment, the segment's light-guiding properties, and the amount of visual pigment and can be nearly doubled through the presence of a light-reflecting tapetum (see Sect. 10.3.3.4). At the physiological level, sensitivity can also be estimated using electroretinogram (ERG) or intracellular recordings in response to light (e.g., Yamashita and Tateda 1976b; Laughlin et al. 1980; Barth et al. 1993; Yamashita and Nakamura 1999).

Some spider groups have evolved adaptations to increase the sensitivity of their inverted secondary eye photoreceptors. For example, convergently in the Sparassidae and Salticidae (Fig. 10.5), cell bodies of the secondary eye photoreceptors have shifted laterally, moving them out of the light path of the rhabdomeres (Homann 1971; Eakin and Brandenburger 1971; Blest 1985; Morehouse 2020). This shift, however, necessarily increases the distance between neighboring photoreceptors, resulting in reductions in visual acuity. In other spider families, the photoreceptor cell bodies have become more transparent, although the effect of these changes on cell physiology is not understood (Morehouse 2020). Many spider species have also increased the width of their secondary eye rhabdomeres to increase sensitivity, again an adaptation that typically comes at a cost to visual acuity. Other spider species increase visual sensitivity by pairing rhabdomeres (even interdigitating microvilli) or reducing pigment granules (shielding) between units, which allows for optical pooling (Cerveira et al. 2021). Microvilli contained within the rhabdomeres can also change size during circadian cycles, allowing for light- and dark-adapted states (Uehara et al. 1993). In some other arthropods, screening pigments in visual cells can migrate in response to light and dark cycles, but this does not appear to happen in the supporting glial cells of spiders to an appreciable extent (Blest 1985). Although jumping spiders are particularly known for their diurnal activity, some species hunt under poor light conditions; for example, *Cyrba algerina* spiders have several adaptations to increase sensitivity, which presumably help them find prey in dark crevices (Cerveira et al. 2021). Thus, improving sensitivity can also be important for some day-active spiders.

10.3.3.4 Secondary Eye Tapeta

The tapetum is a reflective mirror-like layer of material composed of guanine crystals at the base of the retina that returns unabsorbed photons to the rhabdomeres for a second chance at capture, thereby effectively increasing photoreceptor sensitivity. While principal eyes universally lack tapeta, many spider families have tapeta in their secondary eyes. Different types of tapeta have been described in spiders, such as the "primitive" tapetum (which forms a single layer perforated by the passage of retinula axons) of the Theraphosidae and other mygalomorphs, the canoe-shaped tapetum of the Sicariidae and Theridiidae, and the elaborate grate-shaped tapetum of the Oxyopidae and Lycosidae (reviewed in Homann 1971; Land 1985a; Fig. 10.5). In spiders with canoe-shaped tapeta, such as the secondary eyes of the Araneidae, the resolution is poor because the image is focused beneath the retina (i.e., underfocused). In contrast, the grate-shaped tapeta of hunting spiders, such as the Lycosidae, reflect focused light to the photoreceptive segments (Land 1985a). While alternative tapetum morphologies have different effects, they all function to increase sensitivity. However, tapeta do have one downside: they can decrease visual acuity as a result of stray light scattering from the tapetum into neighboring photoreceptive units (Morehouse 2020). Some spider families that primarily hunt during the day therefore lack tapeta, including the Philodromidae, Eresidae, and Salticidae (Fig. 10.5). However, we see different suites of adaptations for hunting at night: some species rely on PM eyes with tapeta for prey capture (e.g., Lycosidae; Rovner 1993), while others lack tapeta but have very large PM eyes with large entrance apertures for light gathering (e.g., Deinopidae; Stafstrom and Hebets 2016).

10.3.3.5 Trade-Off Between Resolution and Sensitivity

The diversity in spider eyes provides excellent examples of the well-known trade-off between resolution and sensitivity (reviewed in Warrant and McIntyre 1993; Land 1985a; Land and Nilsson 2012; Cronin et al. 2014; Meece et al. 2021). For an extreme example, the resolution of the principal (AM) eyes of the diurnal Salticidae is ten times better than the PM eyes of the nocturnal Deinopidae, but deinopid receptors are 2000 times more sensitive (Blest and Land 1977); in fact, deinopid PM eyes are so sensitive that they exhibit strong electrophysiological responses to single photons (Laughlin et al. 1980). The eyes of diurnal animals typically have lower sensitivity and thus require an abundance of light to function optimally. This has broad implications for inter- and intraspecific communication, especially in the Salticidae as these spiders encounter colorful prey in their environments (Taylor et al. 2014), and males often display longer wavelength colors to females during courtship (Taylor and McGraw 2013). Under suboptimal circumstances, such as in the shade or at dusk, the spider is at a disadvantage because they cannot reliably discriminate long wavelength colors, such as reds. Under dim light, spiders show attenuated responses to colorful ornaments used in courtship (Taylor and McGraw 2013; Zurek et al. 2015).

Future work, both across and within families, should endeavor to map variation in eye morphology with different lifestyles. For example, in two surveys across at least 34 families, traits that influence resolution and sensitivity, such as relative lens size, rhabdom length, and interrhabdomeric angles, correlate with the foraging mode (L. Sumner-Rooney, personal communication; N. Morehouse, unpublished data).

10.3.3.6 Specializations of Retinal Anatomy

In addition to variation in the optics and photoreceptor structure of eyes, we also see variation in overall retinal structure. As described previously, salticid principal eyes have boomerang-shaped retinas with a central region of increased photoreceptor density, analogous to the foveal region of the vertebrate retina. While the immovable AL eyes of salticids do not achieve the same spatial acuity as the principal eyes, they do have a forward-facing acute zone with a wider field of view (O'Carroll 1989). The Thomisidae and Lycosidae have an anatomical acute zone as well (Blest and O'Carroll 1989), while comparable specialization is lacking in *Cupiennius salei* (Grusch 1994). In *C. salei*, the retinas are shaped like a hemispherical cup (Land and Barth 1992) that is larger and less narrow than that of the Salticidae. The Lycosidae (Melamed and Trujillo-Cenoz 1966) and Thomisidae (Insausti et al. 2012) also have retinas that are hemispherical. However, the retinas of other spiders, such as the Pisauridae, are trough shaped (Williams 1979). The four-layered retinal tiering of the Salticidae discussed previously is certainly not universal; for example, the principal eyes of the wolf spider *Geolycosa godeffroyi* have two layers (Blest and O'Carroll 1989), and those of *C. salei* have only a single layer (Land and Barth 1992). The distribution of photoreceptors and the overall morphology of each layer undoubtedly have functions for vision that are poorly understood, especially for non-salticid spiders, and the number of retinal layers across many spider groups remains poorly described.

10.3.4 Physiological Specializations of Photoreceptors

While the morphological and optical properties of eyes are important for focusing an image, photoreceptors are the cells that actually respond to light. Across virtually all spiders, the retinas of both the principal and secondary eyes have a mosaic of nonpigmented glial cells, pigmented glial cells, and photoreceptor cells. The light-sensitive portions within the photoreceptor cells are cylindrical structures called rhabdoms (Blest 1985). Contained within the rhabdoms are receptive structures called rhabdomeres, which themselves contain highly folded arrays of membrane called microvilli (which have a comb-like shape). Embedded in these membranes are rhodopsins, composed of an opsin protein (Koyanagi et al. 2008) and a vitamin-A-derived, light-sensitive chromophore called retinal (Barth et al. 1993). The rhodopsin maximally absorbs light of a particular wavelength determined by the opsin; absorption causes the retinal to undergo a conformational change that triggers a G-protein-coupled signal transduction cascade. This ultimately depolarizes the cell and transmits an electrochemical signal to the optic nerve (reviewed in Cronin et al. 2014; Hardie and Juusola 2015).

10.3.4.1 Opsin Evolution

Our understanding of the molecular evolution of genes that underlie phototransduction in spiders lags considerably behind other arthropods, but progress is underway. Using transcriptome assemblies from Bond et al. (2014) and Garrison et al. (2016), work by Morehouse et al. (2017) indicates that the canonical components of the rhabdomeric phototransduction pathway identified in insects are also found in the genomes of spiders, although whether these components all serve the same functions remains to be verified. More is known about opsin evolution and expression. The ancestor of spiders and their kin probably had at least four opsin genes, some of which are expressed in the eyes (ocular), brain (extraocular), or both (Eriksson et al. 2013). Four major clades within the opsin gene family have been identified in spiders: Gq-opsin, c-opsin, xenopsin, and tetraopsin (Porter et al. 2012; Ramirez et al. 2016; see Fig. 5 in Morehouse et al. 2017). While the c-opsin, xenopsin, and tetraopsin clades have important implications for elucidating the evolutionary relationships of eye development across arthropods (see Morehouse et al. 2017), of particular interest for phototransduction is the Gq-opsin clade.

Within the Gq-opsin clade, spiders generally have two long-wavelength-sensitive (LWS) opsins and one ultraviolet-sensitive (UVS) opsin (Koyanagi et al. 2008; Nagata et al. 2010, 2012). These are collectively known as rhabdomeric opsins (r-opsins), and these show different expression patterns between the eye types and retinal layers in the salticid *Hasarius adansoni*. In the principal eye retinas, the LWS opsin *Rh1* (which produces a green-sensitive visual pigment when bound to retinal) is expressed in layers I and II, while the UVS opsin *Rh3* is expressed in layers III and IV (Nagata et al. 2012). In the secondary eyes, the *Rh1* opsin is also expressed in the AL and PL eyes, while the LWS *Rh2* (which forms a blue-sensitive visual pigment when bound to retinal) and UVS *Rh4* are expressed in the PM eyes (Nagata et al. 2012). It was long thought that either mygalomorphs had lower opsin diversity than araneomorphs (Fig. 10.5), or there was an undetectable expression in previous studies. To test this, Foley et al. (2020) scored the presence of opsin genes from transcriptomic data in a comprehensive survey of 25 tarantula genera (family Theraphosidae) and found that all subfamilies possessed the full complement of typical arthropod opsins. Across the spider phylogeny, the number and types of r-opsin proteins are similar, but the specific gene copies vary, and these opsins show interesting patterns of losses and duplications (Morehouse et al. 2017). It is possible that opsin expression is linked to visual ecology (e.g., UVS opsins are retained but not expressed in four nocturnal or crepuscular species) and the diversification of colorful signals in the Salticidae (Morehouse et al. 2017), an area ripe for investigation.

10.3.4.2 Temporal Resolution

Temporal resolution, or the ability to resolve successive events in time, exhibits short-term physiological plasticity with changes in the dark- and light-adapted states and varies across nocturnal and diurnal spiders. A key parameter of temporal resolution is the integration time, which is the time it takes a photoreceptor to sample and respond to incoming light (reviewed in Meece et al. 2021). This is often referred to as the "speed" of vision. Longer integration times increase light capture and the signal-to-noise ratio, thereby increasing sensitivity at the cost of reduced temporal resolution (reviewed in Warrant 1999). In the nocturnal spider *C. salei*, dark-adapted integration times of about 138 ms in the PM eyes and 86 ms in the AM eyes were found using intracellular recordings (Pirhofer-Walzl et al. 2007). In the light-adapted state, *C. salei* had an integration time of about 79 ms in the PM eyes and 44 ms in the AM eyes (Pirhofer-Walzl et al. 2007). Evidence suggests that this plasticity might be mediated in part by efferent inputs to photoreceptors in some spiders, such as in the *Argiope* (Yamashita and Tateda 1981, 1983), although it cannot be required because other families, such as the Deinopidae, still show dark adaptation in their PM eyes despite a lack of efferent innervation (Blest 1985). In contrast to nocturnal spiders, spiders active in the day often exhibit shorter integration times, which increases temporal resolution by allowing photoreceptors to respond to successive events more quickly (reviewed in Warrant 1999). Pirhofer-Walzl et al. (2007) approximated the integration time of the AM eye of the diurnal jumping spider *Phidippus johnsoni* to be 42 ms, which is about half that of a dark-adapted AM eye of *C. salei*. A relatively low temporal resolution (and relatively high spatial resolution) seems to be effective for the nocturnal sit-and-wait hunter *C. salei* (Fenk and Schmid 2010, 2011), while diurnal active jumping spiders likely benefit from relatively high spatiotemporal resolution.

Another useful temporal property of an eye is flicker-fusion frequency, or the frequency at which an intermittent light stimulus is perceived as steady. During jumping spider courtship, in which a male and female are involved in rapid reciprocal interactions at close proximity, quicker perceptions of movement might help a male thwart sexual cannibalism. This is also important experimentally because it determines whether video playback during experiments is perceived as fluid motion or a series of static images. Jumping spiders have an estimated light- and dark-adapted flicker-fusion frequency of 90–110 Hz and 50–60 Hz in the AL eyes, respectively (Zurek 2012; D. Zurek, pers. comm.), while *C. salei* has a much lower estimated behavioral flicker-fusion frequency of less than 9 Hz in the PM eyes (Fenk and Schmid 2011). Playback experiments should always aim to use an appropriate frame rate, but currently flicker-fusion frequencies are not well described in other groups.

10.3.4.3 Spectral Sensitivity

Spectral sensitivity, or the wavelength-specific response of photoreceptors, is also variable across families. Color vision requires the comparison of inputs from a minimum of two photoreceptor types with distinct spectral sensitivities (reviewed in Cronin et al. 2014). Spectral sensitivity can be estimated using ERG or receptor potential recordings in response to specific wavelengths of light (e.g., Yamashita and Tateda 1976a; Tapia et al. 2020), through the microspectrophotometry of individual photoreceptors (e.g., Zurek et al. 2015), or by assessing the retinal expression of visual pigments with known absorbance profiles (e.g., Zopf et al. 2013; Sugihara et al. 2016). Investigations of opsin genes suggest that the ancestor of spiders likely had a complement of four visual r-opsins: two LWS opsins (*Rh1* and *Rh2*), one middle-wavelength-sensitive (MWS) opsin, and one UVS opsin (*Rh3*/*Rh*, Morehouse et al. 2017). As discussed previously, photoreceptors of different spectral sensitivities can segregate to particular retinal layers (e.g., jumping spiders) but can be also heterogeneously dispersed across a single layer (e.g., wolf spiders, DeVoe 1972).

Trichromacy, or three color channels, enables an animal to more fully disentangle brightness from wavelength across the visible range of wavelengths (Osorio and Vorobyev 2008) and has independently evolved in the principal eyes of several salticids, facilitated by opsin duplications and subsequent peak spectral sensitivity shifts (Morehouse et al. 2017) or innovations such as spectral filters (e.g., *Habronattus pyrrithrix*, Zurek et al. 2015). This variation provides an excellent opportunity for investigating the mechanisms that underlie spectral sensitivity shifts, the selective pressures that drive the evolution of color vision, and their impact on visual signals. In contrast to principal eyes, secondary eyes generally exhibit monochromacy, or color blindness, which likely helps increase their sensitivity. However, secondary eyes might provide color vision in some species. For example, crab spiders (family Thomisidae), which wait on flowers for prey, appear to have dichromatic secondary eyes based on electrophysiological data (Defrize et al. 2011). UV-, blue-, and green-sensitive photoreceptors are found in the secondary eyes of *Cupiennius salei* (Walla et al. 1996), although extracellular recordings of retinal muscle activity showed a lack of behavioral response when spiders were presented with moving colored stripes over backgrounds of brightness-matched shades of grey, suggesting that they do not distinguish between colors (Orlando and Schmid 2011). Much work remains in characterizing the spectral sensitivities and color vision of spiders at genetic, retinal, behavioral, and evolutionary levels.

10.3.4.4 Polarization Sensitivity

Polarization sensitivity, or the ability to perceive the E-vector orientation of light (i.e., the electric field of an electromagnetic wave, which vibrates orthogonally to the direction of propagation), is useful for complex tasks like navigation and has been described in at least four spider families. At the basis of this ability are photoreceptors with microvilli consistently oriented parallel to each other and perpendicular to incoming light, which therefore respond most strongly to polarized light with an E-vector orientation that is aligned with the microvillar axis (reviewed in Meece et al. 2021). Thus, by comparing the stimulation of nearby photoreceptors with different microvillar orientations, spiders can perceive and respond to light polarization in their environment, such as polarization patterns in the sky and polarized reflections from objects and other organisms. For example, the ground spider *Drassodes cupreus* (family Gnaphosidae) dedicates a pair of specialized PM eyes to perceive UV skylight polarization (Dacke et al. 1999). Curiously, these eyes have reduced lenses to maximize sensitivity to polarization cues at the expense of resolution (their PM eyes are non-image forming). The PM eyes are so specialized that the tapetum functions as a polarizer when it reflects light, boosting polarization sensitivity further (Mueller and Labhart 2010). The ability to perceive polarized light has also been found in the Lycosidae, which possess a strip-shaped specialized region of aligned rhabdomeres in their principal-eye retinas (Dacke et al. 2001). This specialization is on the ventral retina, which has a receptive field that points to the sky. Similarly, lynx spiders (family Oxyopidae) and *Agelena labyrinthica* (family: Agelenidae) have rhabdomeres of different alignments in the principal-eye retina, some of which are potentially polarization sensitive (Kovoor and Muñoz-Cuevas 1997; Schröer 2017). Interestingly, Schröer (2017) found that the rhabdomeres in other locations of the retina were twisted, which would abolish polarization sensitivity in these nonspecialized regions, potentially improving the signal-to-noise ratio by removing differences in photoreceptor stimulation due to polarization rather than light intensity. In the Salticidae, it has been suggested that a "staircase" pattern of UV-sensitive photoreceptors in Layer IV of the principal-eye retina might support polarization vision (Land 1969a; Eakin and Brandenburger 1971), but *Phidippus* spiders did not appear to use polarization patterns in the sky when pursuing their prey (Hill 1979). Among the species investigated so far that use polarized light cues, polarization vision has been most often supported by the principal eyes, but it is unclear to what extent this applies to other spiders. For a recent review of polarization vision in spiders, see Ortega-Escobar (2017).

10.3.5 Control and Cooperation of Eyes

10.3.5.1 Movable Principal-Eye Retinas

In many families, principal-eye retinas have the ability to move (to our knowledge, eye movement is unstudied in most families). Moveable retinas are particularly useful for spiders, allowing them to alter their gaze direction without alerting potential predators or prey. The masters of eye movements are the jumping spiders. As described in Sect. 10.2.2.2, salticids have six retinal muscles (three pairs, the same number of muscles attached to vertebrate eyes) that direct each eye tube. This suite of muscles enables fine movements, including both translational and rotational movements of the retina in an arc behind the corneal lens. The Ctenidae, Lycosidae, and Thomisidae (Fig. 10.5) have four retinal muscles (two pairs) and are therefore more limited in their repertoire of retinal movements. Like the salticids, these spiders show spontaneous exploratory activity, saccades (up to 15°), tracking to follow objects, and microsaccade twitches (2–4° appearing as rapid quivering) to avoid retinal habituation (i.e., bleaching of photopigments from persistent exposure to the same beam of light that causes the visual image to fade) (reviewed in Morehouse 2020). However, these spiders are unable to use rotational movements to inspect objects further. The principal eyes of *Cupiennius salei* have two muscle pairs, allowing either latero-medial or dorso-ventral movements (Kaps and Schmid 1996), while *Agelena labyrinthica* (family Agelenidae) and many other web-building spiders (Fig. 10.5) have only a single lateromedial pair (Land 1985a; Schröer 2017).

The properties of ocular muscles also determine other movement patterns of the eyes. For example, in *C. salei*, the retinas drift back to the resting position after a saccade because of restoring forces from the opposing stretched muscle (Kaps and Schmid 1996). How tightly synchronized the movements are between eyes also varies. In *C. salei*, a principal-eye retina can be independently directed to a stimulus occurring on its ipsilateral side (Kaps and Schmid 1996), while in salticids, the principal-eye retinas are more tightly coupled, particularly during the active interrogation of a visual stimulus—only during exploratory activity might one retina lead (likely due to lateralization of eye use), after which they "clap" back together to examine an object.

10.3.5.2 Interaction of Eyes

The division of labor between different pairs of eyes has been well studied in only a handful of spider families. The possible interactions between eyes can be usefully divided into two categories: eye pairs that do not share a field of view and eye pairs that do share a field of view.

Eye pairs that do not share a field of view are perhaps more straightforward to understand. For example, as described in Sect. 10.2.2.1, the PL eyes of salticids are oriented toward the rear of the spider. When the PL eyes detect movement, the spider exhibits a rapid turning response and directs its forward-facing AM and AL eyes toward the stimulus. Flattie spiders (family Selenopidae) can very quickly rotate their body using long laterigrade legs and strike prey approaching from any direction with impressive speed and accuracy, possibly using input from their PL eyes to detect prey that is behind them (Zeng and Crews 2018). Thus, in these and other families, the combined field of view of all the secondary eyes gives the spider the ability to monitor a wide area using eyes that individually are quite compact.

Perhaps more interesting are eyes that overlap in field of view, thus providing visual information about the same visual scene from independent visual organs. In many cases, nonmoving secondary eyes may overlap in field of view, either with members of the same pair (e.g., the AL eyes in salticids) or between different sets of secondary eyes. Little is known about how overlapping secondary eye pairs process redundant information. Binocular (or multiocular) overlap might contribute to depth perception, but it is disputed if the eyes of many spiders are spaced far enough apart for this to be a viable strategy. In salticids, Forster (1979) found that jumping spiders with one AL eye masked behaved more or less normally but sometimes slightly misjudged the distance of predatory leaps. The corneal lenses of spider eyes are fixed, and therefore depth perception due to accommodation cannot occur as it does in vertebrate eyes. While retinal tiering and the "stair-case" pattern of photoreceptors in the principal eyes of salticids appear to support depth perception via focus (Blest et al. 1981) or via defocus (Nagata et al. 2012), how this might be accomplished in other spiders is less clear.

For some spiders, we have better insight into how the moveable principal eyes collaborate with the secondary eyes. In jumping spiders and *Cupiennius salei*, the secondary eyes that overlap in field of view (AL and PM eyes, respectively) primarily detect motion and subsequently direct the principal-eye retinas to a target. This collaboration has been verified through electrophysiological recordings of retinal muscles in *Cupiennius* (e.g., Kaps and Schmid 1996; Neuhofer et al. 2009) and eye tracking in salticids (Jakob et al. 2018). It is unclear whether eyes with contiguous and overlapping fields of view can "predict" the trajectory of a moving object, as do the ommatidial facets of a hunting dragonfly compound eye (Wiederman et al. 2017), but recent data suggest that there is an exchange of information between multiple eye pairs in jumping spiders. If a moving object passes through the visual field of the PL eyes to the AL eyes, the principal-eye retinas appear to search for a stimulus with its features (Y. Dolev and X. Nelson, pers. comm.). The presence of moveable principal-eye retinas in other groups (e.g., the active Lycosidae and sit-and-wait Thomisidae) suggests that different eye types share a similar division of labor, but this has yet to be rigorously tested (but see Rovner 1993). The precise targeting of the principal-eye retina seen in jumping spiders may be necessary because the principal-eye retina is so small (see Sect. 10.2.2.2). As described in the previous section, in other spider families, fewer muscles control the principal eyes, and targeting is presumably less precise. However, if the principal eyes have larger retinas and larger fields of view, precise targeting may not be necessary. To understand the evolution of cooperation between moving and stationary eyes, we need phylogenetically informed studies of retinal shape and size, the overlap in field of view between the principal and secondary eyes, the degree of precision in eye movements, the neural pathways between the eyes and the brain (see Sect. 10.3.6), and ecological factors such as hunting strategy. An intriguing hypothesis is that principal-eye retinal movements first evolved to overcome retinal habituation when examining stationary objects. This, along with the unique evolutionary history of each eye type (see Sect. 10.3.2.1), may have set the stage for functional differentiation and the partial release from trade-offs associated with arthropod eye design.

10.3.6 Neurobiology of Vision

After photoreceptors respond to incoming light, electrochemical signals must be processed in the retina or brain for visual perception to occur. Most of what we know about spider brain morphology and function comes from work on jumping spiders and *Cupiennius salei*. In these taxa, the principal and secondary eyes and their neural underpinnings comprise two distinct visual systems (Strausfeld and Barth 1993; Strausfeld et al. 1993). Each eye supplies information to separate brain regions—with some crosstalk between them in salticids—until it is integrated into higher centers (Strausfeld 2012). In salticids, the principal-eye pathways seem to process the features and colors of objects, while the secondary-eye pathways seem to simultaneously process contrast and motion information. Our knowledge about the function of these brain regions is limited by a lack of neurophysiological studies, but hypotheses can be constructed based on their connectivity and our knowledge of relatively similar, yet still quite different, neuroanatomical organization in other arthropods. Here, we provide a brief primer on spider brains and their evolutionary history, after which we describe in more detail how the brain (or synganglion as it is more properly called) is connected to the peripheral visual system.

10.3.6.1 Evolution of Spider Brains

The study of spider brains began with Saint-Remy (1887), who captured unique and diverse structures in elegant drawings. This work was continued by Hanström (1921, 1923, 1935), who applied the Golgi staining technique to spider brains, highlighting their components with great detail. He presumed homology with structures and cell types in insects and crustaceans and thus used similar terminology to describe their visual centers—a (potentially misleading) nomenclatural tradition that persists today. More recent work has largely focused on jumping spiders (e.g., Hill 1975; Duelli 1980; Steinhoff et al. 2017; Long 2021) and *Cupiennius salei* (Babu and Barth 1984; Strausfeld and Barth 1993; Strausfeld et al. 1993), which share similar gross brain anatomy.

The central nervous system of spiders is highly condensed into a mass called the synganglion, which is contained within the cephalothorax, or prosoma. The visual and higher-order centers are found in a region of the synganglion called the protocerebrum, the anterior-most neuromere. The synganglion contains structures called neuropils—composed of synaptic regions and glial processes—which serve as the location of functional integration (Bullock and Horridge 1965; Babu 1965). The cell bodies, or somata, for associated neurons form a rind outside the neuropil. Much like in other arthropods with sophisticated image-forming eyes, the visual system of jumping spiders and *C. salei* consists of successively nested optic neuropils that contain dense networks of stratified interneurons (Strausfeld 2012).

Despite a history of shared terminology, spider brains have a different organization and evolved independently from those of insects and crustaceans (Strausfeld

2012; Lehmann et al. 2015). Many arthropod brains have a midline neuropil called the central body, which is used for complex functions like sensorimotor integration (reviewed in Turner-Evans and Jayaraman 2016). Spiders have an analogous structure called the arcuate body (sometimes also referred to as the central body), which also lies on the protocerebral midline and appears to serve similar functions, although it is part of the principal-eye pathway. Furthermore, spiders lack a dedicated olfactory appendage and its associated processing centers. Many arthropod brains have large areas devoted to olfaction (although some also have very robust optical centers). In insects, olfactory inputs are received by olfactory receptors of the antennae, sent to specific olfactory glomeruli (dense synaptic bundles), then project onto a structure called the mushroom body in the antennal (olfactory) lobe (reviewed in Masse et al. 2009). The mushroom body of insects is also associated with higher-order functions, such as learning (e.g., reviewed in Heisenberg 2003). A structure with similar morphology in spiders, also called the mushroom body, is nested within the secondary-eye pathway but is thought to be primarily used for visual processing (Strausfeld and Barth 1993).

While morphological, developmental, and genetic evidence suggests deep homology of the central and mushroom bodies of insects and their counterparts in spiders (e.g., Homberg 2008; Doeffinger et al. 2010; Wolff and Strausfeld 2015), the debate of homology or convergence has not been fully resolved. Relative to other chelicerates, the mushroom body of spiders (if homologous) has been repurposed for vision (Strausfeld 2012). Such a functional shift from olfactory to visual processing has been documented in the mushroom body of an anosmic water beetle (Lin and Strausfeld 2012). The spider mushroom body does not show immunoreactivity against proteins associated with learning in other arthropods (Wolff and Strausfeld 2015), potentially emphasizing its primary role as a visual center. Neurotransmitter and neuromodulator use in visual centers is best known in the case of *C. salei* (reviewed in Barth 2002), which also appears to be similar to other arthropods (e.g., histaminergic retinula cells).

10.3.6.2 Principal- and Secondary-Eye Pathways

The principal and secondary eyes have separate laminae and medullae, which are the first- and second-order optic neuropils, respectively (Fig. 10.6). The unpaired arcuate body is the third-order optic neuropil of the principal eyes, and the bilaterally symmetric mushroom body is the third-order optic neuropil of the secondary eyes (Strausfeld and Barth 1993; Strausfeld et al. 1993; Fig. 10.6). In jumping spiders and *Cupiennius salei*, all eight eyes innervate distinct laminae and medullae, and each eye of the same type (principal or secondary) shares a similar pathway organization to one another (but see Steinhoff et al. 2020 for a discussion about the PM-eye pathway of salticids). The specific cell types of each pathway are not discussed here in detail as they are described elsewhere (for a review, see Strausfeld 2012; Lehmann et al. 2015).

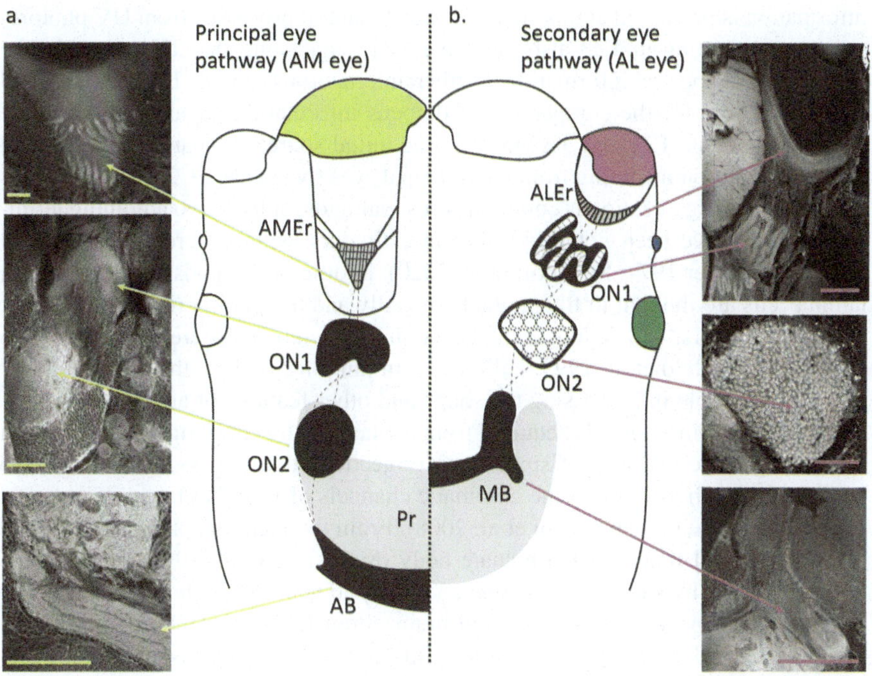

Fig. 10.6 Visual pathways in the brain of *Phidippus audax* depicted in schematic and histological horizontal sections. (**a**) The principal-eye pathway. Shown: AM eye retina (AMEr), lamina (ON1), medulla (ON2), arcuate body (AB), and protocerebrum (Pr). Scale bar represents 50 μm for the AMEr and 200 μm for the rest. (**b**) The secondary-eye pathway. Shown: AL eye retina (ALEr), lamina (ON1), medulla (ON2), and mushroom body (MB). Scale bar represents 50 μm for the ON2 and 200 μm for the rest. The medulla (ON2) of the secondary-eye pathway is comprised of optical glomeruli. Tissue sections were prepared using the methodology described in Long (2018). The diagram is not to scale, and brain structures are not all located in the same plane. The dashed line represents the plane of symmetry. A Nikon Ti confocal microscope (Nikon Instruments Inc., NY) was used for all imaging. Brightness and contrast were adjusted as needed. (Microscope images courtesy of Guilherme Pagoti and the University of Massachusetts Amherst IALS Light Microscopy Core Facility)

In the principal-eye pathway of spiders, projections from the retina enter the rostral edge of the lamina, where they are met by local interneurons (Strausfeld et al. 1993). In salticids, retinula axons are bundled (Oberdorfer 1977) but not pooled (Land 1969a), including those from different retinal layers. The lamina is layered, consisting of four distinct terminal zones in the salticid *Hasarius adansoni* (Nagata et al. 2019). These terminal zones (TZs) correspond to the termination sites of photoreceptor axons from specific retinal layers. For example, green-sensitive photoreceptors from layers I and II terminate in TZ1 and TZ2, respectively. In contrast, lateral UV-sensitive photoreceptors from retinal layers III and IV terminate in TZ3, whereas UV-sensitive central photoreceptors from these layers terminate in TZ4. In all instances, retinotopy is preserved in the TZs, suggesting that spatial information is preserved at this stage. However, lateral processes from UV photoreceptors terminating in TZ3 also innervate TZ1, suggesting that TZ1 may also be able to extract spectral information. Further, interneurons running between TZ1 and TZ2 may allow for the comparison of defocus information postulated to provide depth cues. Thus, TZ1 and TZ2 may be the initial source of spatial, spectral, and depth information incoming from the principal eyes (Nagata et al. 2019). While it is unclear if this organization is common across salticids, at least two adjoining terminal subunits have been noted by other investigators using different species (Hill 1975; Oberdorfer 1977; Steinhoff et al. 2020). In other arthropods, such as insects, lamina cells are thought to filter signals spatially and temporally (e.g., Stöckl et al. 2020). In spiders, projections exit the lamina and target different layers of the medulla (Strausfeld et al. 1993; Hill 2022). In jumping spiders, the medulla likely plays a major role in processing the shape and other features of an object, benefitting from the refined signals acquired from the lamina. It is important to note that in the principal-eye pathway of spiders, no projections that bypass the lamina have been found, such as those in the chromatic channels of flies that target deep layers of the medulla (e.g., Yamaguchi et al. 2008). From the medulla, projections extend to the ipsilateral flange of the arcuate body (Strausfeld et al. 1993). The arcuate body is layered by stratified amacrine and intrinsic neurons, with columnar output neurons that project to areas of the mid-brain (Strausfeld et al. 1993).

In the secondary-eye pathway of salticids and other highly visual hunters, discrete groups of lamina neurons terminate in individual optical glomeruli of the medulla. The axons from the lamina nearly twist 180° before projecting to glomeruli, forming partial chiasmic "chunks" (Strausfeld and Barth 1993; Strausfeld 2012). Perhaps similarly, strepsipteran insects have an organization that subdivides the visual field for "chunk sampling" (Buschbeck et al. 2003). Chiasmic chunking might lend itself to processing adjacent units of visual space at the expense of reduced panoramic image integration. The glomerular organization of the medulla also has analogs in other arthropod brains. Glomerular organization is thought to be efficient given its broad evolutionary convergence for olfaction across phyla (Hildebrand and Shepherd 1997; Eisthen 2002), and some researchers suggest that olfactory and optic glomeruli in arthropods share an organizational ground pattern (Strausfeld et al. 2007; Mu et al. 2012). The optical glomeruli in the lateral protocerebrum of *Drosophila melanogaster* receive nonretinotopic inputs from the lobula yet convey the presence and location of specific visual features to motor centers (Wu et al. 2016). In visually hunting spiders, glomerular chunking may reduce the integration of directional motion across the visual field but enhance the perception of motion in small units. This might help direct principal-eye retinas to a visual target (Strausfeld 2012), facilitated by whole-body turns (fixations) or eye-tube movements (saccades). This organization might function similarly to the small-target motion-detector neurons of predatory insects, such as dragonflies and robber flies, which help them ignore background panoramic motion to pursue a moving target (Buschbeck and Strausfeld 1996; Barnett et al. 2007). In salticids, a mysterious second-order neuropil associated with the secondary eyes (called the "lateral eye neuropil" by Hill 1975, "L2" by Steinhoff et al. 2020, and the "secondary eye lateral neuropil" by Long 2021, the last of which is used hereafter) lacks glomerular organization and bypasses the medulla, which might preserve wide-field panoramic information. However, other investigators have suggested that glomerular organization preserves retinotopy (evidenced by a lack of observable lateral connections), while retinotopy is lost in the secondary eye lateral neuropil, which might enable faster processing of movement (Steinhoff et al. 2020). In the visually guided hunting spiders described in Long (2021), glomeruli then project onto distinct contiguous regions of the ipsilateral mushroom body, which is composed of parallel layers of intrinsic fibers originating from small globuli cells (Strausfeld and Barth 1993; Long 2021). The mushroom bodies are connected by large nerve fibers (the mushroom body "bridge") that cross the midline of the protocerebrum, rostral to the protocerebral commissure. Large output neurons (perhaps similar to those found in the lobula plate of insects) from the mushroom body to the motor centers suggest a role for the mushroom body in behaviors such as prey pursuit (Strausfeld and Barth 1993).

In salticids, the AL eye laminae send direct projections to the arcuate body (Steinhoff et al. 2020), and the principal eye laminae send direct projections to the mushroom body (Strausfeld 2012), affording the opportunity for crosstalk between the two pathways. Centrifugal projections, which originate from higher-order neuropils and target neuropils of a lower order (i.e., so that information can flow in both directions), occur between the medulla and lamina and might allow for the refinement of signals (e.g., accounting for rapid changes in luminance). Secondary-eye optical nerve tracts appear to terminate on the ventral surface of the arcuate body, the probable distal-most location of convergence between the two pathways, although these connections are not well mapped (Weltzien 1988; Long 2021). Extracellular recordings show that, in salticids, a protocerebral region just behind and below the arcuate body responds to inputs from both the secondary and principal eyes (Menda et al. 2014).

10.3.6.3 Variation in Neuromorphology

While brain Bauplans tend to be conserved among closely related taxa, even over geological time (Strausfeld 2019), neural tissue is energetically expensive in both its development and maintenance (Hasenstaub et al. 2010). This means that strong selective pressures can drive divergent brain evolution; specifically, we expect brains to be reduced when they are too costly to maintain and enlarged (or more complex) if fitness benefits outweigh the energetic costs. For example, amphipods of the suborder Hyperiidea have highly variable optic lobe organization and size, correlated with ecological factors, such as the ambient light environment (Lin et al. 2021). Perhaps for similar reasons, visual pathway organization and the percentage of brain volume devoted to visual processing in spiders vary across families (Long 2021).

The principal-eye pathway organization appears to be highly conserved across families (Strausfeld et al. 1993; Kovoor et al. 2005; Nagata et al. 2019; Steinhoff et al. 2020; Long 2021), although there is conspicuous variation in neuropil volume and shape. The arcuate body is relatively robust in most of the species investigated so far (e.g., Weltzien and Barth 1991; Long 2021), likely because of the pivotal role it plays in other functions that would presumably constrain its evolution. However, the arcuate body appears to be somewhat more elaborate in visual hunters (Strausfeld 2012; Long 2021). The volume and shape of the first- and second-order neuropils are highly variable. The volume (or perhaps, more importantly, density) of these regions might be influenced by the extent to which particular information, such as form or color, is important to the spider. Variation in shape might also be influenced by the position and architecture of the retina or eye.

Tremendous variation exists in the secondary-eye pathways across families, as originally noted by Hanström (1935). There are key differences among salticids (Steinhoff et al. 2020), *Cupiennius salei* (Strausfeld and Barth 1993), the less studied orb weavers (Park et al. 2013; Park and Moon 2013), and other spiders. In a recent landmark comparative study, Long (2021) found that the Antrodiaetidae, Hypochilidae, Filistatidae, Scytodidae, and Pholcidae (Fig. 10.5) had the simplest secondary-eye pathways, with underdeveloped laminae, absent or highly reduced medullae, and no apparent mushroom bodies. The Antrodiaetidae was the only mygalomorph represented in this study, and the latter four are relatively basal araneomorphs (Fig. 10.5), all of which are not highly visual (see Sect. 10.3.1). The Araneidae and Deinopidae (Fig. 10.5) had large laminae, absent or highly reduced medullae, and large mushroom bodies. The enlarged mushroom bodies of the Deinopidae were unsurprising, considering their heavy reliance on prey movement cues when hunting at night, but the robust laminae and mushroom bodies of the relatively sedentary orb weavers were an unexpected finding (S. Long, pers. comm.). The Theridiidae, Nephilidae, Amaurobiidae, Agelenidae, and Eutichuridae (formally known as Cheiracanthiidae) (Fig. 10.5) had laminae and some evidence of reduced medullae and mushroom bodies. The Cheiracanthiidae had a simpler secondary-eye pathway than its phylogenetic position would suggest, suggesting a secondary reduction in complexity (Fig. 10.5). Finally, the Ctenidae, Oxyopidae, Pisauridae, Lycosidae, Thomisidae, Philodromidae, and Salticidae (Fig. 10.5) were the most complex, with large laminae, medullae formed from optical glomeruli, and prominent mushroom bodies. We suggest that readers consult Long (2021) for a more detailed discussion of the differences found among groups.

The number of optical glomeruli, when present, varied across families, with the highest density belonging to the Salticidae. The number and connectivity of glomeruli likely relate to the number of visual parameters that can be processed (Strausfeld 2012). Each glomerulus is estimated to have at least a few hundred synapses (about 400 in the salticid *Evarcha arcuata*, Duelli 1980). It is unknown how many retinula cells are represented within each glomerulus and if the ratio is plastic (Long 2021). The degree of chunking (i.e., the number of receptors represented in each glomerulus) would impact their receptive fields and possibly the precision of small-field motion detection and retinal targeting. The presumed ancestral ground pattern of the secondary-eye pathway contains laminae and medullae as first- and second-order neuropils, respectively, so losses are likely to arise from extensive reduction or merging events (Long 2021). Further phylogenetically informed studies should correlate neuromorphology (e.g., Long 2021) and the volume of higher-order centers (e.g., Steinhoff et al. 2018) with different lifestyles and ecological conditions, such as sociality, the complexity of visually guided behaviors, or ambient light environment.

It is presently difficult to draw conclusions about the ecological pressures that drive variation in pathway organization and neuropil shape or volume and the extent to which this variation is related to the diversification of spiders. The homology or convergence of neuromorphological characters remains mysterious without more extensive taxon sampling and ancestral state reconstructions. Chelicerate visual system evolution remains enigmatic, but some investigators have been meticulously conducting new neuroanatomical studies (e.g., Lehmann and Melzer 2021; Brenneis 2022). Recently, methods for sectioning using classical histology, confocal microscopy (Long 2018, 2021), and X-ray microcomputed tomography (Sombke et al. 2015; Steinhoff et al. 2017; Stafstrom et al. 2017) have been refined for use in spiders. Even within investigated families, transmission electron microscopy studies (e.g., Lehmann and Melzer 2018b) are needed to ascertain if unidentifiable neuropils have been lost or fused. Neuronal tracing studies are needed to map the connectivity of the principal- and secondary-eye pathways. Of particular interest is the presence and connectivity of a mushroom body in basal spider groups (i.e., Liphistiidae) as it is still unclear if the mushroom body of spiders is homologous with other arthropods or convergently evolved. If homologous, it is possible that the extensive reduction of the mushroom body in basal groups coincides with the loss of an olfactory appendage when spiders diverged or the mushroom body may have not yet merged into the secondary-eye pathway. The presence of a mushroom-body-like structure, apart from the secondary-eye pathway, has been found in basal mygalomorphs (e.g., Antrodiaetidae in Long 2021; the old-world tarantula *Poecilotheria* in Babu 1965). It would also be of interest to investigate the mushroom bodies of other chelicerates, which are likely olfactory (e.g., amblypygids; Sinakevitch et al. 2021). Electrophysiological studies (e.g., Menda et al. 2014) are required to determine the role of each neuropil (Barth 2002). This work has been neglected due to logistical difficulties associated with pressurized spider prosomas (hydrostatic pressure from hemolymph is used to extend the legs). Spiders rapidly perish if an incision is made in their cuticle, and their brain wobbles with each heartbeat. Fortunately, single- and multiunit extracellular recordings of higher-order centers using thin tungsten microelectrodes have been finally deployed in the jumping spider *Phidippus audax* (Menda et al. 2014; Shamble et al. 2016; A. Winsor, unpubl. data) and the net-casting spider *Deinopis spinosa* (Stafstrom et al. 2020). Across spiders, it is possible that corresponding neuropils are processing different types of information, especially when considering divergent pathway organizations.

10.4 Conclusions and Future Directions

In this section, we wish to highlight a few areas that we think are especially interesting for future work. First, as we have seen, spider visual systems show enormous variation, which is ripe for comparative study. Given that comprehensive, robust, and largely congruent backbone phylogenies are available for spiders (e.g., Bond et al. 2014; Garrison et al. 2016; Wheeler et al. 2017), inferences at the family level or below should be feasible (e.g., Wolff et al. 2022). We currently have a great deal of data on variations in eye arrangement, eye morphology, photoreceptor properties, and neuromorphology for many spider taxa. By strategically adding to this data set and combining it with environmental and behavioral traits, we have many opportunities to understand the selection pressures that led to visual system specialization.

A second, related point is that the distributed visual systems of spiders allow us to investigate how natural selection can act independently on each pair of eyes to overcome functional trade-offs and to understand how the eyes work together. By distributing vision across multiple eyes with different specializations, spiders face relaxed body-size constraints on visual function, which are typically encountered by small animals that mostly rely on a single type of eye (e.g., Rutowski et al. 2009; Warrant and McIntyre 1993). Eye-masking experiments offer the opportunity to study the properties of different visual pathways in isolation and to see how information from multiple eyes is integrated. As elaborated in Sect. 10.3.5.2, our most in-depth knowledge about how eyes divide up tasks is based on only a few families. We especially encourage the study of vision in species that do not appear to rely much on visually based behavior but still have surprisingly robust visual pathways in their central nervous system (e.g., Araneidae) (Long 2021).

Third, the degree to which spider visual processes are flexible is an interesting area. How bottom-up and top-down mechanisms regulate visual cognition is ripe for exploration in invertebrates, especially spiders (reviewed in Winsor et al. 2021). For example, state-dependent modulation of vision, which occurs when changes in physiological states, behavioral states, or environmental conditions impact visual processing (Cheng and Frye 2020), has been virtually unexplored in spiders. How inputs from other sensory modalities influence visual attention is poorly understood outside of salticids and lycosids.

Fourth, how visual information is processed in the spider's brain has just begun to be explored. Researchers have shown that areas in the arcuate body region of the jumping spider brain respond to particular visual images (Menda et al. 2014) and sounds (Shamble et al. 2016). Next, in salticids, recording from more peripheral areas of the separate pathways between the brain and the principal and secondary eyes will help demonstrate exactly where different types of visual information are processed and then integrated. In addition, expanding this work beyond salticids will, of course, be extremely informative, especially given the vast variation in neural architecture across families (Long 2021).

Finally, using spiders for bioinspired engineering solutions is another area of potential research. For example, the computationally efficient construction of a system like that of jumping spiders, which is broadly tuned to lower-resolution, motion-sensitive inputs yet can strategically direct a higher-acuity apparatus to certain stimuli, has already inspired robotic camera systems (Tonet et al. 2008). Similarly, Guo et al. (2019) designed a compact depth sensor inspired by jumping spiders' image-defocusing mechanism.

The visual systems of spiders are both extremely variable and impressively capable. Our knowledge of spider vision is being pushed forward by the collaborative efforts of interdisciplinary researchers worldwide. Much exciting work remains as we attempt to understand how spiders see the world.

Acknowledgments We thank Elke Buschbeck and Michael Bok for the invitation to participate in this volume, as well as for Elke's thorough comments on the manuscript. We are grateful for the thoughtful suggestions provided by Adam Porter, Paul Katz, Skye Long, Lauren Sumner-Rooney, and a very helpful anonymous reviewer. Some of the work described here was supported by NSF IOS-1656714 to EMJ and Ronald R. Hoy and NSF IOS-1557549/IOS-1734291/IOS-1831767 to NIM.

References

Aguilar-Arguello S, Nelson XJ (2021) Jumping spiders: an exceptional group for comparative cognition studies. Learn Behav 49:276–291

Babu KS (1965) Anatomy of the central nervous system of arachnids. Zool Jahrb Abt Ontog Tiere Anat 82:1–154

Babu KS, Barth FG (1984) Neuroanatomy of the central nervous system of the wandering spider, *Cupiennius salei* (Arachnida, Araneida). Zoomorphology 104:344–359

Bacetti B, Bedini C (1964) Research on the structure and physiology of the eyes of a lycosid spider. Arch Ital Biol 102:97–122

Barnett PD, Nordström K, O'Carroll DC (2007) Retinotopic organization of small-field-target-detecting neurons in the insect visual system. Curr Biol 17:569–578

Barth FG (2002) A spider's world: senses and behavior. Springer-Verlag, Berlin/Heidelberg/New York

Barth FG, Nakagawa T, Eguchi E (1993) Vision in the ctenid spider *Cupiennius salei*: spectral range and absolute sensitivity. J Exp Biol 181:63–80

Bartos M, Minias P (2016) Visual cues used in directing predatory strikes by the jumping spider *Yllenus arenarius* (Araneae, Salticidae). Anim Behav 120:51–59

Baudouin-Gonzalez L, Harper A, McGregor AP, Sumner-Rooney L (2022) Regulation of eye determination and regionalization in the spider *Parasteatoda tepidariorum*. Cells 11:631

Bedini C (1967) The fine structure of the eyes of *Euscorpius carpathicus* L. (Arachnida Scorpiones). Arch Ital Biol 105:361–378

Bednarski JV, Taylor P, Jakob EM (2012) Optical cues used in predation by jumping spiders, *Phidippus audax* (Araneae, Salticidae). Anim Behav 84:1221–1227

Bertkau P (1886) Beiträge zur Kenntnis der Sinnesorgane der Spinnen, Die Augen der Spinnen. Arch Mikr Anat 27:589–631

Bleckmann H, Rovner JS (1984) Sensory ecology of a semi-aquatic spider (*Dolomedes triton*): i. Roles of vegetation and wind-generated waves in site selection. Behav Ecol Sociobiol 14:297–301

Blest AD (1985) The fine structure of spider photoreceptors in relation to function. In: Barth FR (ed) Neurobiology of arachnids. Springer-Verlag, Berlin/Heidelberg/New York

Blest AD, Carter M (1987) Morphogenesis of a tiered principal retina and the evolution of jumping spiders. Nature 328:152–155
Blest AD, Land MF (1977) The physiological optics of *Dinopis subrufus* L. Koch: a fish-lens in a spider. Proc R Soc Lond B Biol Sci 196:197–222
Blest AD, O'Carroll D (1989) The evolution of the tiered principal retinae of jumping spiders (Araneae: Salticidae). In: Naresh Singh R, Strausfeld NJ (eds) Neurobiology of sensory systems. Springer, Boston
Blest AD, Price GD (1984) Retinal mosaics of the principal eyes of some jumping spiders (Salticidae: Araneae): adaptations for high visual acuity. Protoplasma 120:172–184
Blest AD, Hardie RC, McIntyre P, Williams DS (1981) The spectral sensitivities of identified receptors and the function of retinal tiering in the principal eyes of a jumping spider. J Comp Physiol A 145:227–239
Blest AD, O'Carroll DC, Carter M (1990) Comparative ultrastructure of Layer I receptor mosaics in principal eyes of jumping spiders: the evolution of regular arrays of light guides. Cell Tissue Res 262:445–460
Bond JE, Garrison NL, Hamilton CA, Godwin RL, Hedin M, Agnarsson I (2014) Phylogenomics resolves a spider backbone phylogeny and rejects a prevailing paradigm for orb web evolution. Curr Biol 24:1765–1771
Brenneis G (2022) The visual pathway in sea spiders (Pycnogonida) displays a simple serial layout with similarities to the median eye pathway in horseshoe crabs. BMC Biol 20:27
Bruce M, Daye D, Long SM, Winsor AM, Menda G, Hoy RR, Jakob EM (2021) Attention and distraction in the modular visual system of a jumping spider. J Exp Biol 224:jeb231035
Bullock T, Horridge GA (1965) Structure and function in the nervous systems of invertebrates. Freeman, San Francisco
Buschbeck EK (2014) Escaping compound eye ancestry: the evolution of single-chamber eyes in holometabolous larvae. J Exp Biol 217:2818–2824
Buschbeck EK, Strausfeld NJ (1996) Visual motion-detection circuits in flies: small-field retinotopic elements responding to motion are evolutionarily conserved across taxa. J Neurosci 16:4563–4578
Buschbeck EK, Ehmer B, Hoy RR (2003) The unusual visual system of the Strepsiptera: external eye and neuropils. J Comp Physiol A 189:617–630
Canavesi C, Long S, Fantone D, Jakob EM, Jackson RR, Harland D, Rolland JP (2011) Design of a retinal tracking system for jumping spiders. Proc SPIE 8129:81290A-8
Carvell GE, Jackson RR, Cross FR (2017) Ontogenetic shift in plant-related cognitive specialization by a mosquito-eating predator. Behav Process 138:105–122
Cerveira AM, Nelson XJ, Jackson RR (2021) Spatial acuity-sensitivity trade-off in the principal eyes of a jumping spider: possible adaptations to a 'blended' lifestyle. J Comp Physiol A 207:437–448
Chan JPY, Lau PR, Tham AJ, Li D (2008) The effects of male—male contests and female eavesdropping on female mate choice and male mating success in the jumping spider, *Thiania bhamoensis* (Araneae: Salticidae). Behav Ecol Sociobiol 62:639–646
Cheng KY, Frye MA (2020) Neuromodulation of insect motion vision. J Comp Physiol A 206:125–137
Clark DL, Morjan CL (2001) Attracting female attention: the evolution of dimorphic courtship displays in the jumping spider *Maevia inclemens* (Araneae: Salticidae). Proc R Soc B 268:2461–2465
Clark DL, Uetz GW (1990) Video image recognition by the jumping spider, *Maevia inclemens* (Araneae: Salticidae). Anim Behav 40:884–890
Clemente CJ, McMaster KA, Fox E, Meldrum L, Stewart T, Main BY (2010) The visual system of the Australian wolf spider *Lycosa leuckartii* (Araneae: Lycosidae): visual acuity and the functional role of the eyes. J Arachnol 38:398–406
Craig CL, Weber RS, Bernard GD (1996) Evolution of predator-prey systems: spider foraging plasticity in response to the visual ecology of prey. Am Nat 147:205–229

Crane J (1949) Comparative biology of salticid spiders at rancho Grande, Venezuela. Part IV. An analysis of display. Zoologica 34:159–214

Cronin TW, Johnsen S, Marshall NJ, Warrant EJ (2014) Visual ecology. Princeton University Press, Princeton

Cross FR, Jackson RR (2006) From eight-legged automatons to thinking spiders. In: Fujita K, Itakura S (eds) Diversity of cognition. Kyoto University Academic Press, Kyoto

Cross FR, Jackson RR (2009) Cross-modality priming of visual and olfactory selective attention by a spider that feeds indirectly on vertebrate blood. J Exp Biol 212:1869–1875

Cross FR, Jackson RR (2010) The attentive spider: search-image use by a mosquito-eating predator. Ethology 116:240–247

Cross FR, Jackson RR (2015) Solving a novel confinement problem by spartaeine salticids that are predisposed to solve problems in the context of predation. Anim Cogn 18:509–515

Cross FR, Jackson RR (2019) Portia's capacity to decide whether a detour is necessary. J Exp Biol 222:jeb203463

Dacke M, Nilsson DE, Warrant EJ, Blest AD, Land MF, O'Carroll DC (1999) Built-in polarizers form part of a compass organ in spiders. Nature 401:470–473

Dacke M, Doan TA, O'Carroll DC (2001) Polarized light detection in spiders. J Exp Biol 204:2481–2490

De Agrò M, Rößler DC, Kim K, Shamble PS (2021) Perception of biological motion by jumping spiders. PLoS Biol 19:e3001172

de Omena PM, Romero GQ (2010) Using visual cues of microhabitat traits to find home: the case study of a bromeliad-living jumping spider (Salticidae). Behav Ecol 21:690–695

Defrize J, Lazzari CR, Warrant EJ, Casas J (2011) Spectral sensitivity of a colour changing spider. J Insect Physiol 57:508–513

DeVoe RD (1972) Dual sensitivities of cells in wolf spider eyes at ultraviolet and visible wavelengths of light. J Gen Physiol 59:247–269

Doeffinger C, Hartenstein V, Stollewerk A (2010) Compartmentalization of the precheliceral neuroectoderm in the spider *Cupiennius salei*: development of the arcuate body, optic ganglia, and mushroom body. J Comp Neurol 518:2612–2632

Dolev Y, Nelson XJ (2014) Innate pattern recognition and categorization in a jumping spider. PLoS One 9:e97819

Drees O (1952) Untersuchungen über die angeborenen Verhaltensweisen bei Springspinnen (Salticidae). Z Tierpsychol 9:169–207

Duelli P (1980) Neuronal organization of the posterior lateral eyes of jumping spiders (Salticidae). Zool Jahrb Abt Ontog Tiere Anat 103:17–40

Dukas R (2002) Behavioural and ecological consequences of limited attention. Philos Trans R Soc Lond Ser B Biol Sci 357:1539–1547

Eakin RM, Brandenburger JL (1971) Fine structure of the eyes of jumping spiders. J Ultrastruct Res 37:618–663

Edwards GB, Jackson RR (1993) Use of prey-specific predatory behaviour by North American jumping spiders (Araneae, Salticidae) of the genus *Phidippus*. J Zool 229:709–716

Eisthen HL (2002) Why are olfactory systems of different animals so similar? Brain Behav Evol 59:273–293

Elias DO, Kasumovic MM, Punzalan D, Andrade MCB, Mason AC (2008) Assessment during aggressive contests between male jumping spiders. Anim Behav 76:901–991

Elias DO, Maddison WP, Peckmezian C, Girard MB, Mason AC (2012) Orchestrating the score: complex multimodal courtship in the *Habronattus coecatus* group of *Habronattus* jumping spiders (Araneae: Salticidae). Biol J Linn Soc 105:522–547

Elliott L (2006) Spider eye arrangements. Iowa State University. Online at https://bugguide.net/node/view/84423. Accessed on 21 July 2021

Eriksson BJ, Fredman D, Steiner G, Schmid A (2013) Characterisation and localisation of the opsin protein repertoire in the brain and retinas of a spider and an onychophoran. BMC Evol Biol 13:186

Fenk LM, Schmid A (2010) The orientation-dependent visual spatial cut-off frequency in a spider. J Exp Biol 213:3111–3117

Fenk LM, Schmid A (2011) Flicker-induced eye movements and the behavioural temporal cut-off frequency in a nocturnal spider. J Exp Biol 214:3658–3663

Fenk LM, Hoinkes T, Schmid A (2010) Vision as a third sensory modality to elicit attack behavior in a nocturnal spider. J Comp Physiol A 196:957–961

Foelix RF (2011) Biology of spiders, 3rd edn. Oxford University Press, Oxford

Foley S, Saranathan V, Piel WH (2020) The evolution of colouration and opsins in tarantulas. Proc R Soc B 287:20201688

Forster LM (1979) Visual mechanisms of hunting behaviour in *Trite planiceps*, a jumping spider (Araneae: Salticidae). N Z J Zool 6:79–93

Forster L (1982) Vision and prey-catching strategies in jumping spiders. Am Sci 70:165–175

Forster L (1985) Target discrimination in jumping spiders (Araneae: Salticidae). In: Barth FG (ed) Neurobiology of arachnids. Springer-Verlag, Berlin/Heidelberg/New York

Friedrich M (2006) Ancient mechanisms of visual sense organ development based on comparison of the gene networks controlling larval eye, ocellus, and compound eye specification in *Drosophila*. Arthropod Struct Dev 35:357–378

Gainett G, Ballesteros JA, Kanzler CR, Zehms JT, Zern JM, Aharon S, Gavish-Regev E, Sharma PP (2020) Systemic paralogy and function of retinal determination network homologs in arachnids. BMC Genomics 21:811

Garrison NL, Rodriguez J, Agnarsson I, Coddington JA, Griswold CE, Hamilton CA, Hedin M, Kocot KM, Ledford JM, Bond JE (2016) Spider phylogenomics: untangling the spider tree of life. PeerJ 4:e1719

Girard MB, Endler JA (2014) Peacock spiders. Curr Biol 24:R588–R590

Girard MB, Elias DO, Kasumovic MM (2015) Female preference for multi-modal courtship: multiple signals are important for male mating success in peacock spiders. Proc R Soc B 282:20152222

Girard MB, Kasumovic MM, Elias DO (2018) The role of red coloration and song in peacock spider courtship: insights into complex signaling systems. Behav Ecol 29:1234–1244

Glenszczyk M, Outomuro D, Gregorič M, Kralj-Fišer S, Schneider JM, Nilsson DE, Morehouse NI, Tedore C (2022) The jumping spider *Saitis barbipes* lacks a red photoreceptor to see its own sexually dimorphic red coloration. Sci Nat 109:6

Goté JT, Butler PM, Zurek DB, Buschbeck EK, Morehouse NI (2019) Growing tiny eyes: how juvenile jumping spiders retain high visual performance in the face of size limitations and developmental constraints. Vis Res 160:24–36

Grenacher H (1879) Untersuchungen tiber das Sehorgan der Arthropoden, insbesondere der Spinnen, Insekten und Crustaceen. Vandenhoek & Ruprecht, Göttingen

Grusch M (1994) Feinstruktur und Retinalgehalt der Augen von *Cupiennius salei* (Araneidae, Ctenidae). Universität Wien, Diplomarbeit

Grusch M, Barth FG, Eguchi E (1997) Fine structural correlates of sensitivity in the eyes of the ctenid spider, *Cupiennius salei* keys. Tissue Cell 29:421–430

Guo Q, Shi Z, Huang Y, Alexander E, Qiu CW, Capasso F, Zickler T (2019) Compact single-shot metalens depth sensors inspired by eyes of jumping spiders. Proc Natl Acad Sci U S A 116:22959–22965

Hanström B (1921) Über die Histologie und vergleichende Anatomie der Sehganglien und Globuli der Araneen. Kungl Svenska Vetenskapsakad Handl 61:1–39

Hanström B (1923) Further notes on the central nervous system of arachnids: scorpions, phalangids, and trap-door spiders. J Comp Neurol 35:249–274

Hanström B (1935) Fortgesetzte Untersuchungen über das Araneengehirn. Zool Jahrb Abt Ontog Tiere Anat 59:455–478

Hardie RC, Juusola M (2015) Phototransduction in *Drosophila*. Curr Opin Neurobiol 34:37–45

Harland DP, Jackson RR (2000) Cues by which *Portia fimbriata*, an araneophagic jumping spider, distinguishes jumping-spider prey from other prey. J Exp Biol 203:3485–3494

Harland DP, Jackson RR (2002) Influence of cues from the anterior medial eyes of virtual prey on *Portia fimbriata*, an araneophagic jumping spider. J Exp Biol 205:1861–1868
Harland DP, Jackson RR (2004) *Portia* perceptions: the Umwelt of an araneophagic jumping spider. In: Prete FR (ed) Complex worlds from simpler nervous systems. MIT Press, Cambridge
Harland DP, Jackson RR, Macnab AM (1999) Distances at which jumping spiders (Araneae: Salticidae) distinguish between prey and conspecific rivals. J Zool 247:357–364
Harland D, Li D, Jackson RR (2012) How jumping spiders see the world. In: Lazareva OF, Shimizu T, Wasserman EA (eds) How animals see the world: comparative behavior, biology, and evolution of vision. Oxford University Press, New York
Hasenstaub A, Otte S, Callaway E, Sejnowskia TJ (2010) Metabolic cost as a unifying principle governing neuronal biophysics. Proc Natl Acad Sci U S A 107:12329–12334
Hazzi NA, Hormiga G (2022) Molecular phylogeny of the tropical wandering spiders (Araneae, Ctenidae) and the evolution of eye conformation in the RTA clade. Cladistics
Hebets EA, Maddison WP (2005) Xenophilic mating preferences among populations of the jumping spider *Habronattus pugillis* Griswold. Behav Ecol 16:981–988
Hebets EA, Uetz GW (1999) Female responses to isolated signals from multimodal male courtship displays in the wolf spider genus *Schizocosa* (Araneae: Lycosidae). Anim Behav 57:865–872
Heil KH (1936) Beiträge zur Physiologie und Psychologie der Springspinnen. Z Vergl Physiol 23:1–25
Heiling AM (1999) Why do nocturnal orb-web spiders (Araneidae) search for light? Behav Ecol Sociobiol 46:43–49
Heiling AM, Cheng K, Chittka L, Goeth A, Herberstein ME (2005) The role of UV in crab spider signals: effects on perception by prey and predators. J Exp Biol 208:3925–3931
Heisenberg M (2003) Mushroom body memoir: from maps to models. Nat Rev Neurosci 4:266–275
Hentschel E (1899) Beiträge zur Kenntnis der Spinnenaugen. Zool Jb Anat 12:509–534
Hildebrand JG, Shepherd GM (1997) Mechanisms of olfactory discrimination: converging evidence for common principles across phyla. Annu Rev Neurosci 20:595–631
Hill DE (1975) The structure of the central nervous system of jumping spiders of the genus *Phidippus* (Araneae: Salticidae). Master's thesis, Oregon State University
Hill DE (1979) Orientation by jumping spiders of the genus *Phidippus* (Araneae: Salticidae) during the pursuit of prey. Behav Ecol Sociobiol 5:301–322
Hill DE (2022) Neurobiology and vision of jumping spiders (Araneae: Salticidae). Peckhamia 255(1):1–81
Hoefler CD, Jakob EM (2006) Jumping spiders in space: movement patterns, nest site fidelity and the use of beacons. Anim Behav 71:109–116
Homann H (1928) Beiträge zur Physiologie der Spinnenaugen. I. Untersuchungsmethoden. II. Das Sehvermögen der Salticiden. Z Vergl Physiol 7:201–269
Homann H (1931) Beiträge zur Physiologie der Spinnenaugen. III. Das Sehvermögen der Lycosiden. Z Vergl Physiol 14:40–67
Homann H (1934) Beiträge zur Physiologie der Spinnenaugen. IV. Das Sehvermögen der Thomisiden. Z Vergl Physiol 20:420–429
Homann H (1951) Die Nebenaugen der Araneen. Zool Jahrb Anat 71:56–144
Homann H (1952) Die Nebenaugen der Araneen. II. Mitteilung. Zool Jahrb Anat 72:345–364
Homann H (1971) Die Augen der Araneae. Anatomie, Ontogenie und Bedeutung für die Systematik. Z Morphol Tiere 69:201–272
Homberg U (2008) Evolution of the central complex in the arthropod brain with respect to the visual system. Arthropod Struct Dev 37:347–362
Hu Z, Liu F, Xu X, Chen Z, Chen A, Li D (2012) Spectral transmission of the principal-eye corneas of jumping spiders: implications for ultraviolet vision. J Exp Biol 215:2853–2859
Hu Z, Xu X, Chen Z, Li H, Wang X, Wu L, Liu F, Chen J, Li D (2014) The spectral transmission of non-salticid spider corneas. J Exp Biol 217:2698–2703
Humphrey B, Helton WS, Bedoya C, Dolev Y, Nelson XJ (2018) Psychophysical investigation of vigilance decrement in jumping spiders: overstimulation or understimulation? Anim Cogn 21:787–794

Humphrey B, Helton WS, Nelson XJ (2019) Caffeine affects the vigilance decrement of *Trite planiceps* jumping spiders (Salticidae). J Comp Psychol 133:551–557

Huseynov EF (2008) Natural prey of the spider *Tibellus macellus* (Araneae, Philodromidae). Bull Br Arachnol Soc 4:206–208

Insausti TC, Defrize J, Lazzari CR, Casas J (2012) Visual fields and eye morphology support color vision in a color-changing crab-spider. Arthropod Struct Dev 41:155–163

Jackson RR, Cross FR (2011) Spider cognition. Adv Insect Physiol 41:115–174

Jackson RR, Harland DP (2009) One small leap for the jumping spider but a giant step for vision science. J Exp Biol 212:2129–2132

Jackson RR, Pollard SD (1996) Predatory behavior of jumping spiders. Annu Rev Entomol 41:287–308

Jackson R, Carter C, Tarsitano M (2001) Trial-and-error solving of a confinement problem by a jumping spider, *Portia fimbriata*. Behaviour 138:1215–1234

Jäger P (2012) Revision of the genus Sinopoda Jäger, 1999 in Laos with discovery of the first eyeless huntsman spider species (Sparassidae: Heteropodinae). Zootaxa 3415:37

Jakob EM, Long SM (2016) How (not) to train your spider: successful and unsuccessful methods for studying learning. N Z J Zool 43:112–126

Jakob EM, Skow CD, Mary PH, Plourde A (2007) Jumping spiders associate food with color cues in a T-maze. J Arachnol 35:487–492

Jakob E, Skow C, Long S (2011) Plasticity, learning and cognition. In: Herberstein ME (ed) Spider behaviour. Cambridge University Press, Cambridge

Jakob EM, Long SM, Harland DP, Jackson RR, Carey A, Searles ME, Porter AH, Canavesi C, Rolland JP (2018) Lateral eyes direct principal eyes as jumping spiders track objects. Curr Biol 28:R1092–R1093

Janeschik M, Schacht MI, Platten F, Turetzek N (2022) It takes two: discovery of spider Pax2 duplicates indicates prominent role in chelicerate central nervous system, eye, as well as external sense organ precursor formation and diversification after neo- and subfunctionalization. Front Ecol Evol 10:810077

Kaps F, Schmid A (1996) Mechanism and possible behavioural relevance of retinal movements in the ctenid spider *Cupiennius salei*. J Exp Biol 199:2451–2458

Kovoor J, Muñoz-Cuevas A (1997) Comparative structure of the visual system of lynx spiders (Oxyopidae) and its relation to habitat and behaviour. Zool Anz 235:133–145

Kovoor J, Muñoz-Cuevas A, Ortega-Escobar J (2005) The visual system of *Lycosa tarentula* (Araneae, Lycosidae): microscopic anatomy of the protocerebral optic centres. Ital J Zool 72:205–216

Koyanagi M, Nagata T, Katoh K, Yamashita S, Tokunaga F (2008) Molecular evolution of arthropod color vision deduced from multiple opsin genes of jumping spiders. J Mol Evol 66:130–137

Lan T, Zhao Y, Zhao F, He Y, Martinez P, Strausfeld NJ (2021) Leanchoiliidae reveals the ancestral organization of the stem euarthropod brain. Curr Biol 31:4397–4404.e2

Land MF (1969a) Structure of the retinae of the principal eyes of jumping spiders (Salticidae: Dendryphantinae) in relation to visual optics. J Exp Biol 51:443–470

Land MF (1969b) Movements of the retinae of jumping spiders (Salticidae: Dendryphantinae) in response to visual stimuli. J Exp Biol 51:471–493

Land MF (1971) Orientation by jumping spiders in the absence of visual feedback. J Exp Biol 54:119–139

Land MF (1972) Stepping movements made by jumping spiders during turns mediated by the lateral eyes. J Exp Biol 57:15–40

Land MF (1974) A comparison of the visual behaviour of a predatory arthropod with that of a mammal. In: Wiersma CAG (ed) Invertebrate neurons and behaviour. MIT Press, Cambridge

Land MF (1981) Optics and vision in invertebrates. In: Autrum H (ed) Handbook of sensory physiology. Springer-Verlag, Berlin/Heidelberg/New York

Land MF (1985a) The morphology and optics of spider eyes. In: Barth FG (ed) Neurobiology of arachnids. Springer-Verlag, Berlin/Heidelberg/New York

Land MF (1985b) Fields of view of the eyes of primitive jumping spiders. J Exp Biol 119:381–384
Land MF, Barth FG (1992) The quality of vision in the ctenid spider *Cupiennius salei*. J Exp Biol 164:227–242
Land MF, Nilsson DE (2012) Animal eyes, 2nd edn. Oxford University Press, Oxford
Landolfa MA, Barth FG (1996) Vibrations in the orb web of the spider *Nephila clavipes*: cues for discrimination and orientation. J Comp Physiol A 179:493–508
Laughlin S, Blest AD, Stowe S (1980) The sensitivity of receptors in the posterior median eye of the nocturnal spider, *Dinopis*. J Comp Physiol A 141:53–65
Lehmann T, Melzer RR (2018a) Also looking like limulus? — retinula axons and visual neuropils of Amblypygi (whip spiders). Front Zool 15:52
Lehmann T, Melzer RR (2018b) A tiny visual system—retinula axons and visual neuropils of *Neobisium carcinoides* (Hermann, 1804) (Chelicerata, Arachnida, Pseudoscorpiones). Zool Anz 273:164–172
Lehmann T, Melzer RR (2021) Outsourcing a visual neuropil—the central visual system of the median eyes of *Galeodes granti* Pocock, 1903 (Arachnida: Solifugae). Arthropod Struct Dev 60:101024
Lehmann T, Melzer RR, Hörnig MK, Michalik P, Sombke A, Harzsch S (2015) Arachnida (excluding Scorpiones). In: Schmidt-Rhaesa A, Harzsch S, Purschke G (eds) Structure and evolution of invertebrate nervous systems. Oxford University Press, Oxford
Levi HW (1994) Heinrich Homann, 1894–1993. J Arachnol 22:87–88
Li D, Lim ML (2005) Ultraviolet cues affect the foraging behaviour of jumping spiders. Anim Behav 70:771–776
Li J, Zhang Z, Liu F, Liu Q, Gan W, Chen J, Lim MLM, Li D (2008) UVB-based mate-choice cues used by females of the jumping spider *Phintella vittata*. Curr Biol 18:699–703
Lin C, Strausfeld NJ (2012) Visual inputs to the mushroom body calyces of the whirligig beetle *Dineutus sublineatus*: modality switching in an insect. J Comp Neurol 520:2562–2574
Lin C, Hoving HT, Cronin TW, Osborn KJ (2021) Strange eyes, stranger brains: exceptional diversity of optic lobe organization in midwater crustaceans. Proc R Soc B 288:20210216
Lohrey AK, Clark DL, Gordon SD, Uetz GW (2009) Antipredator responses of wolf spiders (Araneae: Lycosidae) to sensory cues representing an avian predator. Anim Behav 77:813–821
Long SM (2018) A novel protocol for generating intact, whole-head spider cephalothorax tissue sections. BioTechniques 64:163–169
Long SM (2021) Variations on a theme: morphological variation in the secondary eye visual pathway across the order of Araneae. J Comp Neurol 529:259–280
Long SM, Leonard A, Carey A, Jakob EM (2015) Vibration as an effective stimulus for aversive conditioning in jumping spiders. J Arachnol 43:111–114
Loria SF, Prendini L (2014) Homology of the lateral eyes of Scorpiones: a six-ocellus model. PLoS One 9:e112913
Maddison W, McMahon M (2000) Divergence and reticulation among montane populations of a jumping spider (*Habronattus pugillis* Griswold). Syst Biol 49:400–421
Masse NY, Turner GC, Gregory SXE Jefferis (2009) Olfactory information processing in *Drosophila*. Curr Biol 19:R700–R713
Masta SE, Maddison WP (2002) Sexual selection driving diversification in jumping spiders. Proc Natl Acad Sci U S A 99:4442–4447
McGinley RH, Taylor PW (2016) Video playback experiments support a role for visual assessment of opponent size in male-male contests of *Servaea incana* jumping spiders. Behav Ecol Sociobiol 70:821–829
Meece M, Rathore S, Buschbeck EK (2021) Stark trade-offs and elegant solutions in arthropod visual systems. J Exp Biol 224:jeb215541
Melamed J, Trujillo-Cenóz O (1966) The fine structure of the visual system of *Lycosa* (Araneae: Lycosidae). Z Zellforsch Mikrosk Anat 74:12–31
Melrose A, Nelson XJ, Dolev Y, Helton WS (2019) Vigilance all the way down: vigilance decrement in jumping spiders resembles that of humans. Q J Exp Psychol 72:1530–1538

Menda G, Shamble PS, Nitzany EI, Golden JR, Hoy RR (2014) Visual perception in the brain of a jumping spider. Curr Biol 24:2580–2585
Miether ST, Dunlop JA (2016) Lateral eye evolution in the arachnids. J Arachnol 17:103–119
Morehouse N (2020) Spider vision. Curr Biol 30:R975–R980
Morehouse NI, Buschbeck EK, Zurek DB, Steck M, Porter ML (2017) Molecular evolution of spider vision: new opportunities, familiar players. Biol Bull 233:21–38
Morse DH (2007) Predator upon a flower: life history and fitness in a crab spider. Harvard University Press, Cambridge
Mu L, Ito K, Bacon JP, Strausfeld NJ (2012) Optic glomeruli and their inputs in *Drosophila* share an organizational ground pattern with the antennal lobes. J Neurosci 32:6061–6071
Mueller KP, Labhart T (2010) Polarizing optics in a spider eye. J Comp Physiol A 196:335–348
Nagata T, Koyanagi M, Tsukamoto H, Terakita A (2010) Identification and characterization of a protostome homologue of peropsin from a jumping spider. J Comp Physiol A 196:51
Nagata T, Koyanagi M, Tsukamoto H, Saeki S, Isono K, Shichida Y, Tokunaga F, Kinoshita M, Arikawa K, Terakita A (2012) Depth perception from image defocus in a jumping spider. Science 335:469–471
Nagata T, Arikawa K, Kinoshita M (2019) Photoreceptor projection from a four-tiered retina to four distinct regions of the first optic ganglion in a jumping spider. J Comp Neurol 527:1348–1361
Nakamura T, Yamashita S (2000) Learning and discrimination of colored papers in jumping spiders (Araneae, Salticidae). J Comp Physiol A 186:897–901
Nelson XJ, Jackson RR (2006) Vision-based innate aversion to ants and ant mimics. Behav Ecol 17:676–681
Nelson XJ, Helton WS, Melrose A (2019) The effect of stimulus encounter rate on response decrement in jumping spiders. Behav Process 159:57–59
Neuhofer D, Machan R, Schmid A (2009) Visual perception of motion in a hunting spider. J Exp Biol 212:2819–2823
Nilsson DE (2021) The diversity of eyes and vision. Annu Rev Vis Sci 7:19–41
Niven JE, Laughlin SB (2008) Energy limitation as a selective pressure on the evolution of sensory systems. J Exp Biol 211:1792–1804
Nørgaard T, Henschel JR, Wehner R (2006) The night-time temporal window of locomotor activity in the Namib Desert long-distance wandering spider, *Leucorchestris arenicola*. J Comp Physiol A 192:365–372
O'Carroll DC (1989) An optical assessment of visual performance in the eyes of hunting spiders (Araneae: Labidognatha). Dissertation, Flinders University
Oberdorfer MD (1977) The neural organization of the first optic ganglion of the principal eyes of jumping spiders (Salticidae). J Comp Neurol 174:95–117
Opell BD (1988) Ocular changes accompanying eye loss in the spider family Uloboridae. J Morphol 196:119–126
Opell BD, Cushing PE (1986) Visual fields of orb web and single line web spiders of the family Uloboridae (Arachnida, Araneida). Zoomorphology 106:199–204
Opell BD, Ware AD (1987) Changes in visual fields associated with web reduction in the spider family Uloboridae. J Morphol 192:87–100
Orlando E, Schmid A (2011) Colour blindness of the movement-detecting system of the spider *Cupiennius salei*. J Exp Biol 214:546–550
Ortega-Escobar J (2002) Evidence that the wolf-spider *Lycosa tarentula* (Araneae, Lycosidae) needs visual input for path integration. J Arachnol 30:481–486
Ortega-Escobar J (2017) Polarized-light vision in spiders. Trends Entomol 13:25–34
Osorio D, Vorobyev M (2008) A review of the evolution of animal colour vision and visual communication signals. Vis Res 48:2042–2051
Park YK, Moon MJ (2013) Microstructural organization of the central nervous system in the orb-web spider, *Araneus ventricosus* (Araneae: Araneidae). Appl Microscopy 43:6–74
Park YK, Kim HJ, Kim H, Moon MJ (2013) Fine structure of the CNS ganglia in the geometric spider *Nephila clavata* (Araneae: Nephilidae). J Entomol Res 43:330–343

Paulus H (1979) Eye structure and the monophyly of Arthropoda. In: Gupta AP (ed) Arthropod phylogeny. Van Nostrand Reinhold, New York
Paulus HF (2000) Phylogeny of the Myriapoda—Crustacea—Insecta: a new attempt using photoreceptor structure. J Zool Syst Evol Res 38:189–208
Peckham GW, Peckham EG (1890) Additional observations on sexual selection in spiders of the family Attidae: with some remarks on Mr. Wallace's theory of sexual ornamentation. Natural History Society of Wisconsin, Wisconsin
Peckmezian T, Taylor PW (2015a) A virtual reality paradigm for the study of visually mediated behaviour and cognition in spiders. Anim Behav 107:87–95
Peckmezian T, Taylor PW (2015b) Electric shock for aversion training of jumping spiders: towards an arachnid model of avoidance learning. Behav Process 113:99–104
Peckmezian T, Taylor PW (2017) Place avoidance learning and memory in a jumping spider. Anim Cogn 20:275–284
Persons MH, Uetz GW (1996) The influence of sensory information on patch residence time in wolf spiders (Araneae: Lycosidae). Anim Behav 51:1285–1293
Piacentini LN, Ramírez MJ (2019) Hunting the wolf: a molecular phylogeny of the wolf spiders (Araneae, Lycosidae). Mol Phylogenet Evol 136:227–240
Pirhofer-Walzl K, Warrant E, Barth FG (2007) Adaptations for vision in dim light: impulse responses and bumps in nocturnal spider photoreceptor cells (*Cupiennius salei* Keys). J Comp Physiol A 193:1081–1087
Platnick NI (2020) Spiders of the world: a natural history. Ivy Press, London
Porter ML, Blasic JR, Bok MJ, Cameron EG, Pringle T, Cronin TW, Robinson PR (2012) Shedding new light on opsin evolution. Proc R Soc B 279:3–14
Ramirez MD, Pairett AN, Pankey MS, Serb JM, Speiser DI, Swafford AJ, Oakley TH (2016) The last common ancestor of most bilaterian animals possessed at least nine opsins. Genome Biol Evol 8:3640–3652
Raven PH, Yeates DK (2007) Australian biodiversity: threats for the present, opportunities for the future. Aust J Entomol 46:177–187
Riechert SE (1982) Spider interaction strategies: communication vs. coercion. In: Witt PN, Rovner JS (eds) Spider communication. Princeton University Press, Princeton
Robinson MH, Robinson B (1971) The predatory behavior of the ogre-faced spider *Dinopis longipes* F. Cambridge (Araneae: Dinopidae). Am Midl Nat 85:85–96
Robledo-Ospina LE, Rao D (2022) Dangerous visions: a review of visual antipredator strategies in spiders. Evol Ecol 36(3):163
Rößler DC, De Agrò M, Kim K, Shamble PS (2021) Static visual predator recognition in jumping spiders. Funct Ecol 36:1–11
Rovner JS (1993) Visually mediated responses in the lycosid spider *Rabidosa rabida*: the roles of different pairs of eyes. Mem Queensland Mus 33:635–638
Rovner JS (1996) Conspecific interactions in the lycosid spider *Rabidosa rabida*: the roles of different senses. J Arachnol 24:16–23
Rutowski RL, Gislén L, Warrant EJ (2009) Visual acuity and sensitivity increase allometrically with body size in butterflies. Arthropod Struct Dev 38:91–100
Saint-Remy G (1887) Contribution à l'étude du cerveau chez les arthropodes trachéates. Arch Zool Exp Gen 2:1–274
Samadi L, Schmid A, Eriksson BJ (2015) Differential expression of retinal determination genes in the principal and secondary eyes of *Cupiennius salei* Keyserling (1877). EvoDevo 6:16
Savage SW, Spano LP, Bowers AR (2019) The effects of age and cognitive load on peripheral-detection performance. J Vis 19:1–17
Scheuring L (1913) Die Augen der Arachnoideen. I. Zool lahrb Anat 33:553–636
Scheuring L (1914) Die Augen der Arachnoideen. II. Zool lahrb Anat 37:369–464
Schmid A (1998) Different functions of different eye types in the spider *Cupiennius salei*. J Exp Biol 201:221–225

Schomburg C, Turetzek N, Schacht MI, Schneider J, Kirfel P, Prpic NM, Posnien N (2015) Molecular characterization and embryonic origin of the eyes in the common house spider *Parasteatoda tepidariorum*. EvoDevo 6:15
Schröer W (2017) Fine structure of the anterior median eyes of the funnel-web spider *Agelena labyrinthica* (Araneae: Agelenidae). Arthropod Struct Dev 46:196–214
Shamble PS, Menda G, Golden JR, Nitzany EI, Walden K, Beatus T, Elias DO, Cohen I, Miles RN, Hoy RR (2016) Airborne acoustic perception by a jumping spider. Curr Biol 26:2913–2920
Sinakevitch I, Long SM, Gronenberg W (2021) The central nervous system of whip spiders (Amblypygi): large mushroom bodies receive olfactory and visual input. J Comp Neurol 529:1642–1658
Skow CD, Jakob EM (2005) Jumping spiders attend to context during learned avoidance of aposematic prey. Behav Ecol 17:34–40
Sombke A, Lipke E, Michalik P, Uhl G, Harzsch S (2015) Potential and limitations of X-Ray micro-computed tomography in arthropod neuroanatomy: a methodological and comparative survey. J Comp Neurol 523:1281–1295
Spano L, Long SM, Jakob EM (2012) Secondary eyes mediate the response to looming objects in jumping spiders (*Phidippus audax*, Salticidae). Biol Lett 8:949–951
Stafstrom JA, Hebets EA (2016) Nocturnal foraging enhanced by enlarged secondary eyes in a net-casting spider. Biol Lett 12:20160152
Stafstrom JA, Michalik P, Hebets EA (2017) Sensory system plasticity in a visually specialized, nocturnal spider. Sci Rep 7:46627
Stafstrom JA, Menda G, Nitzany EI, Hebets EA, Hoy RR (2020) Ogre-faced, net-casting spiders use auditory cues to detect airborne prey. Curr Biol 30:5033–5039.e3
Steinhoff POM, Sombke A, Liedtke J, Schneider JM, Harzsch S, Uhl G (2017) The synganglion of the jumping spider *Marpissa muscosa* (Arachnida: Salticidae): insights from histology, immunohistochemistry and microCT analysis. Arthropod Struct Dev 46:156–170
Steinhoff POM, Liedtke J, Sombke A, Uhl G (2018) Early environmental conditions affect the volume of higher-order brain centers in a jumping spider. J Zool 304:182–192
Steinhoff POM, Uhl G, Harzsch S, Sombke A (2020) Visual pathways in the brain of the jumping spider *Marpissa muscosa*. J Comp Neurol 528:1883–1902
Stöckl AL, O'Carroll DC, Warrant EJ (2020) Hawkmoth lamina monopolar cells act as dynamic spatial filters to optimize vision at different light levels. Sci Adv 6:eaaz8645
Strausfeld NJ (2012) Arthropod brains: evolution, functional elegance, and historical significance. The Belknap Press of Harvard University Press, Cambridge
Strausfeld NJ (2019) The divergent evolution of arthropod brains. In: Byrne JH (ed) The Oxford handbook of invertebrate neurobiology. Oxford University Press, Oxford
Strausfeld NJ, Barth FG (1993) Two visual systems in one brain: neuropils serving the secondary eyes of the spider *Cupiennius salei*. J Comp Neurol 328:43–62
Strausfeld NJ, Weltzien P, Barth FG (1993) Two visual systems in one brain: neuropils serving the principal eyes of the spider *Cupiennius salei*. J Comp Neurol 328:63–75
Strausfeld NJ, Sinakevitch I, Okamura JY (2007) Organization of local interneurons in optic glomeruli of the dipterous visual system and comparisons with the antennal lobes. Dev Neurobiol 67:1267–1288
Strausfeld NJ, Ma X, Edgecombe GD, Fortey RA, Land MF, Liu Y, Cong P, Hou X (2016) Arthropod eyes: the early Cambrian fossil record and divergent evolution of visual systems. Arthropod Struct Dev 45:152–172
Sugihara T, Nagata T, Mason B, Koyanagi M, Terakita A (2016) Absorption characteristics of vertebrate non-visual opsin, Opn3. PLoS One 11:e0161215
Tapia F, Olivares J, Schmachtenberg O (2020) The visual spectral sensitivity of the Chilean recluse spider *Loxosceles laeta*. J Exp Biol 223:jeb217133
Tarsitano M (2006) Route selection by a jumping spider (*Portia labiata*) during the locomotory phase of a detour. Anim Behav 72:1437–1442
Tarsitano MS, Jackson RR (1994) Jumping spiders make predatory detours requiring movement away from prey. Behaviour 131:65–73

Tarsitano MS, Jackson RR (1997) Araneophagic jumping spiders discriminate between detour routes that do and do not lead to prey. Anim Behav 53:257–266

Taylor LA, McGraw KJ (2013) Male ornamental coloration improves courtship success in a jumping spider, but only in the sun. Behav Ecol 24:955–967

Taylor PW, Hasson O, Clark DL (2001) Initiation and resolution of jumping spider contests: roles for size, proximity, and early detection of rivals. Behav Ecol Sociobiol 50:403–413

Taylor LA, Maier EB, Byrne KJ, Amin Z, Morehouse NI (2014) Colour use by tiny predators: jumping spiders show colour biases during foraging. Anim Behav 90:149–157

Taylor GJ, Paulk AC, Pearson TWJ, Moore RJD, Stacey JA, Ball D, van Swinderen B, Srinivasan MV (2015) Insects modify their behaviour depending on the feedback sensor used when walking on a trackball in virtual reality. J Exp Biol 218:3118–3127

Taylor LA, Amin Z, Maier EB, Byrne KJ, Morehouse NI (2016) Flexible color learning in an invertebrate predator: *Habronattus* jumping spiders can learn to prefer or avoid red during foraging. Behav Ecol 27:520–529

Tedore C, Johnsen S (2015) Visual mutual assessment of size in male *Lyssomanes viridis* jumping spider contests. Behav Ecol 26:510–518

Tedore C, Johnsen S (2016) Disentangling the visual cues used by a jumping spider to locate its microhabitat. J Exp Biol 219:2396–2401

Terakita A, Nagata T (2014) Functional properties of opsins and their contribution to light-sensing physiology. Zool Sci 31:653–659

Tietjen WJ, Rovner JS (1982) Chemical communication in lycosids and other spiders. In: Witt PN, Rovner JS (eds) Spider communication. Princeton University Press, Princeton

Tonet O, Focacci F, Piccigallo M, Mattei L, Quaglia C, Megali G, Mazzolai B, Dario P (2008) Bioinspired robotic dual-camera system for high-resolution vision. IEEE Trans Robot 24:55–64

Turnbull AL (1973) Ecology of the true spiders (Araneomorphae). Annu Rev Entomol 18:305–348

Turner-Evans DB, Jayaraman V (2016) The insect central complex. Curr Biol 26:R453–R457

Ubick D, Paquin P, Cushing PE, Roth V (eds) (2005) Spiders of North America: an identification manual, 2nd edn. American Arachnological Society, Keene

Uehara A, Uehara K, Ogawa K (1993) Efferent fibers and daily rhabdomal changes in the anteromedial eye of the liphistiid spider, *Heptathela kimurai*. Cell Tissue Res 272:517–522

Vickers ME, Taylor LA (2018) Odor alters color preference in a foraging jumping spider. Behav Ecol 29:833–839

Walla P, Barth FG, Eguchi E (1996) Spectral sensitivity of single photoreceptor cells in the eyes of the ctenid spider *Cupiennius salei* Keys. Zool Sci 13:199–202

Warrant EJ (1999) Seeing better at night: life style, eye design and the optimum strategy of spatial and temporal summation. Vis Res 39:1611–1630

Warrant EJ, McIntyre PD (1993) Arthropod eye design and the physical limits to spatial resolving power. Prog Neurobiol 40:413–461

Weltzien P (1988) Vergleichende Neuroanatomie des Spinnengehirns unter besonderer Berücksichtigung des "Zentralkörpers." Dissertation, Universität Frankfurt am Main

Weltzien P, Barth FG (1991) Volumetric measurements do not demonstrate that the spider brain 'central body' has a special role in web building. J Morphol 208:91–98

Wheeler WC, Coddington JA, Crowley LM, Dimitrov D, Goloboff PA, Griswold CE, Hormiga G, Prendini L, Ramírez MJ, Sierwald P, Almeida-Silva L, Alvarez-Padilla F, Arnedo MA, Benavides Silva LR, Benjamin SP, Bond JE, Grismado CJ, Hasan E, Hedin M, Izquierdo MA, Labarque FM, Ledford J, Lopardo L, Maddison WP, Miller JA, Piacentini LN, Platnick NI, Polotow D, Silva-Dávila D, Scharff N, Szûts T, Ubick D, Vink CJ, Wood HM, Zhang J (2017) The spider tree of life: phylogeny of Araneae based on target-gene analyses from an extensive taxon sampling. Cladistics 33:574–616

Whitcomb WH, Eason R (1965) The mating behavior of *Peucetia viridans* (Araneida: Oxyopidae). Fla Entomol 48:163–167

Widmann E (1907) Der feinere Bau der Augen einiger Spinnen. Zool Anz 31:755–762

Widmann E (1908) Über den feineren Bau der Augen einiger Spinnen. Z Wiss Zool 90:258–312

Wiederman SD, Fabian JM, Dunbier JR, O'Carroll DC (2017) A predictive focus of gain modulation encodes target trajectories in insect vision. elife 6:e26478

Williams DS (1979) The physiological optics of a nocturnal semi-aquatic spider, *Dolomedes aquaticus* (Pisauridae). Z Naturforsch C 34:463–469

Williams DS, McIntyre P (1980) The principal eyes of a jumping spider have a telephoto component. Nature 288:578–580

Winsor AM, Ihle M, Taylor LA (2020) Methods for independently manipulating palatability and color in small insect prey. PLoS One 15:e0231205

Winsor AM, Pagoti GF, Daye DJ, Cheries EW, Cave KR, Jakob EM (2021) What gaze direction can tell us about cognitive processes in invertebrates. Biochem Biophys Res Commun 564:43–54

Wolff GH, Strausfeld NJ (2015) Genealogical correspondence of mushroom bodies across invertebrate phyla. Curr Biol 25:38–44

Wolff JO, Wierucka K, Paterno G, Coddington JA, Hormiga G, Kelly MBJ, Herberstein ME, Ramírez MJ (2022) Stabilized morphological evolution of spiders despite mosaic changes in foraging ecology. Syst Biol 71:1487–1503

World Spider Catalog (2022) World spider catalog version 23.0. Natural History Museum Bern, online at https://wsc.nmbe.ch. Accessed 13 Feb 2022

Wu M, Nern A, Williamson WR, Morimoto MM, Reiser MB, Card GM, Rubin GM (2016) Visual projection neurons in the *Drosophila* lobula link feature detection to distinct behavioral programs. elife 5:e21022

Yamaguchi S, Wolf R, Desplan C, Heisenberg M (2008) Motion vision is independent of color in *Drosophila*. Proc Natl Acad Sci U S A 105:4910–4915

Yamashita S, Nakamura T (1999) Circadian oscillation of sensitivity of spider eyes: diurnal and nocturnal spiders. J Exp Biol 202:2539–2542

Yamashita S, Tateda H (1976a) Spectral sensitivities of jumping spider eyes. J Comp Physiol A 105:29–41

Yamashita S, Tateda H (1976b) Hypersensitivity in the anterior median eye of a jumping spider. J Exp Biol 65:507–516

Yamashita S, Tateda H (1981) Efferent neural control in the eyes of orb weaving spiders. J Comp Physiol A 143:477–483

Yamashita S, Tateda H (1983) Cerebral photosensitive neurons in the orb weaving spiders, *Argiope bruennichii* and *A. amoena*. J Comp Physiol A 150:467–472

Zeng Y, Crews S (2018) Biomechanics of omnidirectional strikes in flat spiders. J Exp Biol 221:jeb166512

Zhou Q, DeSantis DF, Friedrich M, Pignoni F (2016) Shared and distinct mechanisms of atonal regulation in *drosophila* ocelli and compound eyes. Dev Biol 418:10–16

Zhou J, Lai J, Menda G, Stafstrom JA, Miles CI, Hoy RR, Miles RN (2022) Outsourced hearing in an orb-weaving spider that uses its web as an auditory sensor. Proc Natl Acad Sci U S A 199:e2122789119

Zopf LM, Schmid A, Fredman D, Eriksson BJ (2013) Spectral sensitivity of the ctenid spider *Cupiennius salei*. J Exp Biol 216:4103–4108

Zurek DB (2012) The function of the anterior lateral eyes in the modular visual system of jumping spiders (Araneae, Salticidae). Dissertation, Macquarie University

Zurek DB, Nelson XJ (2012a) Hyperacute motion detection by the lateral eyes of jumping spiders. Vis Res 66:26–30

Zurek DB, Nelson XJ (2012b) Saccadic tracking of targets mediated by the anterior-lateral eyes of jumping spiders. J Comp Physiol A 198:411–417

Zurek DB, Taylor AJ, Evans CS, Nelson XJ (2010) The role of the anterior lateral eyes in the vision-based behaviour of jumping spiders. J Exp Biol 213:2372–2378

Zurek DB, Cronin TW, Taylor LA, Byrne K, Sullivan MLG, Morehouse NI (2015) Spectral filtering enables trichromatic vision in colorful jumping spiders. Curr Biol 25:R403–R404

Index

E. Buschbeck, M. Bok (eds.), *Distributed Vision*, Springer Series in Vision Research, https://doi.org/10.1007/978-3-031-23216-9

The manufacturer's authorised representative in the EU is Springer Nature Customer Service Centre GmbH, Europaplatz 3, 69115 Heidelberg, Germany. If you have any concerns regarding our products, please contact ProductSafety@springernature.com

Printed and bound by CPI Group (UK) Ltd, Croydon, CR0 4YY
28/09/2026
02224071-0013